SPON'S
QUARRY
GUIDE
to the British
hard rock industry

SPON'S
QUARRY
GUIDE

to the British
hard rock industry

Compiled by

D. I. E. Jones BSc, ME
H. Gill CEng, MIMinE **and J. L. Watson** MSc

E. & F. N. SPON

An imprint of Chapman and Hall

LONDON • NEW YORK • TOKYO • MELBOURNE • MADRAS

UK   Chapman and Hall, 2–6 Boundary Row, London SE1 8HN

USA   Van Nostrand Reinhold, 115 5th Avenue, New York NY10003

JAPAN   Chapman and Hall Japan, Thomson Publishing Japan, Hirakawacho
      Nemoto Building, 7F, 1-7-11 Hirakawa-cho, Chiyoda-ku, Tokyo 102

AUSTRALIA Chapman and Hall Australia, Thomas Nelson Australia, 480 La Trobe
      Street, PO Box 4725, Melbourne 3000

INDIA   Chapman and Hall India, R Seshadri, 32 Second Main Road, CIT East,
      Madras 600 035

First edition 1991

© 1991 E. & F. N. Spon

Typeset in 8/10pt Helvetica by Mayhew Typesetting, Bristol
Printed in England by Clays Ltd, St. Ives PLC

ISBN 0 419 16710 2   0 442 31330 6 (USA)
ISSN 0960-9377

British Library Cataloguing in Publication Data
available

Library of Congress Cataloging-in-Publication Data
available

Publisher's Note

Whilst great care has been taken in compiling the information in the
Guide, the compilers and the publishers can accept no responsibility
for errors, inaccuracies or omissions or for any consequences arising
therefrom.
  Readers are encouraged to inform the publishers of any such
inaccuracies or omissions so future editions can be corrected.

Contents

Preface

The British quarrying industry is currently experiencing a period of unprecedented activity. Output of all types of stone, crushed rock and products is at an all-time high, and forecasts for the last decade of the century all predict growth in demand. Investment in the country's infrastructure, especially in major civil engineering projects, is expected to be maintained and increased.

Not only is the demand for quarry products greater than ever before but the industry itself is undergoing tremendous changes and realignments as companies are restructured, taken over and merged.

There is clearly a need for all those involved in the commercial, technical and operational aspects of quarrying to have access to reliable, comprehensive and up-to-date information on the industry. We confidently believe that *Spon's Quarry Guide* provides this information in a form that will prove invaluable to those who work in the quarrying industry, supply their products and services to it, purchase or specify its end products, and who make commercial decisions about the companies involved.

To the best of our knowledge, the information in the Guide is not available from a single source. A number of special features should also be noted:

1. Information on holding and operating companies in England, Scotland and Wales is provided.
2. Quarry names, addresses, locations, personnel *etc* are listed and information is given on:
   - geology (rock type, colour, grain size)
   - products
   - drilling and blasting information (hole diameter, bench height *etc*)
   - plant used for drilling, load and haul, and crushing.
3. Comprehensive indexes of companies, quarries, rock types and colours, quarry products and personnel are provided.

Every effort has been made to check and verify the entries in the directory section and the assistance of quarry industry personnel, equipment manufacturers and product end users in this is gratefully acknowledged. With the current rate of change and development in the industry some of the information given may have changed since going to press. If any errors or omissions are found, readers are encouraged to inform the publishers so that future editions can incorporate all necessary changes.

In conjunction with the publishers we are planning to compile and publish new editions of the Guide at regular intervals. Suggestions for ways of improving future editions to make our work an even more valuable information source will be most welcome; a reply-paid card is bound into the Guide for this.

Finally, we should point out that, although contact names and telephone numbers have been given for the quarries, many companies prefer individual enquiries about purchase of products or plant to be referred to a central order/purchasing department. Readers may therefore find it best to contact the head office in the first instance.

D Jones
H Gill
J Watson

How to use this Guide

The Guide is in two main parts: the Directory of Quarry Companies and Quarries (pages 1 – 316) and the Indexes (pages 317 – 480).

The Directory

This is arranged in alphabetical order of quarry company. The quarries owned by each company in England, Scotland and Wales are listed alphabetically. The larger quarry companies operate separate regional companies or divisions and their quarries are listed under the appropriate *operating* company, *not* the corporate or holding company.

Each quarry has a unique reference number, which appears in bold type on the right of each entry, to aid identification and indexing.

The name given to each quarry is that by which it is commonly known. Where quarries have alternative names these are also given, both in the Directory and in the Indexes.

Quarries which are not being worked at present are identified as 'Quarries on stand'.

The Indexes

For easy reference to the Directory, ten indexes are provided. These cover: companies, quarries, rock type and colour, products and personnel. The indexes for quarries (2 and 3), rock type (4 and 5), rock colour (6 and 7) and products (8 and 9) are presented in two ways:

- in a single combined listing for Britain (England, Scotland and Wales)
- in separate listings for England, Scotland and Wales, county by county.

This is to enable quarries, sources or products to be identified on a countrywide or local scale, as appropriate.

Index 1, Companies, refers to the page numbers of the book; Indexes 2–9 refer to the unique quarry numbers; and Index 10 refers to page numbers (for head office personnel) and quarry numbers (for quarry personnel).

Introduction

The British hard rock industry

The practice of quarrying rock is nearly as old as mankind itself. We are aware that during the Neolithic period in Great Britain a form of organized quarrying was already taking place; the disused flint workings in parts of Norfolk and Sussex are visible witness of these primitive industries which were concerned with the making of flint tools and weapons.

Building with stone was also practised during this early period. The Neolithic stone built dwellings at Skara Brae on Orkney show how the early buildings used slabs of local flaggy sandstone found readily on the nearby sea shore and shore line outcrops.

The extraction of a stone from an exposed rock face developed as the centuries progressed. Stone was quarried to build the Roman cities of Roman Britain and Hadrian's Wall, for the later castles of Saxon and Norman nobles and for the building of the monasteries, minsters and cathedrals of the Middle Ages.

The uses to which stone is put have increased greatly since these early times. The dimension and monumental stone quarrying industry has been overtaken in size of operation by the requirements for crushed stone, tar-macadam, aggregates for concrete and concrete products for the road construction industry; for broken gritstones and igneous rocks for railway ballast; for crushed limestone and chalk for the chemical, cement, sugar and food processing industries; for crushed silica sandstones and limestone for glass making industries, to name but a few.

Hard rock quarries exist today in many locations and sizes, depending on the type of stone being quarried and the use to which that stone is to be put.

The location of hard rock quarries in Great Britain

The vast majority of hard rock quarries in Great Britain are located west and north-west of a line from Exeter, in Devon, north-eastwards to the mouth of The Wash in Lincolnshire. Eastward and south of this line hard rock quarries are very few indeed and are usually scattered at wide distances from each other.

We need to know why this is so, since while a very large tonnage of hard rock is consumed each year in the densely populated south-east of England — in the form of crushed stone, tar-macadam, railway ballast, ready-mixed concrete, concrete products, slates, stone cladding and paving — the rock raw material is quarried many miles away in south-west England, Wales, the Midlands, northern England and Scotland, then transported by road, rail and sea to its final destination.

While existing and potential market locations, the nature and requirements of the market and their related transport routes do have influence on the size and the local siting of the quarries, it is the geology — the type, nature and characteristics of the rock being quarried or required to be quarried — which determine the geographical location of each quarry.

Geology

Whilst this Guide is not intended to be a geology textbook some readers may find it a useful reference to the geological sequence of the sedimentary system of rocks in Great Britain and the influence these have on the location of the quarries. Each of the following geological eras and periods is to be found represented within Great Britain.

Geological eras	Geological periods	Age (millions of years)
Cenozoic	Quaternary	
	Pleiocene	
	Miocene	
	Oligocene	
	Eocene	
Mesozoic	Cretaceous	100
	Jurassic	
	Triassic	200
Palaeozoic	Permian	
	Carboniferous	300
	Devonian	400
	Silurian	
	Ordovician	500
	Cambrian	600
	Pre Cambrian	

Rocks of economic importance of the various geological periods

1. PRE-CAMBRIAN
   The igneous rocks of the Charnwood area of Leicestershire, the schists, gneisses and Torridonian sandstones of north-west Scotland, also the Dalradian limestones which occur in various parts of Scotland.

2. CAMBRIAN
   The slates of north Wales and the metamorphosed sandstones of the English midlands around Nuneaton.

3. ORDOVICIAN
   The slates of Borrowdale and Skiddaw in Cumbria. The igneous rocks of south-west Wales, north Wales, Shropshire, the Lake District and the southern uplands of Scotland.

4. SILURIAN
   The limestones of east Wales and Shropshire, also the great thickness of greywackes in west and central Wales, the Lake District and the southern

uplands of Scotland. There is a virtual absence of volcanic rocks in this geological period.

5. DEVONIAN
The sandstones of south-west England and north-east Scotland.
The igneous rocks of the Ochil Hills and Sidlaw Hills, also the Midland Valley, all situated in Scotland and the Cheviots on the border between north-east England and Scotland.

6. CARBONIFEROUS
The widespread limestones of the Pennine areas of Northumberland, Durham, Cumbria, Lancashire, Yorkshire, Derbyshire and Staffordshire and the Mendip Hills of Avon and Somerset, also parts of Gloucestershire, north and south Wales.
The Millstone Grits and the coal measures sandstones of west Yorkshire, east Lancashire and parts of Derbyshire also the Forest of Dean and the Pennant sandstones of south Wales are very important sources of good quality dimension and building stone.
The igneous rocks of the Midland Valley in Scotland including the Clyde Plateau lavas, Stirling Castle rock and Arthur's Seat, Edinburgh. Also the dolorite Whin Sill of north-east England.

7. PERMIAN
The sandstones of Dumfries and Galloway in south-west Scotland also the Penrith area of Cumbria. In south-west England the igneous rocks of the Dartmoor and Bodmin Moor granites. The magnesian limestones of County Durham and the Doncaster to Wetherby area of Yorkshire.

8. TRIASSIC
The Bunter sandstones and conglomerates of Staffordshire and Nottinghamshire.

9. JURASSIC
The important building stones of the Cotswold Hills stretching through Northamptonshire, into Lincolnshire also the Portland stone of Dorset.

10. CRETACEOUS
The chalk of the Chiltern Hills in Buckinghamshire and Bedfordshire, the Warminster Downs of Wiltshire, parts of Sussex, Kent, Essex and Suffolk; also the southern edge of the Yorkshire Wolds to the west of Hull is quarried for cement making and other industrial uses.

11. TERTIARY
Hard rock from this period (also known as the Cenozoic Era) is of little importance from a commercial point of view.

Plant and equipment

The large tonnages of rock consumed by the British market each year combined with the various characteristics of the rocks which occur in Great Britain have led to the development of a major industry with a wide variety of products, making a valuable contribution to the nation's economy and well-being.

The quarry industry sells its products into a very competitive market and, in common with many industries, it has been forced to replace the labour-intensive methods of yesteryear with a smaller but better trained and equipped workforce, the degree of mechanisation at many quarries, both large and small, being quite impressive.

The safety and health of the workforce is constantly being monitored and reviewed, leading to continuing improvements in the working conditions to be found at the majority of quarries. These improved conditions extend to the provision of heated, air-conditioned and sound-suppressed cabs for the operators of loaders, dumpers, trucks, graders and the large self-contained drill rigs. In the Guide we have listed wherever possible the production equipment in use in the quarries.

At certain quarries it can be seen that the nature of the rock allows it to be lifted from the surrounding strata by hand or by mechanical means. At the majority of quarries, however, the rock has to be prepared by blasting which requires the drilling of holes into the strata and into which explosives are placed.

The drillability of a rock depends on the hardness of the minerals included in the rock and the grain size of the minerals. Quartz is one of the common rock forming minerals and, since it is very hard, a rock with a high quartz content is difficult to drill and is also very abrasive, leading to wearing of the surfaces of not only the drilling equipment but also the loaders, trucks, dumpers, rock crushing and classifying equipment, and the tyres of all the wheel-mounted vehicles in use.

Rocks which are by comparison softer and less abrasive, such as chalk and the softer limestones, have a high calcium content and little or no silica. These rocks are easier to drill and are much less wearing on the quarry equipment and machinery. It can be seen that several factors, including these important rock characteristics, need to be considered in addition to the annual output required from a quarry before decisions can be made regarding the plant and equipment to be employed.

Directory of
British quarry companies
and their quarries

BARMAC DUOPACTOR

ROCK-ON-ROCK CRUSHER

9600 and 6900 Single and Dual Feed Models

With Barmac's exclusive high capacity Cascade Feed System

TIDCO CROFT LIMITED

23 BREWMASTER BUILDINGS, CHARLTON TRADING ESTATE,
SHEPTON MALLET, SOMERSET BA4 5QE, UNITED KINGDOM

Tel:	National	(0749) 344321
	International	+44 749 344321
Fax:	National	(0749) 345117
	International	+44 749 345117
Telex:	444132 BARMAC G	

A S QUARRIES LTD
Head Office *43–53 Trafalgar Road, Greenwich, London SE10 9TT*
Phone *081 858 5161* **Fax** *081 293 4103*
Personnel *D W Poultney, Director*
M J Poultney, Director
A W Northam, Director
Quarries *Bowyers, Independent (Admiralty)*

A S QUARRIES LTD

BOWYERS QUARRY

1

Wide Street, Easton, Portland, Dorset DT5 2JP
Phone 0305 823704
OS Map 194 **OS Grid Reference** SY686725
Personnel P Saunders, Unit Manager

PRODUCT DATA	**Rock type**	Jurassic Oolitic Limestone
	Colour	Grey
	Grain	Fine
	Product	Dressed or Lump Stone

DRILLING DATA		Hand-held equipment

A S QUARRIES LTD

INDEPENDENT QUARRY (ADMIRALTY QUARRY)

2

Grove Road, Easton, Portland, Dorset DT5 1DD
Phone 0305 860369 **Fax** 0305 860379
OS Map 194 **OS Grid Reference** SY692718
Personnel P Saunders, Unit Manager

PRODUCT DATA	**Rock type**	Jurassic Oolitic Limestone
	Colour	Greyish white
	Grain	Fine
	Products	Dressed or Lump Stone
		Crushed Stone

DRILLING DATA		Hand-held equipment

ABBEY QUARRIES LTD
Head Office *Abbey Quarry, Quarry Lane, Linby, Hucknall, Notts NG15 8GA*
Phone *0602 630760*
Personnel *T Barfield, Owner*
Quarries *Abbey*

ABBEY QUARRIES LTD

ABBEY QUARRY

3

Quarry Lane, Linby, Hucknall, Notts NG15 8GA
Phone 0602 630760
OS Map 120 **OS Grid Reference** SK536518
Personnel T Barfield, Owner and Unit Manager

entry continues overleaf

PRODUCT DATA	**Rock type**	Permian Magnesian Limestone
	Colour	Dark buff to orangey red
	Grain	Fine to medium
	Product	Dressed or Lump Stone

DRILLING DATA		Bench Height 2.5m
		Hand-held equipment
		Contractor occasionally

| COMPRESSORS | | CompAir |

| LOAD AND HAUL | **Trucks** | None |
| | **Loaders** | Massey Ferguson |

ABER STONE LTD
Head Office Aberstrecht Quarry, Moelfre, Anglesey, Gwynedd LL72 8NN
Phone 024888 644
Personnel T Kellet, Owner
Quarries Aberstrecht

ABER STONE LTD

ABERSTRECHT QUARRY   4

Moelfre, Anglesey, Gwynedd LL72 8NN
Phone 024888 644
OS Map 114   **OS Grid Reference** SH508867
Personnel T Kellet, Owner and Unit Manager

PRODUCT DATA	**Rock type**	Carboniferous Limestone
	Colour	Light grey
	Grain	Fine
	Products	Dressed or Lump Stone
		Crushed Stone

| DRILLING DATA | | Hand-held equipment |

| LOAD AND HAUL | **Loaders** | Caterpillar |

ACRESFORD SAND AND GRAVEL LTD
Head Office High Street, Syston, Leicester, Leics LE7 8GS
Phone 0533 609666
Personnel J Remington, Managing Director
Quarries Huncote

ACRESFORD SAND AND GRAVEL LTD

HUNCOTE QUARRY   5

Forest Wood, Huncote, Leics LE7 8SD
Phone 0533 864871
OS Map 140   **OS Grid Reference** SP503969
Personnel B Cook, Unit Manager

PRODUCT DATA	**Rock type**	Triassic Sandstone
	Colour	Dull red
	Grain	Medium
	Product	Dressed or Lump Stone

DRILLING DATA		Hand-held equipment

AGGETTS LTD
Head Office Knowle Quarry, North Road, Okehampton, Devon EX20 1RQ
Phone 0837 52609
Personnel T Barkwell, Owner
Quarries Knowle

AGGETTS LTD

KNOWLE QUARRY   6

North Road, Okehampton, Devon EX20 1RQ
Phone 0837 52609
OS Map 191 **OS Grid Reference** SX595963
Personnel T Barkwell, Owner and Production Director

PRODUCT DATA	**Rock type**	Carboniferous Sandstone
	Colour	Medium to dark grey
	Grain	Medium
	Products	Dressed or Lump Stone
		Crushed Stone

DRILLING DATA		Hole Diameter 105mm
		Bench Height 12–15m
		Contractor Devon Rock Services

SECONDARY BREAKING		Dropball on Ruston-Bucyrus 22-RB

LOAD AND HAUL	**Trucks**	Aveling Barford RD40
	Loaders	Caterpillar 966C

D G AITKENS
Head Office Fornham Park, Fornham St Martin, Bury St Edmunds, Suffolk IP28 6TT
Phone 028484 432
Personnel D G Aitken, Owner
Quarries Ingham

D G AITKENS

INGHAM QUARRY   7

The Folly, Ingham, Bury St Edmunds, Suffolk IP28 6UX
Phone 028484 432
OS Map 155 **OS Grid Reference** TL856717
Personnel D G Aitken, Owner and Unit Manager

PRODUCT DATA	**Rock type**	Jurassic Oolitic Limestone
	Colour	White
	Grain	Fine
	Product	Crushed Stone

RIPPABLE STRATA

ANTRON HILL GRANITE LTD
Head Office Meadowbank, Lestraynes Road, Rame, Penryn, Cornwall TR10 9EL
Phone 0209 860672
Personnel B Brook, Managing Director
Quarries Antron Hill, Burnthouse

ANTRON HILL GRANITE LTD
ANTRON HILL QUARRY 8
Meadow Bank, Lestraynes Lane, Rame, Penryn, Cornwall TR10 9EL
Phone 0209 860672
OS Map 204 **OS Grid Reference** SW760335

QUARRY ON STAND

ROCK DATA		
	Rock type	Granite
	Colour	Silver grey
	Grain	Fine

ANTRON HILL GRANITE LTD
BURNTHOUSE QUARRY 9
Mabe, Penryn, Cornwall TR10 9HH
Phone 0209 860672
OS Map 204 **OS Grid Reference** SW768341
Personnel B Brook, Managing Director and Unit Manager

PRODUCT DATA		
	Rock type	Granite
	Colour	Silver grey
	Grain	Fine
	Product	Dressed or Lump Stone

DRILLING DATA	
	Hand-held equipment

ARC LTD (HANSON plc)
Corporate Head Office The Ridge, Chipping Sodbury, Bristol, Avon BS17 6AY
Phone 0454 316000 **Fax** 0454 325161 **Telex** 449353
Personnel A R Cotton, Executive Chairman
I Menzies-Gow, Chief Executive
P T Turner, Director
A J H Dougal, Director
Companies ARC Central Ltd, ARC Northern Ltd, ARC Powell Duffryn Ltd,
ARC Southern Ltd, ARC South West Ltd, Bath and Portland Stone Ltd

---

ARC CENTRAL LTD (ARC LTD)
Head Office Ashby Road East, Shepshed, Loughborough, Leics LE12 9BU
Phone 0509 503161 **Fax** 0509 504120
Personnel G Fisk, Managing Director
P Wood, Quarrying Director
C James, Commercial Manager
D Steeples, Production Manager
M Hardy, Production Manager
Quarries Blodwell, Charnwood, Clee Hill, Criggion, Edwin Richards, Groby, Judkins, Whitwick

ARC CENTRAL LTD (ARC LTD)

BLODWELL QUARRY

10

Llynclys, Nr Oswestry, Shrops SY10 8LY
Phone 0691 81343/81430
OS Map 126   **OS Grid Reference** SJ258235
Personnel I P Cross, Unit Manager

PRODUCT DATA	**Rock type**	Basalt
	Colour	Medium grey
	Grain	Fine
	Products	Crushed Stone
		Asphalt or Tarmacadam
DRILLING DATA		Hole Diameter 105mm
		Bench Height 12m
	Drills	Holman Voltrak
COMPRESSORS		CompAir 600
SECONDARY BREAKING		Dropball on Ruston-Bucyrus 22-RB
LOAD AND HAUL	**Trucks**	Terex
	Loaders	Caterpillar 966C
		Caterpillar 980C
CRUSHING PLANT	**Primary**	Hadfield 42 × 36 Jaw (Double Toggle)
	Secondary	Kennedy 14 Cone
	Tertiary	(7)Kennedy Cone
		Boliden Allis Cone

ARC CENTRAL LTD (ARC LTD)

CHARNWOOD QUARRY

11

Ashby Road East, Shepshed, Nr Loughborough, Leics LE12 9BU
Phone 0509 503161   **Telex** 34365
OS Map 129   **OS Grid Reference** SK486182
Personnel A S L Jones, Unit Manager

PRODUCT DATA	**Rock type**	Diorite
	Colour	Speckled black and white
	Grain	Coarse
	Products	Crushed Stone
		Asphalt or Tarmacadam
		Ready-mixed Concrete
DRILLING DATA		Hole Diameter 105mm
		Bench Height 10–12m
	Drills	Halco 410C
		Halco 450H
COMPRESSORS		Atlas Copco XRH350
SECONDARY BREAKING		Hydraulic Hammer
		Dropball on Ruston-Bucyrus 22-RB

entry continues overleaf

LOAD AND HAUL	**Trucks**	(2)Terex R35S
		Aveling Barford RD40
	Loaders	Poclain 300
		Caterpillar 980C
		Caterpillar 988B
		Volvo BM4600
		Terex 72-31B
CRUSHING PLANT	**Primary**	Kue Ken 60 × 48 Jaw (Double Toggle)
	Secondary	Nordberg 5.5' Cone
	Tertiary	Nordberg 4.25' Cone
	Quaternary	Nordberg 2' Cone
		Parker 900 Cone
		Parker 1200 Cone

ARC CENTRAL LTD (ARC LTD)

CLEE HILL QUARRY 12

Clee Hill, Ludlow, Shrops SY8 3QA
Phone 0584 890516
OS Map 126   **OS Grid Reference** SO598762
Personnel Vacant, Unit Manager

PRODUCT DATA	**Rock type**	Basalt
	Colour	Bluish black
	Grain	Fine
	Products	Crushed Stone
		Asphalt or Tarmacadam
DRILLING DATA		Hole Diameter 105mm
		Bench Height 8–20m
	Drills	Halco 410C
SECONDARY BREAKING		Dropball on Ruston-Bucyrus 22-RB
LOAD AND HAUL	**Trucks**	Terex
	Loaders	Poclain 220
CRUSHING PLANT	**Primary**	Baxter DT50 × 36 Jaw (Double Toggle)
	Secondary	Kue Ken 36 × 24 Jaw (Double Toggle)
	Tertiary	Symons 3' Cone
		(3)Symons 2' Cone

ARC CENTRAL LTD (ARC LTD)

CRIGGION QUARRY 13

Criggion, Shrewsbury, Shrops SY5 9BA
Phone 0938 74215   **Telex** 35162
OS Map 126   **OS Grid Reference** SJ290145
Personnel G Brown, Unit Manager

PRODUCT DATA	**Rock type**	Dolerite
	Colour	Sea green
	Grain	Medium
	Products	Crushed Stone
		Asphalt or Tarmacadam

DRILLING DATA		Hole Diameter 110mm
		Bench Height 12m
	Drills	Halco 450H

SECONDARY BREAKING		Dropball on Ruston-Bucyrus 22-RB

LOAD AND HAUL	**Trucks**	Terex
	Loaders	Caterpillar 966C
		Caterpillar 980C

CRUSHING PLANT	**Primary**	Kue Ken 48 × 42 Jaw (Double Toggle)
	Secondary	Babbitless 16″ Gyratory
	Tertiary	Allis Chalmers 3′ Cone
		(2)Babbitless 704 Cone
		Babbitless 703 Cone

ARC CENTRAL LTD (ARC LTD)

EDWIN RICHARDS QUARRY 14

Portway Road, Rowley Regis, Warley, W Midlands B65 9DN
Phone 0215 599881
OS Map 139 **OS Grid Reference** SO969883
Personnel M Hardy, Unit Manager

PRODUCT DATA	**Rock type**	Rowley Rag Basalt
	Colour	Very dark grey
	Grain	Fine
	Products	Crushed Stone
		Asphalt or Tarmacadam

DRILLING DATA		Hole Diameter 105mm
		Bench Height 30m
	Drills	Halco 410C
		Holman Voltrak

COMPRESSORS		CompAir RO60–170S

SECONDARY BREAKING		Dropball on Ruston-Bucyrus 22-RB

LOAD AND HAUL	**Face shovel**	O&K RH40
	Trucks	(2)Aveling Barford RD55
	Loaders	Caterpillar

CRUSHING PLANT	**Primary**	Hazemag SAP5 Broad Impactor
	Secondary	Hazemag APK1313 Impactor
	Tertiary	Allis Chalmers 345 34″ Cone

ARC CENTRAL LTD (ARC LTD)

GROBY QUARRY 15

Newtown Linford Lane, Groby, Leics LE6 0HF
Phone 0533 876161 **Telex** 341040
OS Map 140 **OS Grid Reference** SK523083
Personnel M Manning, Unit Manager

entry continues overleaf

PRODUCT DATA	**Rock type**	Granophyric Diorite
	Colour	Speckled black and white
	Grain	Coarse
	Products	Crushed Stone
		Asphalt or Tarmacadam
		Ready-mixed Concrete
DRILLING DATA		Hole Diameter 105mm
		Bench Height 12m
		Contractor GTS Drilling
CRUSHING PLANT	**Primary**	Pegson 46 × 36 Jaw (Mobile)
	Secondary	(2)Pegson 1200MF Autocone (Mobile)
	Tertiary	Barmac 9600 MkIII Duopactor (Mobile)

ARC CENTRAL LTD (ARC LTD)

JUDKINS QUARRY 16

Tuttle Hill, Nuneaton, Warwicks CV10 4JQ
Phone 0203 348231
OS Map 140 **OS Grid Reference** SP352926
Personnel H P Gibbs, Unit Manager

PRODUCT DATA	**Rock type**	Cambrian Hartshill Quartzitic Sandstone
	Colour	Bluish grey
	Grain	Coarse
	Products	Crushed Stone
		Asphalt or Tarmacadam
		Pre-cast Concrete Products
DRILLING DATA		Hole Diameter 127–130mm
		Bench Height 12m
		Contractor Blastrite
	Drills	Halco 410C
SECONDARY BREAKING		Dropball on Ruston-Bucyrus 22-RB
LOAD AND HAUL	**Trucks**	Aveling Barford RD40
	Loaders	Caterpillar
		Komatsu
CRUSHING PLANT	**Primary**	Pegson 36 × 6 Jaw
	Secondary	Nordberg 4′ Cone (Standard)
		Nordberg 4′ Cone (Shorthead)
	Tertiary	Nordberg 3′ Gyratory
		Allis Chalmers 2′ Cone

ARC CENTRAL LTD (ARC LTD)

WHITWICK QUARRY 17

Leicester Road, Whitwick, Coalville, Leics LE6 3GR
Phone 0530 36671 **Telex** 341040
OS Map 129 **OS Grid Reference** SK449155
Personnel K T Chambers, Unit Manager

PRODUCT DATA	**Rock type**	Andesite
	Colour	Grey
	Grain	Fine
	Products	Crushed Stone
		Asphalt or Tarmacadam
		Pre-cast Concrete Products

DRILLING DATA		Hole Diameter 105mm
		Bench Height 18m
	Drills	Halco 450H
		Holman Voltrak

COMPRESSORS	Atlas Copco XRH350
	Ingersoll-Rand DXL700H

SECONDARY BREAKING	Dropball on Ruston-Bucyrus 22-RB

LOAD AND HAUL	**Face shovel**	Poclain 380
	Trucks	(2)Aveling Barford RD55
	Loaders	Volvo BM 4500
		Volvo BM 4600

CRUSHING PLANT	**Primary**	Pegson 60 × 48 Jaw (Double Toggle)
	Secondary	Nordberg 5.5′ Cone
	Tertiary	Pegson 1200 Autocone
	Quaternary	(3)Pegson 900 Autocone
		Barmac Duopactor

ARC NORTHERN LTD (ARC LTD)

Head Office *Clifford House, Wetherby Business Park, York Road, Wetherby, W Yorks LS22 4NS*
Phone *0937 61977* **Fax** *0937 61610*
Personnel *J Hopkins, Managing Director*
J Brooks, Operations Director
N Vaughan, Trading Manager
J Adams, Production Manager
M A C Gale, Production Manager
K Heywood, Production Manager
Quarries *Auchinleck, Caer Glaw, Dumfries, East (Appley Bridge), Goldmire, Houghton, Ingleton, Kelhead,*
Llanddulas, Montcliffe, Parkhead, Penmaenmawr, Round 'O', Shap Beck, Silverdale, Stranraer,
Swinburne, Wigtown

ARC NORTHERN LTD (ARC LTD)

AUCHINLECK QUARRY   18

Auchinleck Station Yard, Cumnock, Strathclyde KA18 2R0
Phone 0937 61977
OS Map 71   **OS Grid Reference** NS555220

QUARRY ON STAND

ROCK DATA	**Rock type**	Carboniferous Limestone
	Grain	Medium

ARC NORTHERN LTD (ARC LTD)

CAER GLAW QUARRY   **19**

Gwalchmai, Sandrygran, Llanerchymedd, Anglesey, Gwynedd ML65 4PW
Phone 0407 720292
OS Map 114 **OS Grid Reference** SH418841
Personnel E Evans, Unit Manager

PRODUCT DATA	**Rock type**	Granite
	Colour	White, light grey
	Grain	Coarse
	Product	Crushed Stone

DRILLING DATA		Hole Diameter 105mm
		Bench Height 12m
		Contractor K&H Rockdrillers

| SECONDARY BREAKING | | Dropball on Priestman |

LOAD AND HAUL	**Trucks**	Aveling Barford
		Terex
	Loaders	Poclain

CRUSHING PLANT	**Primary**	Kue Ken 95 36×24 Jaw (Double Toggle)
	Secondary	Allis Chalmers 436 Cone
		Kue Ken 70 10×24 Jaw
	Tertiary	Allis Chalmers 22″ Hydrocone

ARC NORTHERN LTD (ARC LTD)

DUMFRIES QUARRY   **20**

Dargavel, Lockerbie Road, Dumfries, Dumfries & Galloway, DG1 3PG
Phone 0937 61977
OS Map 84 **OS Grid Reference** NX979760

QUARRY ON STAND

| ROCK DATA | **Rock type** | Carboniferous Limestone |

ARC NORTHERN LTD (ARC LTD)

EAST QUARRY (APPLEY BRIDGE QUARRY)   **21**

Appley Lane North, Appley Bridge, Wigan, Lancs WN6 9AF
Phone 0937 61977
OS Map 108 **OS Grid Reference** SD525098

QUARRY ON STAND

ROCK DATA	**Rock type**	Carboniferous Sandstone
	Colour	Greenish grey
	Grain	Medium

ARC NORTHERN LTD (ARC LTD)

GOLDMIRE QUARRY

22

Thwaites Flatt, Dalton-in-Furness, Cumbria LA14 4QG
Phone 0229 65502
OS Map 90 **OS Grid Reference** SD220749
Personnel S Train, Unit Manager

PRODUCT DATA	**Rock type**	Carboniferous Limestone
	Colour	Dark grey
	Grain	Medium
	Products	Crushed Stone
		Asphalt or Tarmacadam
		Ready-mixed Concrete
		Pre-cast Concrete Products
DRILLING DATA		Hole Diameter 95mm
		Bench Height 30m
		Contractor Peak Drilling
	Drills	Halco 400
SECONDARY BREAKING		Dropball on Ruston-Bucyrus 22-RB
LOAD AND HAUL	**Trucks**	Terex 3305B
		Foden
	Loaders	Aveling Barford 35B
		Caterpillar 966C
CRUSHING PLANT	**Primary**	Kue Ken 120S 48 × 36 Jaw
	Secondary	BJD Impactor (reversible)
	Tertiary	BJD 2024 Pulverizer

ARC NORTHERN LTD (ARC LTD)

HOUGHTON QUARRY

23

Newbottle Street, Houghton-le-Spring, Tyne & Wear DH4 4AU
Phone 0783 842205/841831/842845
OS Map 88 **OS Grid Reference** NZ339506
Personnel J Whitfield, Unit Manager

PRODUCT DATA	**Rock type**	Permian Dolomitic Limestone
	Colour	Cream to grey
	Grain	Coarse
	Products	Crushed Stone
		Ready-mixed Concrete
		Agricultural Lime
DRILLING DATA		Hole Diameter 110mm
		Bench Height 15m
	Drills	Halco 410C
COMPRESSORS		CompAir RO60HP
SECONDARY BREAKING		Dropball
LOAD AND HAUL	**Trucks**	(3)Aveling Barford RD40
	Loaders	(2)Caterpillar 966C
CRUSHING PLANT	**Primary**	Pegson 34 × 42 Jaw
	Secondary	Parker Kubitizer

ARC NORTHERN LTD (ARC LTD)

INGLETON QUARRY

24

Ingleton, Via Carnforth, Lancs LA6 3AW
Phone 0468 41264
OS Map 98 **OS Grid Reference** SD707739
Personnel C H Lis, Unit Manager

PRODUCT DATA	**Rock type**	Carboniferous Sandstone (Gritstone)
	Colour	Greenish grey
	Grain	Medium
	Product	Crushed Stone
DRILLING DATA		Hole Diameter 105–155mm
		Bench Height 15–30m
	Drills	Halco 425HP
		Halco 620C
SECONDARY BREAKING		Dropball on Ruston-Bucyrus 22-RB
LOAD AND HAUL	**Trucks**	Aveling Barford RD35
		Terex
	Loaders	Volvo 4500
		Michigan 275C
CRUSHING PLANT	**Primary**	Pegson 48 × 42 Jaw (Single Toggle)
	Secondary	Nordberg 4.25′ Cone
	Tertiary	Allis Chalmers 436 Cone
	Quaternary	Pegson 900 Autocone
		Barmac Duopactor

ARC NORTHERN LTD (ARC LTD)

KELHEAD QUARRY

25

Kinmount, Annan, Dumfries & Galloway, DG12 5RH
Phone 046 17432
OS Map 85 **OS Grid Reference** NY145696
Personnel R Potts, Unit Manager

PRODUCT DATA	**Rock type**	Pre-Cambrian Dalradian Limestone
	Colour	Not commercially significant
	Product	Crushed Stone
DRILLING DATA		Hole Diameter 105mm
		Bench Height 12m
		Contractor R & J Blasting
SECONDARY BREAKING		Dropball
LOAD AND HAUL	**Trucks**	Terex
	Loaders	Terex
CRUSHING PLANT	**Primary**	Goodwin Barsby Jaw
	Secondary	Hazemag APK60
	Tertiary	None

ARC NORTHERN LTD (ARC LTD)

LLANDDULAS QUARRY

26

Llanddulas, Abergele, Clwyd LL22 8HP
Phone 0492 518202
OS Map 116   **OS Grid Reference** SH900789
Personnel M Ripley, Unit Manager

PRODUCT DATA	**Rock type**	Carboniferous Limestone
	Colour	White, grey
	Grain	Medium
	Product	Crushed Stone
DRILLING DATA		Hole Diameter 105mm
		Bench Height 10–20m
		Contractor A Jones Rock Drillers
SECONDARY BREAKING		Dropball on Ruston-Bucyrus 22-RB
LOAD AND HAUL	**Face shovel**	Caterpillar 269
	Trucks	(3)Aveling Barford RD50
	Loaders	Caterpillar 980C
CRUSHING PLANT	**Primary**	Jaw
	Secondary	Cone
	Tertiary	Impactor

NO FURTHER INFORMATION PROVIDED

ARC NORTHERN LTD (ARC LTD)

MONTCLIFFE QUARRY

27

Georges Lane, Horwich, Lancs BL6 6RS
Phone 0204 696171
OS Map 109   **OS Grid Reference** SD656123
Personnel S Lamond, Unit Manager

PRODUCT DATA	**Rock type**	Carboniferous Sandstone (Gritstone)
	Colour	Not commercially significant
	Grain	Medium
	Product	Crushed Stone
DRILLING DATA		Hole Diameter 105mm
		Bench Height 12m
		Contractor Peak Drilling
LOAD AND HAUL	**Trucks**	Terex 3307
	Loaders	Poclain
CRUSHING PLANT	**Primary**	Kue Ken 42×36 Jaw
		(2)Goodwin Barsby Goliath 42×24 Jaw
	Secondary	Pegson 16B 4' Cone
	Tertiary	Hazemag APK1013 Impactor

ARC NORTHERN LTD (ARC LTD)

PARKHEAD QUARRY 28

Caldbeck, Cumbria CA5 7HL
Phone 06998 324/383
OS Map 90 **OS Grid Reference** NY341408
Personnel R Hardy, Unit Manager

PRODUCT DATA		
	Rock type	Carboniferous Limestone
	Colour	Dark grey
	Products	Dressed or Lump Stone
		Crushed Stone

RIPPABLE STRATA

ARC NORTHERN LTD (ARC LTD)

PENMAENMAWR QUARRY 29

Bangor Road, Penmaenmawr, Gwynedd LL34 5NA
Phone 0492 622256
OS Map 115 **OS Grid Reference** SH710755
Personnel B Waldron, Unit Manager

PRODUCT DATA		
	Rock type	Microdiorite
	Colour	Grey to dark grey
	Grain	Medium
	Products	Crushed Stone
		Asphalt or Tarmacadam
		Ready-mixed Concrete

DRILLING DATA	
	Hole Diameter 110mm
	Bench Height 18m
	Contractor A Jones Rock Drillers

SECONDARY BREAKING	
	Dropball on Ruston-Bucyrus 22-RB

LOAD AND HAUL		
	Face shovel	Poclain
	Trucks	(2)Aveling Barford RD55
	Loaders	Caterpillar 980C

CRUSHING PLANT		
	Primary	Allis Chalmers 36×25 Gyratory
		Allis Chalmers 15/12-60 Hydrocone
		Allis Chalmers 236 Cone

ARC NORTHERN LTD (ARC LTD)

ROUND 'O' QUARRY 30

Cobbs Brow Lane, Lathom, Ormskirk, Lancs L40 6JJ
Phone 0257 63569
OS Map 108 **OS Grid Reference** SD488078
Personnel T Aspinal, Unit Manager

PRODUCT DATA		
	Rock type	Carboniferous Sandstone (Gritstone)
	Colour	Dark grey
	Grain	Medium
	Product	Crushed Stone

DRILLING DATA		Hole Diameter 105mm
		Bench Height 13m
	Drills	Ingersoll-Rand CM351

COMPRESSORS		Ingersoll-Rand DXL700H

SECONDARY BREAKING		(2)Dropball on Ruston-Bucyrus 22-RB

LOAD AND HAUL	**Trucks**	Volvo
	Loaders	Caterpillar
		Volvo

CRUSHING PLANT	**Primary**	Kue Ken 20 × 42 Jaw (semi-Mobile)
	Secondary	Pegson 4′ Gyrasphere
	Tertiary	None

ARC NORTHERN LTD (ARC LTD)

SHAP QUARRY (SHAP BECK QUARRY)   **31**

Shap, Penrith, Cumbria CA10 3NX
Phone 09316 241
OS Map 90   **OS Grid Reference** NY557084
Personnel B W Buckman, Unit Manager

PRODUCT DATA	**Rock type**	Granite
	Colour	Pinkish grey
	Grain	Coarse
	Products	Crushed Stone
		Asphalt or Tarmacadam

DRILLING DATA		Hole Diameter 105mm
		Bench Height 25m
		Contractor Rocklift

SECONDARY BREAKING		Dropball on Ruston-Bucyrus 22-RB

LOAD AND HAUL	**Trucks**	Terex
	Loaders	Komatsu WA250
		Komatsu WA300

CRUSHING PLANT	**Primary**	Kue Ken 160 Jaw (Double Toggle)
	Secondary	Hazemag AP4 Hammermill
	Tertiary	Nordberg 2′ Cone

ARC NORTHERN LTD (ARC LTD)

SILVERDALE QUARRY   **32**

Silverdale, Nr Carnforth, Lancs LA5 0UH
Phone 0524 701375/701155/701634
OS Map 97   **OS Grid Reference** SD469770
Personnel T Atkinson, Unit Manager

PRODUCT DATA	**Rock type**	Carboniferous Great Scar Limestone
	Colour	Medium grey
	Products	Crushed Stone
		Asphalt or Tarmacadam

entry continues overleaf

DRILLING DATA		Hole Diameter 105mm Bench Height 20m Contractor Peak Drilling
SECONDARY BREAKING		Hydraulic Hammer Dropball
LOAD AND HAUL	**Face shovel** **Trucks** **Loaders**	Poclain 220 Terex Caterpillar 966C Caterpillar 988B
CRUSHING PLANT	**Primary** **Secondary**	Lokomotrac 125B Jaw (Mobile) Mansfield No 4 Hammermill Mansfield No 5 Hammermill

ARC NORTHERN LTD (ARC LTD)

STRANRAER QUARRY

33

Cults Store, Castle Kennedy, Stranraer, Dumfries & Galloway, DG9 8SH
Phone 0937 61977
OS Map 82   **OS Grid Reference** NX060608

QUARRY ON STAND

ROCK DATA	**Rock type**	Carboniferous Limestone

ARC NORTHERN LTD (ARC LTD)

SWINBURNE QUARRY

34

Colwell, Hexham, Northumb NE38 3DN
Phone 043 481616/481617/481618
OS Map 87   **OS Grid Reference** NY942762
Personnel J Austin, Unit Manager

PRODUCT DATA	**Rock type** **Colour** **Grain** **Products**	Quartzitic Dolerite Very dark grey Coarse Crushed Stone Asphalt or Tarmacadam Pre-cast Concrete Products
DRILLING DATA	 **Drills**	Hole Diameter 105-125mm Bench Height 13-17m Contractor A Jones Rock Drillers Halco 410C
SECONDARY BREAKING		Hydraulic Hammer Dropball on Ruston-Bucyrus 22-RB
LOAD AND HAUL	**Face shovel** **Trucks** **Loaders**	Ruston-Bucyrus Terex R25T Terex 33-05B Caterpillar 950E Volvo 44-100

CRUSHING PLANT	**Primary**	Kue Ken 1205 42 × 36 Jaw (Double Toggle)
	Secondary	Allis Chalmers 1336 Gyratory
	Tertiary	Symons 3′ Cone (Standard)
		Symons 3′ Cone (Shorthead)
	Quaternary	Breakring 200 Impactor

ARC NORTHERN LTD (ARC LTD)

WIGTOWN QUARRY 35

Baldon Store, Wigtown, Dumfries & Galloway, DG8 9AF
Phone 0937 61977
OS Map 83   **OS Grid Reference** NX435554

QUARRY ON STAND

| ROCK DATA | **Rock type** | Silurian Limestone |

ARC POWELL DUFFRYN LTD (ARC LTD)
Head Office *Canal Road, Cwmbach, Aberdare, Mid Glam CF44 0AG*
Phone *0685 884444* **Fax** *0685 884444 Ext 227*
Personnel *F M Edwards, Managing Director*
M Hobbs, Commercial Manager
N Eastwood, Production Manager
Quarries *Builth Wells, Craig-y-Hesg, Cwmleyshon, Ifton, Lithalun, Livox, Machen, Penderyn, Penhow, Risca, Vaynor*

ARC POWELL DUFFRYN LTD (ARC LTD)

BUILTH WELLS QUARRY 36

Builth Wells, Powys LB2 3UB
Phone 0982 553608
OS Map 148   **OS Grid Reference** SO048519
Personnel L Tomlinson, Unit Manager

PRODUCT DATA	**Rock type**	Dolerite
	Colour	Light greenish grey
	Grain	Coarse
	Products	Crushed Stone
		Asphalt or Tarmacadam
DRILLING DATA		Hole Diameter 105mm
		Bench Height 20–23m
	Drills	Halco 410C
COMPRESSORS		CompAir 650HE
SECONDARY BREAKING		Dropball on Ruston-Bucyrus 30-RB
LOAD AND HAUL	**Face shovel**	Caterpillar 235
	Trucks	Terex 3307
CRUSHING PLANT	**Primary**	Kue Ken 42 × 36 Jaw
		Kue Ken 90 (Mobile)
	Secondary	Sheepbridge Kennedy 14
	Tertiary	Nordberg 3′ Cone
		(2)Nordberg 2′ Cone

ARC POWELL DUFFRYN LTD (ARC LTD)

CRAIG-YR-HESG QUARRY 37

Berw Road, Pontypridd, Mid Glam CF37 3EG
Phone 0443 403078/404984/405750
OS Map 170 **OS Grid Reference** ST079915
Personnel A Brown, Unit Manager

PRODUCT DATA	**Rock type**	Carboniferous Pennant Sandstone
	Colour	Light to medium grey
	Grain	Medium
	Products	Crushed Stone
		Asphalt or Tarmacadam
		Ready-mixed Concrete
		Pre-cast Concrete Products

DRILLING DATA		Hole Diameter 105mm
		Bench Height 12–15m
	Drills	Halco 410C

| COMPRESSORS | | CompAir 700HE |

| SECONDARY BREAKING | | Dropball on Ruston-Bucyrus 22-RB |

LOAD AND HAUL	**Trucks**	Terex R24
		Terex R25
	Loaders	Caterpillar 980C
		Volvo BM L120

| CRUSHING PLANT | **Primary** | Kue Ken Jaw (Double Toggle) |

NO FURTHER INFORMATION PROVIDED

ARC POWELL DUFFRYN LTD (ARC LTD)

CWMLEYSHON QUARRY 38

Rudry, Caerphilly, Mid Glam CF8 3EB
Phone 0685 884444
OS Map 171 **OS Grid Reference** ST212868

QUARRY ON STAND

ROCK DATA	**Rock type**	Carboniferous Dolomitic Limestone
	Colour	Light grey
	Grain	Coarse

ARC POWELL DUFFRYN LTD (ARC LTD)

IFTON QUARRY 39

Rogiet Caldicot, Gwent NP6 4LR
Phone 0685 884444
OS Map 171 **OS Grid Reference** ST346188

QUARRY ON STAND

ROCK DATA	**Rock type**	Carboniferous Dolomitic Limestone
	Colour	Light grey
	Grain	Coarse

ARC POWELL DUFFRYN LTD (ARC LTD)

LITHALUN QUARRY

40

Ewenny, Nr Bridgend, Mid Glam CF35 5AN
Phone 0656 653786
OS Map 170   **OS Grid Reference** SS899765
Personnel A Kenwood, Unit Manager

PRODUCT DATA	**Rock type**	Carboniferous Limestone
	Colour	Brownish grey
	Product	Crushed Stone
DRILLING DATA		Hole Diameter 105mm
		Bench Height 12m
		Contractor Rees Blasting
SECONDARY BREAKING		Dropball on Ransomes & Rapier NCK-Rapier 305B
LOAD AND HAUL	**Trucks**	Aveling Barford RD40
	Loaders	Caterpillar 966C
CRUSHING PLANT	**Primary**	Goodwin Barsby 48×36 Jaw
	Secondary	Hazemag APK50 Impactor
	Tertiary	Lanway No 3
		Lanway No 1 Hammermill

ARC POWELL DUFFRYN LTD (ARC LTD)

LIVOX QUARRY

41

St Arvans Road, Tintern, Chepstow, Gwent MP6 6HD
Phone 02912 3603/3492
OS Map 162   **OS Grid Reference** ST542975
Personnel M Deakin, Unit Manager

PRODUCT DATA	**Rock type**	Carboniferous Dolomitic Limestone
	Colour	Light grey
	Grain	Medium
	Products	Crushed Stone
		Asphalt or Tarmacadam
		Agricultural Use
DRILLING DATA		Hole Diameter 105mm
		Bench Height 15m
	Drills	Holman Voltrak
COMPRESSORS		Ingersoll-Rand DXL600P
SECONDARY BREAKING		Dropball on Ruston-Bucyrus 22-RB
LOAD AND HAUL	**Trucks**	Aveling Barford RD40
	Loaders	Caterpillar 966C
CRUSHING PLANT	**Primary**	Pegson 48×44 Jaw
	Secondary	Hazemag APK60 Impactor
		(2)Hazemag APKM0805 Impactor

ARC POWELL DUFFRYN LTD (ARC LTD)

MACHEN QUARRY

42

Machen, Newport, Gwent NP1 8YP
Phone 0633 441111/441144
OS Map 171 **OS Grid Reference** ST220890
Personnel G Rhydderch, Unit Manager

PRODUCT DATA	**Rock type**	Carboniferous Dolomitic Limestone
	Colour	Medium brownish grey
	Grain	Coarse
	Products	Crushed Stone
		Asphalt or Tarmacadam
		Ready-mixed Concrete
		Agricultural Lime
DRILLING DATA		Hole Diameter 127mm
		Bench Height 25m
	Drills	Halco MPD60–25T
SECONDARY BREAKING		Dropball on Ruston-Bucyrus 30-RB
LOAD AND HAUL	**Trucks**	Aveling Barford
		(3)Terex 2366/64
	Loaders	Terex 72–31B
		Terex 72–51
		Caterpillar 988B
CRUSHING PLANT	**Primary**	Goodwin Barsby 42×30 Jaw
	Secondary	Hazemag APK50 Impactor
	Tertiary	Lanway No 1 Hammermill
		Lanway No 3 Hammermill

ARC POWELL DUFFRYN LTD (ARC LTD)

PENDERYN QUARRY

43

Penderyn, Aberdare, Mid Glam CF44 0TX
Phone 0685 811431
OS Map 160 **OS Grid Reference** SN949088
Personnel R Thomas, Unit Manager

PRODUCT DATA	**Rock type**	Carboniferous Limestone
	Colour	Light brownish grey
	Grain	Medium
	Products	Crushed Stone
		Ready-mixed Concrete
DRILLING DATA		Hole Diameter 105mm
		Bench Height 15m
	Drills	Holman Voltrak
COMPRESSORS		CompAir 650HE
SECONDARY BREAKING		Dropball on Ruston-Bucyrus 22-RB
LOAD AND HAUL	**Trucks**	Terex
	Loaders	Caterpillar 980C
		Volvo BM L120

CRUSHING PLANT	**Primary**	Hazemag SAP5 Impactor
	Secondary	Hazemag APK50 Impactor
	Tertiary	Hazemag APKM0805 Impactor
	Quaternary	BJD Hammermill
		Hazemag V10/05

ARC POWELL DUFFRYN LTD (ARC LTD)

PENHOW QUARRY 44

Penhow, Newport, Gwent NP6 3YD
Phone 0633 400211
OS Map 171   **OS Grid Reference** ST423912
Personnel D McIntosh, Unit Manager

PRODUCT DATA	**Rock type**	Carboniferous Limestone
	Colour	Light grey
	Grain	Medium
	Products	Dressed or Lump Stone
		Crushed Stone
		Industrial Limestone
DRILLING DATA		Hole Diameter 105mm
		Bench Height 17m
	Drills	Halco 410C
COMPRESSORS		CompAir 650HE
SECONDARY BREAKING		Dropball on Ruston-Bucyrus 22-RB
LOAD AND HAUL	**Trucks**	Aveling Barford RD40
		(2)Volvo
	Loaders	(2)Terex
CRUSHING PLANT	**Primary**	Kue Ken 150 48×42 Jaw
	Secondary	Hazemag APK60 Impactor
	Tertiary	Hazemag APKM0805 Impactor

ARC POWELL DUFFRYN LTD (ARC LTD)

RISCA QUARRY 45

Risca, Gwent NP1 6HD
Phone 0685 884444
OS Map 171   **OS Grid Reference** ST237913

QUARRY ON STAND	— crushing and coating plant only
ROCK DATA	**Rock type**	Carboniferous Dolomitic Limestone
	Colour	Medium brownish grey
	Grain	Medium to coarse

ARC POWELL DUFFRYN LTD (ARC LTD)

VAYNOR QUARRY 46

Cefn Coed, Merthyr Tydfil, Mid Glam CF48 2LA
Phone 0685 723355
OS Map 160   **OS Grid Reference** SO050093
Personnel D Clifford, Unit Manager

entry continues overleaf

PRODUCT DATA	**Rock type**	Carboniferous Limestone
	Colour	Light grey
	Grain	Medium
	Products	Crushed Stone
		Asphalt or Tarmacadam
		Ready-mixed Concrete
		Agricultural Lime
DRILLING DATA		Hole Diameter 110mm
		Bench Height 15m
	Drills	Holman Voltrak
COMPRESSORS		CompAir 650H
SECONDARY BREAKING		Dropball on Ruston-Bucyrus 22-RB
LOAD AND HAUL	**Trucks**	Terex
	Loaders	Caterpillar 980C
		Caterpillar 988B
		Volvo BM L120
CRUSHING PLANT	**Primary**	Kue Ken 150 Jaw (Double Toggle)
	Secondary	Mansfield No 5 Hammermill
	Tertiary	Hazemag APK40

ARC SOUTHERN LTD (ARC LTD)

Head Office Stoneleigh House, Frome, Somerset BA11 2HB
Phone 0373 63211 **Fax** 0373 65843
Personnel J Draper, Managing Director
C Edwards, Quarrying Director
D Rogers, Commercial Manager
T Jones, Production Manager
M Lowthian, Production Manager
T Reed, Production Manager
Quarries Allington, Batts Combe, Chipping Sodbury, Cromhall, Daglingworth, Drybrook, Guiting, Offham, Tytherington, Whatley, White Hill

ARC SOUTHERN LTD (ARC LTD)

ALLINGTON QUARRY   **47**

London Road, Allington, Maidstone, Kent ME16 0LT
Phone 0622 679461
OS Map 188   **OS Grid Reference** TQ746576
Personnel D Brown, Unit Manger

PRODUCT DATA	**Rock type**	Lower Cretaceous Hythe Beds Greensand (Ragstone)
	Colour	Light greenish grey
	Grain	Fine
	Products	Crushed Stone
		Asphalt or Tarmacadam
		Ready-mixed Concrete

RIPPABLE STRATA

ARC SOUTHERN LTD (ARC LTD)

BATTS COMBE QUARRY

48

Cheddar, Somerset BS27 3LR
Phone 0934 742733 **Telex** 44391
OS Map 182 **OS Grid Reference** ST459546
Personnel N Lewis, Unit Manager

PRODUCT DATA	**Rock type**	Carboniferous Limestone
	Colour	Light grey
	Products	Crushed Stone
		Asphalt or Tarmacadam
		Industrial Limestone
		Burnt Limestone
		Agricultural Lime
DRILLING DATA		Hole Diameter 105–115mm
		Bench Height Up to 50m
	Drills	Halco 410C
		Halco 625H
COMPRESSORS		Ingersoll-Rand DXL600P
SECONDARY BREAKING		Dropball on Ruston-Bucyrus 22-RB
LOAD AND HAUL	**Trucks**	Terex
	Loaders	Caterpillar 966C
CRUSHING PLANT	**Primary**	Allis Chalmers 42 × 65 Gyratory
	Secondary	Nordberg 5.5′ Cone
		Hazemag APKM0805 Impactor

ARC SOUTHERN LTD (ARC LTD)

CHIPPING SODBURY QUARRY

49

Chipping Sodbury, Avon BS17 6AY
Phone 0454 314400
OS Map 172 **OS Grid Reference** ST725842
Personnel G Wooley, Unit Manager

PRODUCT DATA	**Rock type**	Carboniferous Limestone
	Colour	Light brownish grey
	Products	Crushed Stone
		Asphalt or Tarmacadam
		Ready-mixed Concrete
		Pre-cast Concrete Products
		Industrial Limestone
		Agricultural Lime
DRILLING DATA		Hole Diameter 125mm
		Bench Height 17m
	Drills	Halco HPD90-35T
		Halco 410C
		Halco 650H
		Holman Voltrak
SECONDARY BREAKING		Hydraulic Hammer
		Talisker on Poclain 220CK

entry continues overleaf

LOAD AND HAUL	**Loaders**	Volvo
CRUSHING PLANT	**Primary**	Hazemag AP5 Impactor
	Secondary	Hazemag APK107 Impactor
	Tertiary	Nordberg Gyradisc
		Lightning Impactor

ARC SOUTHERN LTD (ARC LTD)

CROMHALL QUARRY

50

Wotton-under-Edge, Glos GL12 8AA
Phone 0454 260347
OS Map 172   **OS Grid Reference** ST705915
Personnel R Webb, Unit Manager

PRODUCT DATA	**Rock type**	Carboniferous Dolomitic Limestone
	Colour	Light brownish grey
	Grain	Coarse
	Products	Crushed Stone
		Asphalt or Tarmacadam
		Industrial Limestone
		Agricultural Lime
DRILLING DATA		Hole Diameter 105–127mm
		Bench Height 17m
	Drills	Halco 625H
LOAD AND HAUL	**Trucks**	Euclid
		Aveling Barford
		Caterpillar
	Loaders	Caterpillar 969D
		Caterpillar 980C
CRUSHING PLANT	**Primary**	BJD 55 Impactor
	Secondary	BJD 44 Impactor
	Tertiary	Nordberg 3' Cone (Shorthead)

ARC SOUTHERN LTD (ARC LTD)

DAGLINGWORTH QUARRY

51

Gloucester Road, Cirencester, Glos GL7 7JB
Phone 0285 655961/655962
OS Map 163   **OS Grid Reference** SP640180
Personnel D Smith, Unit Manager

PRODUCT DATA	**Rock type**	Jurassic Limestone
	Colour	Pinkish grey
	Grain	Fine to medium
	Products	Dressed or Lump Stone
		Crushed Stone
		Ready-mixed Concrete
		Industrial Limestone
DRILLING DATA		Hole Diameter 105mm
		Bench Height 8m
		Contractor Railside Engineering

SECONDARY BREAKING		Dropball on Ruston-Bucyrus 22-RB
LOAD AND HAUL	**Face shovel**	Poclain 220CK
	Trucks	Aveling Barford RD30
	Loaders	Caterpillar
CRUSHING PLANT	**Primary**	O&K 140/150 Impactor (Mobile)
	Secondary	BJD 48″ Swing-hammer
	Tertiary	Christy & Norris Swing-hammer

ARC SOUTHERN LTD (ARC LTD)

DRYBROOK QUARRY

52

Drybrook, Glos GL17 9BT
Phone 0594 542291
OS Map 162 **OS Grid Reference** SO644181
Personnel F G Treasure, Unit Manager

PRODUCT DATA	**Rock type**	Carboniferous Dolomitic Limestone
	Colour	Light reddish brown
	Grain	Coarse
	Products	Crushed Stone
		Agricultural Lime
DRILLING DATA		Hole Diameter 105mm
		Bench Height 10m
	Drills	Halco 410C
COMPRESSORS		CompAir 700HE
SECONDARY BREAKING		Dropball on Ruston-Bucyrus 22-RB
LOAD AND HAUL	**Trucks**	Caterpillar
	Loaders	Caterpillar
CRUSHING PLANT	**Primary**	Pegson 48 × 42 Jaw
	Secondary	Mansfield No 4 Hammermill
		Mansfield No 5 Hammermill
	Tertiary	BJD Hammermill

ARC SOUTHERN LTD (ARC LTD)

GUITING QUARRY

53

Temple Guiting, Cheltenham, Glos GL54 5SB
Phone 038673 285
OS Map 150 **OS Grid Reference** SP079302
Personnel I Southgate, Unit Manager

PRODUCT DATA	**Rock type**	Jurassic Inferior Oolitic Limestone
	Colour	Light to dark yellowish/orangey cream
	Grain	Fine
	Products	Dressed or Lump Stone
		Crushed Stone
		Industrial Limestone

entry continues overleaf

DRILLING DATA		Hole Diameter 105mm Bench Height 6–8m Hand-held equipment Contractor Railside Engineering
COMPRESSORS		Holman
SECONDARY BREAKING		Dropball on Ruston-Bucyrus 22-RB
LOAD AND HAUL	**Trucks**	Aveling Barford RD40
	Loaders	Caterpillar
CRUSHING PLANT	**Primary**	Kue Ken 120 36 × 24 Jaw (Double Toggle)
	Secondary	(2)Parker 36 × 10 Jaw (Double Toggle)
	Tertiary	Mansfield No 3 Hammermill

ARC SOUTHERN LTD (ARC LTD)

OFFHAM QUARRY   54

West Malling, Kent ME19 5PT
Phone 0732 843071
OS Map 188   **OS Grid Reference** TQ654575
Personnel P Barker, Unit Manager

PRODUCT DATA	**Rock type**	Cretaceous Lower Hythe Beds Greensand (Ragstone)
	Colour	Light greenish grey
	Product	Crushed Stone

RIPPABLE STRATA

ARC SOUTHERN LTD (ARC LTD)

TYTHERINGTON QUARRY   55

Wotton-under-Edge, Glos GL12 8UW
Phone 0454 416161
OS Map 172 **OS Grid Reference** ST659889
Personnel M J Bishop, Unit Manager

PRODUCT DATA	**Rock type**	Carboniferous Limestone
	Colour	Light brownish grey
	Grain	Medium
	Products	Crushed Stone
		Asphalt or Tarmacadam
		Ready-mixed Concrete
		Industrial Limestone
		Agricultural Lime
DRILLING DATA		Hole Diameter 110–125mm
		Bench Height 15m
	Drills	Halco 410C
		Ingersoll-Rand Drillmaster 25
COMPRESSORS		Compair 650HE

SECONDARY BREAKING		Dropball on Ruston-Bucyrus 22-RB
LOAD AND HAUL	**Trucks**	Caterpillar 769C
	Loaders	Caterpillar 988B
CRUSHING PLANT		No information provided

ARC SOUTHERN LTD (ARC LTD)

WHATLEY QUARRY

56

Frome, Somerset BA11 3LF
Phone 0373 52515 **Telex** 444522
OS Map 183 **OS Grid Reference** ST733480
Personnel P Hovergill, Unit Manager

PRODUCT DATA	**Rock type**	Carboniferous Limestone
	Colour	Light brownish grey
	Grain	Medium
	Products	Crushed Stone
		Asphalt or Tarmacadam
		Industrial Limestone
		Agricultural Lime
DRILLING DATA		Hole Diameter 165mm
		Bench Height 17m
		Contractor W C D Sleaman
	Drills	Halco 620C
		(2) Halco MPD115-35T
SECONDARY BREAKING		Dropball on Ruston-Bucyrus 22-RB
LOAD AND HAUL	**Face shovel**	O&K RH90
	Trucks	(4)Caterpillar 777B
	Loaders	Caterpillar 992C
CRUSHING PLANT	**Primary**	Hazemag APK1030 Impactor
	Secondary	Hazemag APK800 Impactor
	Tertiary	Nordberg Omnicone

ARC SOUTHERN LTD (ARC LTD)

WHITE HILL QUARRY

57

Witney Road, Burford, Oxon OX8 4EU
Phone 0373 63211
OS Map 163 **OS Grid Reference** SP245125

QUARRY ON STAND

ROCK DATA	**Rock type**	Jurassic Great Oolitic Limestone
	Colour	Honey brown
	Grain	Fine

ARC SOUTH WEST LTD (ARC LTD)
Head Office Grace Road, Marsh Barton, Trading Estate, Exeter, Devon EX2 8PU
Phone 0392 215353 **Fax** 0392 51849
Personnel P E Gilbert, Managing Director
        D Sharman, Operations Director
        M Bews, Trading Manager
        Vacant, Production Manager
Quarries Beer, Charnage, Hingston Down, Long Bredy, Portland, Trusham, Whitecleaves, Yalberton Tor

---

ARC SOUTH WEST LTD (ARC LTD)

BEER QUARRY

58

Beer, Seaton, Devon EX12 38E
Phone 0297 80214
OS Map 192 **OS Grid Reference** SY215895
Personnel L Davis, Unit Manager

PRODUCT DATA	**Rock type**	Cretaceous Middle Chalk
	Colour	White, light brown
	Grain	Fine
	Products	Crushed Stone
		Industrial Limestone

RIPPABLE STRATA

ARC SOUTH WEST LTD (ARC LTD)

CHARNAGE QUARRY

59

Mere, Warminster, Wilts BA12 6AP
Phone 0747 860210
OS Map 183 **OS Grid Reference** ST830330
Personnel L Davis, Unit Manager

PRODUCT DATA	**Rock type**	Cretaceous Middle Upper Chalk
	Colour	White
	Grain	Fine
	Products	Crushed Stone
		Industrial Limestone
RIPPABLE STRATA		Ripper Caterpillar D8
LOAD AND HAUL	**Trucks**	Terex
	Loaders	Michigan
		Caterpillar

ARC SOUTH WEST LTD (ARC LTD)

HINGSTON DOWN QUARRY

60

Gunnislake, Cornwall PL18 9AU
Phone 0822 832271
OS Map 201 **OS Grid Reference** SX409720
Personnel T Williams, Unit Manager

PRODUCT DATA	**Rock type**	Hingston Down Granite
	Colour	Light speckled grey
	Grain	Fine
	Products	Crushed Stone
		Asphalt or Tarmacadam
		Ready-mixed Concrete
		Pre-cast Concrete Products

DRILLING DATA		Hole Diameter 110mm
		Bench Height 15m
	Drills	Halco 410C
		Holman Voltrak

| COMPRESSORS | | Ingersoll-Rand DXL700H |

| SECONDARY BREAKING | | Dropball on Ruston-Bucyrus 22-RB |

| LOAD AND HAUL | **Trucks** | (2)Aveling Barford RD40 |
| | **Loaders** | Caterpillar 988B |

CRUSHING PLANT	**Primary**	Pegson 60 × 48 Jaw (Single Toggle)
	Secondary	Nordberg 5.5′ Cone
	Tertiary	Nordberg 3′ Cone

ARC SOUTH WEST LTD (ARC LTD)

LONG BREDY QUARRY 61

Long Bredy, Dorchester, Dorset DT2 9EF
Phone 03083 391
OS Map 194 **OS Grid Reference** SY560905
Personnel L Davis, Unit Manager

PRODUCT DATA	**Rock type**	Cretaceous Chalk
	Colour	White
	Grain	Fine
	Products	Crushed Stone
		Industrial Limestone

RIPPABLE STRATA

ARC SOUTH WEST LTD (ARC LTD)

PORTLAND QUARRY 62

Bumpers Lane, Easton, Portland, Dorset DT5 1HY
Phone 0305 820207
OS Map 194 **OS Grid Reference** SY695716
Personnel J W Reay, General Manager
E Maidment, Unit Manager

PRODUCT DATA	**Rock type**	Jurassic Portland Limestone
	Colour	Buff
	Grain	Shelly
	Products	Dressed or Lump Stone
		Crushed Stone

| DRILLING DATA | | Hand-held equipment |

entry continues overleaf

| COMPRESSORS | | Broomwade static |
| LOAD AND HAUL | **Loaders** | Liebherr R974 |

COMPRESSORS		Broomwade static
SECONDARY BREAKING		Hydraulic Hammer - Rammer S86
LOAD AND HAUL	**Loaders**	Liebherr R974

ARC SOUTH WEST LTD (ARC LTD)

TRUSHAM QUARRY

63

Trusham, Newton Abbott, Devon TQ13 0NX
Phone 0626 852115
OS Map 191   **OS Grid Reference** SX850810
Personnel P Hughes, Unit Manager

PRODUCT DATA	**Rock type**	Dolerite
	Colour	Medium greenish grey
	Grain	Medium
	Products	Crushed Stone
		Asphalt or Tarmacadam

DRILLING DATA		Hole Diameter 105mm
		Bench Height 13m
	Drills	Halco 400C

| COMPRESSORS | | CompAir 650HE |

| SECONDARY BREAKING | | Dropball on Ruston-Bucyrus 22-RB |

| LOAD AND HAUL | **Trucks** | Volvo |
| | **Loaders** | Caterpillar |

CRUSHING PLANT	**Primary**	Hadfield 48 × 36 Jaw (Double Toggle)
	Secondary	Kue Ken 90 Jaw
	Tertiary	Symons 3′ Cone (Shorthead)
	Quaternary	Symons 3′ Cone (Standard)

ARC SOUTH WEST LTD (ARC LTD)

WHITECLEAVES QUARRY

64

Buckfastleigh, Devon TQ11 0EA
Phone 0364 43017
OS Map 202   **OS Grid Reference** SX738656
Personnel M German, Unit Manager

PRODUCT DATA	**Rock type**	Basalt
	Colour	Dark grey
	Grain	Fine
	Products	Crushed Stone
		Asphalt or Tarmacadam

DRILLING DATA		Hole Diameter 105mm
		Bench Height 12m
		Contractor Celtic Rock Services

| SECONDARY BREAKING | | Dropball on Ruston-Bucyrus 22-RB |

LOAD AND HAUL	**Face shovel**	Poclain
	Trucks	Heathfield
	Loaders	Fiatallis

CRUSHING PLANT	**Primary**	Nordberg VB1048 Jaw (Single Toggle)
	Secondary	Allis Chalmers 11 × 36 Gyratory
	Tertiary	Nordberg 3′ Cone (Shorthead)
	Quaternary	Allis Chalmers 436 Hydrocone

ARC SOUTH WEST LTD (ARC LTD)

YALBERTON TOR QUARRY 65

Yalberton Road, Paignton, Devon TQ4 7RJ
Phone 0803 559491
OS Map 202 **OS Grid Reference** SX869592
Personnel J Parker, Unit Manager

PRODUCT DATA	**Rock type**	Devonian Limestone
	Colour	Light grey
	Products	Crushed Stone
		Industrial Limestone

DRILLING DATA		Hole Diameter 108mm
		Bench Height 11–15m
		Contractor Celtic Rock Services

| SECONDARY BREAKING | | Hydraulic Hammer |
| | | Dropball on Ruston-Bucyrus 22-RB |

LOAD AND HAUL	**Face shovel**	O&K
	Trucks	Heathfield
	Loaders	Caterpillar 966C

CRUSHING PLANT	**Primary**	Parker 40 × 40 (Rotary)
	Secondary	Hazemag APK40
	Tertiary	None

C D ASHBRIDGE

Head Office *Pigeon Cote Farm, Malton Road, York, N Yorks YO3 9TD*
Phone *0904 425301*
Personnel *C D Ashbridge, Owner*
Quarries *Hovingham*

C D ASHBRIDGE

HOVINGHAM QUARRY 66

Hovingham, York, N Yorks YO6 4LB
Phone 065382 594
OS Map 105 **OS Grid Reference** SE674749
Personnel C D Ashbridge, Owner and Unit Manager

PRODUCT DATA	**Rock type**	Jurassic Oolitic Limestone
	Colour	Creamy white
	Grain	Fine
	Product	Dressed or Lump Stone

| DRILLING DATA | | Hand-held equipment |

J & J ASHCROFT LTD
Head Office *Little Quarries, Whittle-le-Woods, Chorley, Lancs PR6 7QR*
Phone *02572 62137*
Personnel *J R Ashcroft, Managing Director*
Quarries *Little*

J & J ASHCROFT LTD
LITTLE QUARRY 67
Whittle-le-Woods, Chorley, Lancs PR6 7QR
Phone 02572 62137
OS Map 103 **OS Grid Reference** SD586226
Personnel J R Ashcroft, Managing Director and Unit Manager

PRODUCT DATA	**Rock type**	Carboniferous Sandstone
	Colour	Various browns, greys, blues
	Grain	Coarse
	Product	Crushed Stone
DRILLING DATA	**Drills**	Halco 410C
COMPRESSORS		Ingersoll-Rand
SECONDARY BREAKING		Hydraulic Hammer
LOAD AND HAUL	**Trucks**	Aveling Barford
	Loaders	Caterpillar
CRUSHING PLANT	**Primary**	Pegson 25 × 40 Jaw (Single Toggle)
	Secondary	Goodwin Barsby 36 × 10 Jaw
	Tertiary	None

ASKERNISH QUARRY COMPANY LTD
Head Office *Askernish Quarry, Hillside Garage, Lochboisdale, South Uist, Western Isles, PA81 5TS*
Phone *08784 278*
Personnel *D J Macaulay, Owner*
Quarries *Askernish*

ASKERNISH QUARRY COMPANY LTD
ASKERNISH QUARRY 68
Hillside Garage, Lochboisdale, South Uist, Western Isles, PA81 5TS
Phone 08784 278
OS Map 31 **OS Grid Reference** NF795198
Personnel D J Macaulay, Owner and Unit Manager

PRODUCT DATA	**Rock type**	Basalt
	Colour	Black to greyish black
	Grain	Fine
	Product	Crushed Stone
DRILLING DATA		No drilling
LOAD AND HAUL	**Trucks**	Aveling Barford
	Loaders	Caterpillar

D S ATKINSON AND PARTNERS
Head Office *10 Royal Crescent, Whitby, N Yorks YO21 1RT*
Phone *0947 810936*
Personnel *D S Atkinson, Owner*
Quarries *Blue Bank*

D S ATKINSON AND PARTNERS

BLUE BANK QUARRY   69

Sleights, Whitby, N Yorks YO22 5EU
Phone 0274 832379
OS Map 94   **OS Grid Reference** NZ868060

QUARRY ON STAND

ROCK DATA	**Rock type**	Jurassic Sandstone
	Colour	Cream to brown
	Grain	Coarse

R E ATKINSON
Head Office *42 Garden Road, Brighouse, Halifax, W Yorks HD6 2ES*
Phone *0274 832379*
Personnel *R E Atkinson, Partner*
Quarries *Friendly, Ten Yards Lane*

R E ATKINSON

FRIENDLY QUARRY   70

Burnley Road, Friendly, Halifax, W Yorks HD5 8RQ
Phone 0274 832379
OS Map 104   **OS Grid Reference** SE080250
Personnel R E Atkinson, Partner and Unit Manager

PRODUCT DATA	**Rock type**	Carboniferous Sandstone (Flagstone)
	Colour	Buff, cream
	Grain	Fine
	Product	Dressed or Lump Stone

DRILLING DATA		Hand-held equipment

R E ATKINSON

TEN YARDS LANE QUARRY   71

Ten Yards Lane, Thornton, N Yorks BD13 3SE
Phone 0274 832379
OS Map 104   **OS Grid Reference** SE105327
Personnel R E Atkinson, Partner and Unit Manager

PRODUCT DATA	**Rock type**	Carboniferous Sandstone
	Colour	Cream to brown
	Grain	Fine
	Product	Dressed or Lump Stone

DRILLING DATA		Hand-held equipment

ATTWOODS plc
Head Office *The Pickeridge, Stoke Common Road, Fulmer, Bucks SL3 6HA*
Phone *0281 62700* **Fax** *0281 62464*
Personnel *M K Foreman, Chairman and Chief Executive*
T Pensold, Managing Director
M A Ashcroft, Director
A Pontin, Director
Companies *DRINKWATER SABEY LTD* — see separate entry

ALLEN BAILEY
Head Office *Hainworth Shaw Quarry, Winfield, Shaw Lane, Hainworth Shaw, Keighley, W Yorks BD21 5QR*
Phone *0535 604340*
Personnel *A Bailey, Owner*
Quarries *Hainworth Shaw*

ALLEN BAILEY

HAINWORTH SHAW QUARRY   72
Winfield, Shaw Lane, Hainworth Shaw, Keighley, W Yorks BD21 5QR
Phone 0535 604340
OS Map 104   **OS Grid Reference** SE066388
Personnel A Bailey, Owner and Unit Manager

PRODUCT DATA		
	Rock type	Carboniferous Millstone Grit
	Colour	Grey to black
	Grain	Fine
	Product	Dressed or Lump Stone

RIPPABLE STRATA

BAIRD AND STEVENSON (QUARRYMASTERS) LTD
Head Office *Locharbriggs Quarry, Locharbriggs, Dumfries, Dumfries & Galloway DG1 1QS*
Phone *0387 710237*
Personnel *W A Simon, Director*
Quarries *Locharbriggs*

BAIRD AND STEVENSON (QUARRYMASTERS) LTD

LOCHARBRIGGS QUARRY   73
Locharbriggs, Dumfries, Dumfries & Galloway DG1 1QS
Phone 0387 710237
OS Map 85   **OS Grid Reference** NX990805
Personnel N Rennie, Unit Manager

PRODUCT DATA		
	Rock type	Permian Sandstone
	Colour	Dull reddish pink
	Grain	Medium
	Product	Dressed or Lump Stone

DRILLING DATA	
	Hand-held equipment

S J BAKER
Head Office *Trevillet Quarry, Tintagel, Cornwall PL34 0HL*
Phone *0840 770659*
Personnel *S J Baker, Owner*
Quarries *Trevillet*

S J BAKER

TREVILLET QUARRY   74

Tintagel, Cornwall PL34 0HL
Phone 0840 770659
OS Map 200   **OS Grid Reference** SX050882
Personnel S J Baker, Owner and Unit Manager

PRODUCT DATA	**Rock type**	Devonian Slate
	Colour	Red to brown
	Grain	Medium
	Product	Dressed or Lump Stone

DRILLING DATA		Hand-held equipment

BALFOUR BEATTY LTD
Corporate Head Office *7 Mayday Road, Thornton Heath, Croydon, Surrey CR4 7XA*
Phone *081 684 6922* **Fax** *081 684 6597* **Telex** *264042*
Personnel *D Cawthraw, Executive Director*
Companies *Scottish Natural Stones Ltd* — see separate entry

BARDON QUARRIES LTD
Head Office *Bardon Quarry, Bardon Hill, Coalville, Leics LE6 2TL*
Phone *0530 510066* **Fax** *0530 35780* **Telex** *341017*
Personnel *P W G Tom, Managing Director*
Quarries *Bardon*

BARDON QUARRIES LTD

BARDON QUARRY   75

Bardon Hill, Coalville, Leics LE6 2TL
Phone 0530 510066   **Telex** 341017
OS Map 129   **OS Grid Reference** SK450130
Personnel A Noble, Unit Manager

PRODUCT DATA	**Rock type**	Andesite
	Colour	Greyish green and pink
	Grain	Fine
	Products	Crushed Stone
		Asphalt or Tarmacadam
		Pre-cast Concrete Products

DRILLING DATA		Hole Diameter 135mm
		Bench Height 21m
	Drills	(2)Tamrock DHA1000

COMPRESSORS	(2)Atlas Copco XRH350

SECONDARY BREAKING	Hydraulic Hammer
	Dropball on Ruston-Bucyrus 22-RB

entry continues overleaf

LOAD AND HAUL	**Face shovel**	RH90
	Trucks	(3)Caterpillar 777
	Loaders	None

CRUSHING PLANT	**Primary**	Fuller 54″ Gyratory (Mobile)
	Secondary	Nordberg 5.5′ Cone
	Tertiary	(3)Nordberg 4′ Cone (Shorthead)

BARHAM BROS LTD
Head Office *Station Quarry, Charlton Mackrell, Somerton, Somerset TA11 6AG*
Phone *045822 3249*
Personnel *B R Wellsead, Owner*

BARHAM BROS LTD
STATION QUARRY   **76**
Charlton Mackrell, Somerton, Somerset TA11 6AG
Phone 045822 3249
OS Map 183   **OS Grid Reference** ST528284
Personnel B R Wellsead, Owner and Unit Manager

PRODUCT DATA	**Rock type**	Carboniferous Limestone
	Colour	Greyish blue, light brown
	Product	Dressed or Lump Stone

| DRILLING DATA | | Hand-held equipment |

BARLAND QUARRY LTD
Head Office *Barland Quarry, Pennard Road, Bishopston, Swansea, W Glam SA3 3JE*
Phone *044128 2284/2434*
Personnel *E Gooch, Managing Director*
Quarries *Barland*

BARLAND QUARRY LTD
BARLAND QUARRY   **77**
Pennard Road, Bishopston, Swansea, W Glam SA3 3JE
Phone 044128 2284
OS Map 159   **OS Grid Reference** SS575895
Personnel D Eves, General Manager
          P Lewis, Unit Manager

PRODUCT DATA	**Rock type**	Carboniferous Limestone
	Colour	Not commercially significant
	Product	Crushed Stone

DRILLING DATA		Hole Diameter 110mm
		Bench Height 12m
		Contractor Drill Quip UK

| SECONDARY BREAKING | | Dropball on Ruston-Bucyrus 22-RB |

LOAD AND HAUL	**Trucks**	(4)Aveling Barford 017
	Loaders	JCB
		Yale
		Caterpillar 966C

CRUSHING PLANT	**Primary**	Baxter 32 × 22 Jaw (Single Toggle)
	Secondary	Ajax AB Hammermill
	Tertiary	None

BARNEY'S QUARRY DIVISION LTD
Head Office *Woodeaton, Oxford, Oxon OX3 9TP*
Phone *0865 52646*
Personnel *J E Barney, Managing Director*
    J A Barney, Director
Quarries *Woodeaton*

BARNEY'S QUARRY DIVISION LTD
WOODEATON QUARRY
Woodeaton, Oxford, Oxon OX3 9TP
Phone 0865 52646
OS Map 164   **OS Grid Reference** SP534122
Personnel J E Barney, Managing Director and Unit Manager

78

PRODUCT DATA	**Rock type**	Jurassic Great Oolitic Limestone
	Colour	White
	Grain	Fine, fossiliferous
	Product	Crushed Stone
DRILLING DATA		Hole Diameter 105mm
		Bench Height 7-8m
	Drills	Halco 150
COMPRESSORS		CompAir 37HP
LOAD AND HAUL	**Face shovel**	Caterpillar 225
		Akermans H16
	Trucks	(4)Volvo 180
	Loaders	Volvo BM L1240
		Zettelmeyer ZL2002
CRUSHING PLANT	**Primary**	Parker 5702 rotary (Mobile)
		Parker Track Ranger (Mobile)
	Secondary	None
	Tertiary	None

W & J BARR AND SONS LTD
Head Office *Tongland Quarry, Heathfield, Ayr, Strathclyde KA8 9SX*
Phone *0292 281311*
Personnel *W Barr, Owner*
    W Barr (Jnr), Director
Quarries *Barlockhart, Cairnryan, Tongland, Tormitchell*

W & J BARR AND SONS LTD
BARLOCKHART QUARRY
Glen Luce, Dumfries & Galloway DG8 0JQ
Phone 05813 329
OS Map 82   **OS Grid Reference** NX210560
Personnel H M McClurg, Unit Manager

79

entry continues overleaf

PRODUCT DATA	**Rock type**	Silurian Greywacke
	Colour	Grey to black
	Grain	Fine
	Product	Crushed Stone

DRILLING DATA		Hole Diameter 100mm
		Bench Height 18m
	Drills	Tamrock DHA600

| SECONDARY BREAKING | | Dropball on Smiths |

| LOAD AND HAUL | **Trucks** | (2)Aveling Barford RD30 |
| | **Loaders** | Caterpillar 950 |

CRUSHING PLANT	**Primary**	Baxter 36×30 Jaw
	Secondary	Goodwin Barsby 24×10 Jaw (Single Toggle)
	Tertiary	Nordberg Omnicone

W & J BARR AND SONS LTD

CAIRNRYAN QUARRY

80

Cairnryan, Stranraer, Dumfries & Galloway DG9 8OX
Phone 0292 281311
OS Map 82 **OS Grid Reference** NX064685

QUARRY ON STAND

| ROCK DATA | **Rock type** | Silurian Sandstone |

W & J BARR AND SONS LTD

TONGLAND QUARRY

81

Heathfield, Ayr, Strathclyde KA8 9SX
Phone 046 587239/587250
OS Map 84 **OS Grid Reference** NY695545
Personnel W Barr (Jnr), Director and Unit Manager

PRODUCT DATA	**Rock type**	Silurian Greywacke
	Colour	Grey to black
	Grain	Fine
	Product	Crushed Stone

DRILLING DATA		Hole Diameter 100mm
		Bench Height 16m
	Drills	Tamrock DHA600

| SECONDARY BREAKING | | Dropball on Smiths |

| LOAD AND HAUL | **Trucks** | (2)Aveling Barford RD30 |
| | **Loaders** | Caterpillar 950 |

CRUSHING PLANT	**Primary**	Kue Ken 36×42 Jaw
	Secondary	Nordberg 4.25' Cone
		Nordberg 3' Cone

W & J BARR AND SONS LTD

TORMITCHELL QUARRY

82

Tormitchell, Girvan, Strathclyde KA26 0TT
Phone 046587 251
OS Map 76 **OS Grid Reference** NX235944
Personnel J Wilson, Unit Manager

PRODUCT DATA	**Rock type**	Ordovician Gritstone
	Colour	Not commercially significant
	Product	Crushed Stone

DRILLING DATA		Hole Diameter 100mm
		Bench Height 16m
	Drills	Tamrock DHA600

| SECONDARY BREAKING | | Dropball |

| LOAD AND HAUL | **Trucks** | Aveling Barford RD40 |
| | **Loaders** | Caterpillar |

| CRUSHING PLANT | **Primary** | Baxter 40 × 32 Jaw |
| | **Secondary** | (2)Nordberg 3′ Cone |

BATH AND PORTLAND STONE LTD (ARC LTD)
Head Office *Moor Park House, Moor Green, Corsham, Wilts SN13 9SE*
Phone *0225 810456* **Fax** *0225 811234*
Personnel *P Highfield, Managing Director*
　　　　　J E Williams, Production Director
Quarries *Doulting, Monks Park Mine, Westwood Ground*

BATH AND PORTLAND STONE LTD (ARC LTD)

DOULTING QUARRY

83

Doulting, Shepton Mallett, Somerset BA4 4QE
Phone 0225 810367
OS Map 183 **OS Grid Reference** ST648434
Personnel D W Selby, Unit Manager

PRODUCT DATA	**Rock type**	Jurassic Inferior Oolitic Limestone
	Colour	Light brown
	Grain	Fine to coarse
	Product	Dressed or Lump Stone

| DRILLING DATA | | Hand-held equipment |

BATH AND PORTLAND STONE LTD (ARC LTD)

MONKS PARK MINE

84

Corsham, Chippenham, Wilts SN13 9PJ
Phone 0225 810367
OS Map 173 **OS Grid Reference** ST881683
Personnel D W Selby, Unit Manager

entry continues overleaf

PRODUCT DATA	**Rock type**	Jurassic Great Oolitic Limestone
	Colour	Light cream
	Grain	Fine
	Product	Dressed or Lump Stone

| DRILLING DATA | | Hand-held equipment |

BATH AND PORTLAND STONE LTD (ARC LTD)

WESTWOOD GROUND QUARRY 85

Bradford-on-Avon, Trowbridge, Wilts BA15 2AR
Phone 0225 8866645
OS Map 183 **OS Grid Reference** ST803591
Personnel D W Selby, Unit Manager

PRODUCT DATA	**Rock type**	Jurassic Great Oolitic Limestone
	Colour	Greyish white
	Grain	Fine
	Product	Dressed or Lump Stone

| DRILLING DATA | | Hand-held equipment |

BATH STONE CO LTD
Head Office Midford Lane, Limpley Stoke, Bath, Avon BA2 5SA
Phone 022122 3792
Personnel B Marson, Director
Quarries Hayes Wood Mine

BATH STONE CO LTD

HAYES WOOD MINE 86

Midford Lane, Limpley Stoke, Bath, Avon BA2 5SA
Phone 022122 3792
OS Map 183 **OS Grid Reference** ST778603
Personnel B Marson, Director and Unit Manager

PRODUCT DATA	**Rock type**	Jurassic Great Oolitic Limestone
	Colour	Light buff
	Grain	Fine, fossiliferous
	Product	Dressed or Lump Stone

| DRILLING DATA | | Hand-held equipment |

BECKSTONE QUARRY LTD
Head Office Burford Quarry, Burford Road, Brize Norton, Oxon OX8 8NN
Phone 0993 842391
Personnel R Becket, Director
Quarries Burford

BECKSTONE QUARRY LTD

BURFORD QUARRY 87

Burford Road, Brize Norton, Oxon OX8 8NN
Phone 0993 842391
OS Map 163 **OS Grid Reference** SP277095
Personnel K Deakin, Unit Manager

PRODUCT DATA	**Rock type**	Jurassic Great Oolitic Limestone
	Colour	Honey brown
	Grain	Fossiliferous
	Products	Dressed or Lump Stone
		Crushed Stone
DRILLING DATA		Hole Diameter 105mm
		Bench Height 18m
		Contractor Railside Engineering
LOAD AND HAUL	**Trucks**	(3)Heathfield
	Loaders	(3)JCB
CRUSHING PLANT	**Primary**	Jaw
	Secondary	None
	Tertiary	None

THE BELLISTON QUARRY COMPANY LTD
Head Office *Sconnie Park, Leven, Fife KY8 4TD*
Phone *0333 26841*
Personnel *A Cook, Managing Director*
Quarries *Belliston*

THE BELLISTON QUARRY COMPANY LTD

BELLISTON QUARRY   88
Colinsburgh, Leven, Fife KY8 4TD
Phone 0333 8223
OS Map 59   **OS Grid Reference** NO499063
Personnel J McOwen, Unit Manager

PRODUCT DATA	**Rock type**	Dolerite Sill
	Colour	Black
	Grain	Medium
	Product	Crushed Stone
DRILLING DATA		Hole Diameter 105mm
		Bench Height 23m
		Contractor Drilling & Blasting
SECONDARY BREAKING		Dropball on Priestman
LOAD AND HAUL	**Trucks**	Aveling Barford RD30
	Loaders	Furukawa
CRUSHING PLANT	**Primary**	Kue Ken 42×27 Jaw (Double Toggle)
		Parker 24×10 Jaw (Mobile)
	Secondary	None
	Tertiary	None

BEN BENNETT JNR LTD
Head Office *Lisle Road, Rotherham, S Yorks S60 2RL*
Phone *0709 382251*
Personnel *B Bennett, Managing Director*
Quarries *Grange Mill*

entry continues overleaf

BEN BENNETT JNR LTD
GRANGE MILL QUARRY

89

Nr Wirksworth, Derbys DE4 4HD
Phone 0629 85334
OS Map 119 **OS Grid Reference** SK241574
Personnel A Rowson, Unit Manager

PRODUCT DATA	**Rock type**	Carboniferous Bee Low Limestone
	Colour	White
	Grain	Medium
	Products	Crushed Stone
		Industrial Limestone
		Agricultural Lime
DRILLING DATA		Hole Diameter 110mm
		Bench Height 20m
	Drills	Halco 410C
COMPRESSORS		CompAir 650HE
SECONDARY BREAKING		Dropball on Ruston-Bucyrus 22-RB
LOAD AND HAUL	**Trucks**	(3)Iveco
	Loaders	Caterpillar 966C
		Liebherr 551
CRUSHING PLANT	**Primary**	Goodwin Barsby 36 × 24 Jaw
		Goodwin Barsby 30 × 18 Jaw
	Secondary	None
	Tertiary	None

PETER BENNIE LTD

Head Office Oxwich Close, Brackmills Ind Estate, Northampton, Northants NN4 0BH
Phone 0604 765252
Personnel J Goodjohn, Managing Director
        D Warwick, Quarry Director
Quarries Alkerton, Boughton, Pitsford, Shennington, Shotley

PETER BENNIE LTD
ALKERTON QUARRY

90

Edge Hill, Alkerton, Oxon OX15 6HZ
Phone 0295 87505
OS Map 151 **OS Grid Reference** SP395429
Personnel C Neville, Unit Manager

PRODUCT DATA	**Rock type**	Jurassic Middle Lias Banbury Ironstone
	Colour	Greenish blue, brown
	Grain	Fine
	Products	Dressed or Lump Stone
		Crushed Stone
DRILLING DATA		Hole Diameter 75mm
		Bench Height 4–5m
	Drills	Dando RO60 Rotary

LOAD AND HAUL	**Loaders**	Caterpillar 950E
		Komatsu
		Poclain

CRUSHING PLANT	**Primary**	Sheepbridge Impactor
	Secondary	None
	Tertiary	None

PETER BENNIE LTD

BOUGHTON QUARRY 91

Brampton Lane, Chapel Brampton, Northampton, Northants NN6 8AA
Phone 0604 847814
OS Map 152 **OS Grid Reference** SP728662
Personnel Vacant, Unit Manager

PRODUCT DATA	**Rock type**	Jurassic Sandstone
	Colour	Greyish brown
	Grain	Fine
	Product	Dressed or Lump Stone

RIPPABLE STRATA

PETER BENNIE LTD

PITSFORD QUARRY 92

Pitsford, Northampton, Northants NN2 8SR
Phone 0604 765252
OS Map 152 **OS Grid Reference** TQ756677

QUARRY ON STAND

| ROCK DATA | **Rock type** | Jurassic Inferior Oolitic Limestone |

PETER BENNIE LTD

SHENNINGTON QUARRY 93

Shenington, Banbury, Oxon OX15 6NT
Phone 0295 87505
OS Map 151 **OS Grid Reference** SP359434
Personnel C Neville, Unit Manager

PRODUCT DATA	**Rock type**	Jurassic Middle Lias Banbury Ironstone
	Colour	Brown, blue
	Grain	Fine
	Products	Dressed or Lump Stone
		Crushed Stone

DRILLING DATA		Hole Diameter 75mm
		Bench Height 4-5m
	Drills	Dando RO60 Rotary

LOAD AND HAUL	**Loaders**	Caterpillar 950E
		Komatsu
		Poclain

entry continues overleaf

CRUSHING PLANT	**Primary**	Babbitless Rolls
	Secondary	None
	Tertiary	None

PETER BENNIE LTD

SHOTLEY QUARRY 94

Spanhoe Airfield, Harringworth, Corby, Northants NN17 3AG
Phone 0780 85317
OS Map 141   **OS Grid Reference** SP933968
Personnel D York, Unit Manager

PRODUCT DATA	**Rock type**	Permian Limestone
	Colour	White
	Grain	Medium
	Products	Dressed or Lump Stone
		Crushed Stone

| DRILLING DATA | | Hole Diameter 108mm |
| | **Drills** | Halco SPD43-22 |

| CRUSHING PLANT | **Primary** | Sheepbridge 24 × 42 Jaw |
| | **Secondary** | Goodwin Barsby 30 × 42 Jaw |

BERWYN SLATE QUARRY LTD

Head Office Clogau Quarry, Horse Shoe Pass, Llangollen, Clwyd LL20 8BS
Phone 0792 851468
Personnel L Rees, Owner
Quarries Clogau

BERWYN SLATE QUARRY LTD

CLOGAU QUARRY 95

Horse Shoe Pass, Llangollen, Clwyd LL20 8BS
Phone 0792 851468
OS Map 116   **OS Grid Reference** SJ185466
Personnel L Rees, Owner and Unit Manager

PRODUCT DATA	**Rock type**	Silurian Slate
	Colour	Deep blue
	Grain	Fine
	Product	Dressed or Lump Stone

| DRILLING DATA | | Contractor |

NO FURTHER INFORMATION

BILLOWN LIME QUARRIES LTD

Head Office Balladoole Quarry, Malew, Ballasalla, Isle of Man
Phone 0624 822527
Personnel A J Broome, Managing Director
Quarries Balladoole

BILLOWN LIME QUARRIES LTD

BALLADOOLE QUARRY

96

Malew, Ballasalla, Isle of Man
Phone 0624 822527
OS Map 95   **OS Grid Reference** SC282700
Personnel A Kelly, Unit Manager

PRODUCT DATA	**Rock type**	Ordovician Limestone
	Products	Crushed Stone
		Agricultural Lime
DRILLING DATA		Hole Diameter 105mm
		Bench Height 18m
		Contractor A Jones Rock Drillers
SECONDARY BREAKING		Dropball
CRUSHING PLANT	**Primary**	Goodwin Barsby 36 × 18 Jaw
	Secondary	Marsden Rotary
	Tertiary	Marsden Rotary

BINNS BROS

Head Office *Southowram, Halifax, W Yorks HX3 9QH*
Phone *0422 50700*
Personnel *J Binns, Owner*
Quarries *Watson*

BINNS BROS

WATSON QUARRY

97

Southowram, Halifax, W Yorks HX3 9QH
Phone 0422 715146
OS Map 104   **OS Grid Reference** SE108237
Personnel J Binns, Owner and Unit Manager

PRODUCT DATA	**Rock type**	Carboniferous Sandstone (Gritstone)
	Colour	Light buff
	Grain	Fine
	Product	Dressed or Lump Stone
DRILLING DATA		Hand-held equipment

D A BIRD LTD

Head Office *Pury End Quarry, Bugbrooke, Northampton, Northants NN7 3PH*
Phone *0604 830455*
Personnel *D A Bird, Managing Director*
Quarries *Pury End*

D A BIRD LTD

PURY END QUARRY (CAMP HILL QUARRY)

98

Bugbrooke, Northampton, Northants NN7 3PH
Phone 0604 830455
OS Map 152   **OS Grid Reference** SP680574
Personnel D A Bird, Managing Director and Unit Manager

entry continues overleaf

| PRODUCT DATA | **Rock type** | Jurassic Oolitic Limestone |
| | **Product** | Dressed or Lump Stone |

RIPPABLE STRATA

| LOAD AND HAUL | **Face shovel** | Akermans |

BLUE CIRCLE INDUSTRIES plc
Corporate Head Office *Eccleston Square, London SW1V 1PX*
Phone *071 828 3456* **Fax** *071 245 8169* **Telex** *849939*
Personnel *D A R Poole, Group Managing Director*

BLUE CIRCLE INDUSTRIES plc (CEMENT WORKS)
Head Office *12th Floor, Churchill Plaza, Churchill Way, Basingstoke, Hants RG21 1QU*
Phone *0256 844500* **Fax** *0256 470608*
Personnel *D A R Poole, Managing Director*
Quarries *Aberthaw, Cauldon, Dunbar Northwest, Eastgate, Great Blakenham, Holborough, Hope, Masons, Melton, Norman, Northfleet, Plymstock, Ruthin, Shoreham, Westbury*

BLUE CIRCLE INDUSTRIES plc (CEMENT WORKS)

ABERTHAW QUARRY   99
Rhoose, Barry, S Glam CF6 9ZR
Phone 0446 750204 **Telex** 497040
OS Map 170 **OS Grid Reference** ST038672
Personnel G L Reid, General Manager
A Dauncey, Unit Manager

PRODUCT DATA	**Rock type**	Jurassic Lower Lias Limestone
	Colour	Grey
	Grain	Fine
	Products	Crushed Stone
		Industrial Limestone

DRILLING DATA		Hole Diameter 105mm
		Bench Height 20m
		Contractor EMK Drilling & Blasting

LOAD AND HAUL	**Trucks**	(2)Aveling Barford RD40
		(2)Aveling Barford RD55
	Loaders	(2)Caterpillar 966C
		Volvo

CRUSHING PLANT	**Primary**	GEC Double Rotor
	Secondary	None
	Tertiary	None

BLUE CIRCLE INDUSTRIES plc (CEMENT WORKS)

CAULDON QUARRY   100
Waterhouses, Stoke-on-Trent, Staffs ST10 3EQ
Phone 0538 308000
OS Map 119 **OS Grid Reference** SK085492
Personnel D Beetham, Unit Manager

PRODUCT DATA	**Rock type**	Carboniferous Milldale Limestone
	Colour	Light creamy grey
	Products	Crushed Stone
		Asphalt or Tarmacadam
		Industrial Limestone
		Agricultural Lime
DRILLING DATA		Hole Diameter 135mm
		Bench Height 20m
		Contractor EMK Drilling & Blasting
SECONDARY BREAKING		Hydraulic Hammer
LOAD AND HAUL	**Trucks**	(4)Aveling Barford RD50
	Loaders	Caterpillar 992C
		Caterpillar 996C
CRUSHING PLANT	**Primary**	Nordberg 74-75 Gyratory
	Secondary	MMD Mineral Sizer
	Tertiary	None

BLUE CIRCLE INDUSTRIES plc (CEMENT WORKS)

DUNBAR NORTHWEST QUARRY   **101**

Dunbar, Lothian EH42 1SL
Phone 0368 63371
OS Map 67   **OS Grid Reference** NT705770
Personnel C J Johnson, General Manager
       A Ruxton, Unit Manager

PRODUCT DATA	**Rock type**	Carboniferous Lower Limestone
	Colour	Not commercially significant
	Products	Crushed Stone
		Industrial Limestone
DRILLING DATA		Hole Diameter 110mm
		Bench Height 7.5m
		Contractor EMK Drilling & Blasting
	Drills	Drilltec 25
LOAD AND HAUL	**Trucks**	Aveling Barford RD50
		Aveling Barford RD55
	Loaders	Caterpillar 988B
		Demag H121
CRUSHING PLANT	**Primary**	Hazemag AP7 Impactor
	Secondary	Hazemag AP13 Impactor
	Tertiary	None

BLUE CIRCLE INDUSTRIES plc (CEMENT WORKS)

EASTGATE QUARRY   **102**

Eastgate, Stanhope, Bishop Auckland, Durham DL13 2AA
Phone 0388 517286
OS Map 92   **OS Grid Reference** NY940375
Personnel Dr E Dack, Unit Manager

entry continues overleaf

PRODUCT DATA	**Rock type**	Carboniferous Great Limestone
	Colour	Grey
	Products	Crushed Stone
		Industrial Limestone

DRILLING DATA		Hole Diameter 115mm
		Bench Height 7–12m
	Drills	Halco SPD43–22

| SECONDARY BREAKING | | Dropball on Ruston-Bucyrus 22-RB |

LOAD AND HAUL	**Face shovel**	O&K RH40
	Trucks	(2)Aveling Barford RD225
		(2)Aveling Barford RD255
		(3)Aveling Barford RD40
		Terex 3307
	Loaders	Caterpillar 988B

CRUSHING PLANT	**Primary**	Babbitless 42″ Twinroll
	Secondary	None
	Tertiary	None

BLUE CIRCLE INDUSTRIES plc (CEMENT WORKS)

GREAT BLAKENHAM QUARRY 103

Claydon, Ipswich, Suffolk IP6 0JX
Phone 0473 830213
OS Map 155 **OS Grid Reference** TM111500
Personnel R Gnatt, Unit Manager

PRODUCT DATA	**Rock type**	Cretaceous Upper Chalk
	Colour	White
	Grain	Fine
	Product	Crushed Stone

RIPPABLE STRATA

BLUE CIRCLE INDUSTRIES plc (CEMENT WORKS)

HOLBOROUGH QUARRY 104

Snodland, Kent ME6 5PH
Phone 0256 844500
OS Map 178 **OS Grid Reference** TQ695624

QUARRY ON STAND

ROCK DATA	**Rock type**	Cretaceous Lower Middle Chalk
	Colour	White
	Grain	Fine

BLUE CIRCLE INDUSTRIES plc (CEMENT WORKS)

HOPE QUARRY 105

Hope, Sheffield, S Yorks S30 2RP
Phone 0433 20317
OS Map 110 **OS Grid Reference** SK159820
Personnel S Joyce, General Manager
P Dumenil, Unit Manager

PRODUCT DATA	**Rock type**	Carboniferous Monsal Dale Limestone
	Colour	Not commercially significant
	Products	Crushed Stone
		Industrial Limestone

DRILLING DATA		Hole Diameter 160mm
		Bench Height 12.5-20m
		Contractor Peak Drilling
	Drills	Halco HPD900-35
		Ingersoll-Rand DM45

| SECONDARY BREAKING | | Dropball on Ruston-Bucyrus 22-RB |

LOAD AND HAUL	**Face shovel**	O&K RH40
	Trucks	(2)Aveling Barford RD50
		(5)Aveling Barford RD55
	Loaders	(3)Caterpillar 988D

CRUSHING PLANT	**Primary**	Fuller 54″ Gyratory (Mobile)
	Secondary	(2)Cone
	Tertiary	(2)Nordberg Cone

BLUE CIRCLE INDUSTRIES plc (CEMENT WORKS)

MASONS QUARRY 106

Claydon, Nr Ipswich, Suffolk IP6 0JX
Phone 0473 830213
OS Map 169   **OS Grid Reference** TM136498
Personnel M Watson, General Manager
R J Gant, Unit Manager

PRODUCT DATA	**Rock type**	Cretaceous Middle Chalk
	Colour	White
	Grain	Fine
	Product	Crushed Stone

RIPPABLE STRATA

LOAD AND HAUL	**Face shovel**	Marion 101
		Ruston-Bucyrus 54-RB
	Trucks	None
	Loaders	None

BLUE CIRCLE INDUSTRIES plc (CEMENT WORKS)

MELTON QUARRY 107

Melton Whiting Works, North Ferriby, Humbers HU14 3HW
Phone 0482 633351
OS Map 106   **OS Grid Reference** SE966278
Personnel J G E Nichols, Unit Manager

PRODUCT DATA	**Rock type**	Cretaceous Ferriby Welton Chalk
	Colour	White
	Products	Industrial Limestone
		Burnt Limestone

RIPPABLE STRATA

BLUE CIRCLE INDUSTRIES plc (CEMENT WORKS)

NORMAN QUARRY

108

Sawston, Cambs CB2 4HE
Phone 0256 844500
OS Map 154   **OS Grid Reference** TL480510

QUARRY ON STAND

ROCK DATA	**Rock type**	Cretaceous Lower Chalk
	Colour	White
	Grain	Fine

BLUE CIRCLE INDUSTRIES plc (CEMENT WORKS)

NORTHFLEET QUARRY

109

The Shore, Northfleet, Dartford, Kent DA11 9AN
Phone 0474 64355
OS Map 177   **OS Grid Reference** TQ590735
Personnel M J Boyce, General Manager

PRODUCT DATA	**Rock type**	Cretaceous Middle Chalk
	Colour	White
	Grain	Fine
	Product	Crushed Stone

RIPPABLE STRATA

LOAD AND HAUL	**Face shovel**	Bucyrus-Erie 280B
		Liebherr 994
	Trucks	None
	Loaders	None
CRUSHING PLANT	**Primary**	(2)Weserhutte Titan-Miag (Mobile)
		(2)Babbitless (Mobile)
	Secondary	None
	Tertiary	None

BLUE CIRCLE INDUSTRIES plc (CEMENT WORKS)

PLYMSTOCK QUARRY

110

Plymstock, Plymouth, Devon PL9 7JA
Phone 0752 402121
OS Map 202   **OS Grid Reference** SX509544
Personnel G Davies, Unit Manager

PRODUCT DATA	**Rock type**	Devonian Limestone
	Colour	Medium grey
	Products	Crushed Stone
		Industrial Limestone
DRILLING DATA		Hole Diameter 108mm
		Bench Height 10m
	Drills	Halco SPD43-22
SECONDARY BREAKING		Hydraulic Hammer

LOAD AND HAUL	**Face shovel**	O&K
	Trucks	(3)Aveling Barford RD40
	Loaders	(2)Clark Michigan L275
CRUSHING PLANT	**Primary**	BJD Hammermill
	Secondary	Hammermill

BLUE CIRCLE INDUSTRIES plc (CEMENT WORKS)

RUTHIN QUARRY   111

St Mary Hill, Bridgend, Mid Glam CF35 5DY
Phone 0656 860207
OS Map 170   **OS Grid Reference** SS973793
Personnel A Dauncey, Unit Manager

PRODUCT DATA	**Rock type**	Carboniferous Limestone
	Colour	Medium grey
	Grain	Medium
	Products	Crushed Stone
		Industrial Limestone

DRILLING DATA		Hole Diameter 105mm
		Bench Height 12-17m
		Contractor EMK Drilling & Blasting

| SECONDARY BREAKING | | Hydraulic Hammer |

LOAD AND HAUL	**Face shovel**	Caterpillar 225
	Trucks	Aveling Barford RD40
	Loaders	Caterpillar 988C

CRUSHING PLANT	**Primary**	Hazemag SAP5 Impactor
	Secondary	None
	Tertiary	None

BLUE CIRCLE INDUSTRIES plc(CEMENT WORKS)

SHOREHAM QUARRY   112

Beeding, Steyning, W Sussex BN4 3TX
Phone 0273 452266
OS Map 198   **OS Grid Reference** TQ205088
Personnel V Langley, Unit Manager

PRODUCT DATA	**Rock type**	Cretaceous Upper Chalk
	Colour	White
	Grain	Fine
	Product	Industrial Limestone

RIPPABLE STRATA

BLUE CIRCLE INDUSTRIES plc (CEMENT WORKS)

WESTBURY QUARRY   113

Trowbridge Road, Westbury, Wilts BA13 4LX
Phone 0373 822481
OS Map 183   **OS Grid Reference** ST880500
Personnel B Capon, General Manager
               D Beatty, Unit Manager

entry continues overleaf

PRODUCT DATA	**Rock type**	Cretaceous Chalk
	Colour	Medium grey
	Grain	Fine
	Products	Crushed Stone
		Industrial Limestone

RIPPABLE STRATA

| LOAD AND HAUL | **Trucks** | Terex |
| | **Loaders** | Caterpillar 966C |

BODFARI QUARRIES LTD
Head Office Midland Bank Chambers, Holywell, Clwyd CH8 7TH
Phone 0352 712581
Personnel G P Gibson, Director and General Manager
Quarries Burley Hill

BODFARI QUARRIES LTD
BURLEY HILL QUARRY 114
Pant Du Nercwys, Mold, Clwyd CH8 4EH
Phone 0352 85551
OS Map 117 **OS Grid Reference** SJ204596
Personnel M Wilkon, Unit Manager

PRODUCT DATA	**Rock type**	Carboniferous Limestone
	Colour	Not commercially significant
	Product	Crushed Stone

DRILLING DATA		Hole Diameter 125mm
		Bench Height 15m
	Drills	Hausherr HBM60

| SECONDARY BREAKING | | Hydraulic Hammer |

LOAD AND HAUL	**Trucks**	Aveling Barford RD30
		Aveling Barford RD40
	Loaders	Caterpillar 950
		Caterpillar 980C
		Caterpillar 988B
		Komatsu WA210

CRUSHING PLANT	**Primary**	SBM Wagoner
	Secondary	Mansfield No 5 Hammermill
	Tertiary	None

BORDER HARDCORE AND ROCKERY STONE CO LTD
Head Office Middletown Quarry, Middletown, Welshpool, Powys SY21 8DJ
Phone 093874 253
Personnel D W Hilditch, Director
Quarries Middletown

BORDER HARDCORE AND ROCKERY STONE CO LTD

MIDDLETOWN QUARRY

115

Middletown, Welshpool, Powys SY21 8DJ
Phone 093874 253
OS Map 126 **OS Grid Reference** SJ288119
Personnel D W Hilditch, Director and Unit Manager

PRODUCT DATA	**Rock type**	Dolerite
	Colour	Not Commercially Significant
	Grain	Medium
	Product	Crushed Stone

| DRILLING DATA | | No drilling |

BORDERS REGION COUNCIL
Head Office Council Offices, Newtown St Boswells, Melrose, Borders TD6 0SA
Phone 0835 23301 **Fax** 0835 22145
Personnel R I Hill, Technical Director
Quarries Edstone

BORDERS REGION COUNCIL

EDSTONE QUARRY

116

Peebles, Borders EH45 8NW
Phone 0721 20129
OS Map 73 **OS Grid Reference** NT227397
Personnel W Birchel, Unit Manager

PRODUCT DATA	**Rock type**	Silurian Greywacke
	Colour	Grey to black
	Grain	Fine
	Product	Crushed Stone

DRILLING DATA		Hole Diameter 105mm
		Bench Height 12-15m
		Contractor Rockblast

| SECONDARY BREAKING | | Dropball on Ransomes & Rapier NCK-Rapier 305B |

| LOAD AND HAUL | **Trucks** | Aveling Barford RD30 |
| | **Loaders** | Caterpillar |

| CRUSHING PLANT | | No information provided |

BOSTON SPA MACADAMS LTD
Head Office Inglebank Quarry, Tadcaster Road, Boston Spa, Wetherby, W Yorks LS23 6BZ
Phone 0937 843493
Personnel M Kirby, Managing Director
Quarries Sandfold

BOSTON SPA MACADAMS LTD

SANDFOLD QUARRY

117

Barr Lane, Clifford, Wetherby, W Yorks LS23 4RZ
Phone 0937 843493
OS Map 105 **OS Grid Reference** SE428443

entry continues overleaf

QUARRY ON STAND

ROCK DATA   **Rock type**   Carboniferous Sandstone

BOTHEL LIMESTONE AND BRICK COMPANY
Head Office *Moota Quarry, Cockermouth, Cumbria CA13 0QE*
Phone *0965 202234*
Personnel *M Milbourn, Managing Director*
Quarries *Moota*

BOTHEL LIMESTONE AND BRICK COMPANY

MOOTA QUARRY   **118**
Cockermouth, Cumbria CA13 0QE
Phone 0965 202234
OS Map 90   **OS Grid Reference** NY140363
Personnel K Stevenson, Works Manager

PRODUCT DATA   **Rock type**   Carboniferous Limestone
              Colour   Bluish grey
              Products   Crushed Stone
                            Asphalt or Tarmacadam

DRILLING DATA   Hole Diameter 105mm
                              Bench Height 12m
              Drills   Holman Tractor Vole Mkl

COMPRESSORS   Holman on board

SECONDARY BREAKING   Dropball on Ruston-Bucyrus 22-RB

LOAD AND HAUL   **Trucks**   Terex
                Loaders   Furakawa

CRUSHING PLANT   **Primary**   Baxter Jaw
                Secondary   Mansfield No 2 Hammermill
                               Mansfield No 3 Hammermill
                Tertiary   None

H V BOWEN AND SONS LTD
Head Office *Tan-y-Foel Quarry, Cefncoch, Llanfair Caereinion, Welshpool, Powys SY21 0AN*
Phone *0691 830642*
Personnel *H V Bowen, Director*
Quarries *Tan-y-Foel*

H V BOWEN AND SONS LTD

TAN-Y-FOEL QUARRY   **119**
Cefncoch, Llanfair Caereinion, Welshpool, Powys SY21 0AN
Phone 0691 830642
OS Map 136   **OS Grid Reference** SJ015015
Personnel H V Bowen, Director and Unit Manager

PRODUCT DATA	**Rock type**	Silurian Sandstone
	Colour	Not commercially significant
	Grain	Fine
	Products	Crushed Stone
		Asphalt or Tarmacadam
		Ready-mixed Concrete
DRILLING DATA		Hole Diameter 110mm
		Bench Height 17m
		Contractor Powys Drilling
SECONDARY BREAKING		Dropball
LOAD AND HAUL	**Loaders**	Volvo
CRUSHING PLANT	**Primary**	Kue Ken 42×27 Jaw (Double Toggle)
	Secondary	Kue Ken 36×10 Jaw
	Tertiary	Allis Chalmers 436 Cone

BOYNE BAY LIME COMPANY LTD
Head Office *54 Park Road, Aberdeen, Grampian AB2 1PA*
Phone *0224 632281*
Personnel *V Michie, Managing Director*
Quarries *Boyne Bay*

BOYNE BAY LIME COMPANY LTD

BOYNE BAY QUARRY

120

Boyndie, Banff, Grampian AB4 2AE
Phone 0261 7466
OS Map 29 **OS Grid Reference** NJ615659
Personnel A Innes, Unit Manager

PRODUCT DATA	**Rock type**	Pre-Cambrian Dalradian Limestone
	Colour	Not commercially significant
	Product	Agricultural Lime
DRILLING DATA		Hole Diameter 90mm
		Bench Height 15m
	Drills	Halco 17HR Wagon-drill
SECONDARY BREAKING		Dropball on Ruston-Bucyrus 19-RB
LOAD AND HAUL	**Trucks**	None
	Loaders	Ford A62
		Ford A64
CRUSHING PLANT	**Primary**	Baxter 32×22 Jaw
	Secondary	Kue Ken 24×10 Jaw
	Tertiary	Kue Ken 36×10 Jaw

BREEDON plc
Head Office *Breedon Hill Quarry, Breedon on the Hill, Derby, Leics DE7 1AP*
Phone *0332 862254* **Fax** *0332 863149*
Personnel *J G Shields, Managing Director*
Quarries *Breedon Hill*

entry continues overleaf

BREEDON plc

BREEDON HILL QUARRY

121

Breedon on the Hill, Derby, Leics DE7 1AP
Phone 0332 862254
OS Map 129   **OS Grid Reference** SK406233
Personnel B Cowan, Unit Manager

PRODUCT DATA	**Rock type**	Carboniferous Limestone
	Colour	Reddish grey
	Product	Crushed Stone

DRILLING DATA		Hole Diameter 115mm
		Bench Height 12.5m
	Drills	Halco SPD43-22

| SECONDARY BREAKING | | Hydraulic Hammer - Montabert |

LOAD AND HAUL	**Trucks**	(2)Euclid R501
	Loaders	Michigan L270
		Volvo

CRUSHING PLANT	**Primary**	Pegson 42 × 48 Jaw (Double Toggle)
	Secondary	Hazemag APK1313 Impactor
		Pegson 48″ Gyrasphere
	Tertiary	None

R A BRIGGS AND COMPANY LTD

Head Office *Woodside Quarry, Ring Road, Leeds, W Yorks LS16 6QG*
Phone *0532 582214*
Personnel *R P Spragg, Director*
         H D Walker, Director
Quarries *Woodside*

R A BRIGGS AND COMPANY LTD

WOODSIDE QUARRY

122

Ring Road, Leeds, W Yorks LS16 6QG
Phone 0532 582214
OS Map 104   **OS Grid Reference** SE256383
Personnel J Hirst, Manager

PRODUCT DATA	**Rock type**	Carboniferous Millstone Grit
	Colour	White to yellow
	Grain	Fine
	Product	Dressed or Lump Stone

| DRILLING DATA | | Contractor |

NO FURTHER INFORMATION

BRIGHOUSE BRICK, TILE AND STONE CO LTD

Corporate Head Office *Tower Farm, Hollins Lane, Steeton, Keighley, W Yorks BD20 6LY*
Phone *0535 654864*
Personnel *L Evans, Managing Director*
         K Atkinson, Director
Companies *Cornish Roadstones* — see separate entry

BRITISH GYPSUM LTD
Head Office *Ruddington Hall, Ruddington, Notts NG11 6LX*
Phone *0602 844844* **Fax** *0602 846644* **Telex** *377292*
Personnel *A Brook, Chairman*
A Clark, Managing Director
A Thompson, Production Director
B Breem, Chief Mining Engineer
Quarries *Bantycock, Staple Farm*

BRITISH GYPSUM LTD

BANTYCOCK QUARRY

123

Beacon Hill Works, Newark, Notts NG24 2JO
Phone 0636 703351
OS Map 121   **OS Grid Reference** SK810500
Personnel T Baldwin, General Manager
B Adams, Unit Manager

PRODUCT DATA	**Rock type**	Triassic Mercia Mudstone
	Colour	Not commercially significant
	Grain	Fine
	Product	Industrial Use
DRILLING DATA		Hole Diameter 50mm
		Bench Height 5.5–8m
	Drills	(4)Turmag-Vol
LOAD AND HAUL	**Face shovel**	Ruston-Bucyrus
		(3)O&K RH75
	Trucks	Caterpillar 773B
		(2)Volvo BM A25
		(8)Volvo BM A35

BRITISH GYPSUM LTD

STAPLE FARM QUARRY

124

Beacon Hill Works, Newark, Notts NG24 2JO
Phone 0636 703351
OS Map 121   **OS Grid Reference** SK805500
Personnel T Baldwin, General Manager
B Adams, Unit Manager

PRODUCT DATA	**Rock type**	Triassic Mercia Mudstone
	Colour	Not commercially significant
	Grain	Fine
	Product	Industrial Use
DRILLING DATA		Hole Diameter 50mm
		Bench Height 5.5–8m
	Drills	(4)Turmag-Vol
LOAD AND HAUL	**Face shovel**	Ruston-Bucyrus
		(3)O&K RH75
	Trucks	Caterpillar 773B
		(2)Volvo BM A25
		(8)Volvo BM A35

BRITISH RAIL
Head Office *Meldon Quarry, Okehampton, Devon EX20 4LT*
Phone *0837 2749*
Personnel *A Dumpleton, General Manager*
Quarries *Meldon*

BRITISH RAIL

MELDON QUARRY

125

Okehampton, Devon EX20 4LT
Phone 0837 2749
OS Map 191 **OS Grid Reference** SX570925
Personnel G Norris, Unit Manager

PRODUCT DATA	**Rock type**	Hornfels Schist
	Colour	Light speckled grey
	Grain	Fine
	Product	Crushed Stone
DRILLING DATA		Hole Diameter 115mm
		Bench Height 14–17m
	Drills	Halco MPD60-25
		Holman Voltrak
SECONDARY BREAKING		Hydraulic Hammer
		Talisker on Poclain 90P
LOAD AND HAUL	**Face shovel**	Ruston-Bucyrus 38-RB
		Liebherr
	Trucks	(4)Aveling Barford RD40
		(2)Foden 27t
		(3)Foden 35t
	Loaders	Aveling Barford TS500
		Komatsu WA500
CRUSHING PLANT	**Primary**	Pegson 28B Gyratory
	Secondary	(2)Nordberg 5.5′ Cone
	Tertiary	Barmac 6900 Duopactor

BRITISH STEEL plc
Corporate Head Office *20 Inch Mill, Brenda Road, Hartlepool, Cleveland TS25 2EG*
Phone *0429 266611* **Fax** *0429 266611* **Telex** *341561*
Personnel *F W Wallace, Managing Director*

BRITISH STEEL plc (QUARRY PRODUCTS)
Head Office *Hardendale Quarry, Shap, Penrith, Cumbria CA10 3QG*
Phone *09316 647*
Personnel *J G Edwards, General Manager*
Quarries *Hardendale*

BRITISH STEEL plc (QUARRY PRODUCTS)

HARDENDALE QUARRY

126

Shap, Penrith, Cumbria CA10 3QG
Phone 09316 647
OS Map 91 **OS Grid Reference** NY585140
Personnel A Anderson, Unit Manager

PRODUCT DATA	**Rock type**	Carboniferous Great Scar Limestone
	Colour	Grey
	Products	Crushed Stone
		Industrial Limestone

DRILLING DATA		Hole Diameter 110mm
		Bench Height 15m
		Contractor Ritchies Equipment

| SECONDARY BREAKING | | Dropball on Ruston-Bucyrus 22-RB |

| LOAD AND HAUL | **Trucks** | Terex |
| | **Loaders** | Caterpillar 920 |

CRUSHING PLANT	**Primary**	Morgardshammar ER65×48 Hammermill
	Secondary	Babbitless double roll
	Tertiary	(5)BJD Pulverizer

BRODEN CONSTRUCTION (NE) LTD
Head Office Industrial Estate, New Herrington, Houghton-le-Spring, Tyne and Wear DH4 7BG
Phone Not available
   Not Trading
Quarries Copp Grag, Prudham

BRODEN CONSTRUCTION (NE) LTD
COPP GRAG QUARRY 127
Redesdale Camp, Byrness, Northumb NE19 1TD
Phone 1234
OS Map 8004   **OS Grid Reference** NY822995

QUARRY ON STAND

| ROCK DATA | **Rock type** | Carboniferous Limestone |
| | **Colour** | Brown to white |

BRODEN CONSTRUCTION (NE) LTD
PRUDHAM QUARRY 128
Fourstones, Hexham, Northumb NE47 4DL
Phone 1234
OS Map 87   **OS Grid Reference** NY885688

QUARRY ON STAND

ROCK DATA	**Rock type**	Carboniferous Limestone
	Colour	Cream to buff
	Grain	Fine

BRONTE STONE COMPANY
Head Office Coldedge, Withens, Halifax, W Yorks HX3 5LL
Phone 0422 240909
Personnel J Brown, Partner
Quarries Fly Flatts, Hunters Hill

entry continues overleaf

FLY FLATTS QUARRY

129

Coldedge, Withens, Halifax, W Yorks HX3 5LL
Phone 0422 240909
OS Map 104 **OS Grid Reference** SD049302
Personnel J Brown, Partner and Unit Manager

PRODUCT DATA	**Rock type**	Carboniferous Sandstone
	Colour	Bluish grey
	Grain	Medium
	Product	Dressed or Lump Stone

| DRILLING DATA | | No drilling |

BRONTE STONE COMPANY
HUNTERS HILL QUARRY

130

Coldedge, Withens, Halifax, W Yorks HX3 5LL
Phone 0422 240909
OS Map 104 **OS Grid Reference** SD052293
Personnel J Brown, Partner and Unit Manager

PRODUCT DATA	**Rock type**	Carboniferous Sandstone
	Colour	Bluish grey
	Grain	Medium
	Product	Dressed or Lump Stone

| DRILLING DATA | | No drilling |

BROWN BROS (LONGRIDGE) LTD
Head Office *Altham Quarry, Whinney Hill Road, Accrington, Lancs BB5 6NR*
Phone *0254 871049*
Personnel *C McMalus, Director*
Quarries *Altham, Leeming*

BROWN BROS (LONGRIDGE) LTD
ALTHAM QUARRY

131

Whinney Hill Road, Accrington, Lancs BB5 6NR
Phone 0254 871049
OS Map 103 **OS Grid Reference** SD767332
Personnel C McMalus, Director and Unit Manager

PRODUCT DATA	**Rock type**	Carboniferous Sandstone
	Products	Dressed or Lump Stone
		Crushed Stone

RIPPABLE STRATA

BROWN BROS (LONGRIDGE) LTD
LEEMING QUARRY

132

Aighton, Stony Hurst, Blackburn, Lancs BB6 9HY
Phone 0254 86457
OS Map 103 **OS Grid Reference** SD646278
Personnel D McGrath, Unit Manager

PRODUCT DATA	**Rock type**	Carboniferous Millstone Grit
	Grain	Medium
	Products	Dressed or Lump Stone
		Crushed Stone

DRILLING DATA		Hole Diameter 110mm
		Bench Height 5-10m
	Drills	Halco 410C

| COMPRESSORS | | CompAir 650HE |

| SECONDARY BREAKING | | Dropball on Ruston-Bucyrus 19-RB |
| | | Dropball on Ruston-Bucyrus 22-RB |

| LOAD AND HAUL | **Face shovel** | Ruston-Bucyrus 22-RB |
| | **Trucks** | Scammel |

| CRUSHING PLANT | | None |

BRUCE PLANT LTD
Head Office *Spring Works, High Street, Inverbervie by Montrose, Grampian DD10 0RH*
Phone *0561 61265*
Personnel *I Bruce, Managing Director*
Quarries *Black Hills, Longside*

BRUCE PLANT LTD
BLACK HILLS QUARRY
Longhaven, Peterhead, Grampian AB4 7NU
Phone 0779 74507
OS Map 30   **OS Grid Reference** NK115372
Personnel R Stephen, Unit Manager

133

PRODUCT DATA	**Rock type**	Peterhead Granite
	Colour	Pink
	Grain	Coarse
	Product	Dressed or Lump Stone

| DRILLING DATA | | Hand-held equipment |

BRUCE PLANT LTD
LONGSIDE QUARRY
Peterhead, Grampian AB4 7TX
Phone 0561 61265
OS Map 30   **OS Grid Reference** NK027483
Personnel L Skea, Unit Manager

134

PRODUCT DATA	**Rock type**	Peterhead Granite
	Colour	Pink
	Grain	Coarse
	Product	Crushed Stone

DRILLING DATA		Hole Diameter 105mm
		Bench Height 22m
		Contractor Albion Drilling Services

entry continues overleaf

SECONDARY BREAKING		Hydraulic Hammer - Talisker
LOAD AND HAUL	**Face shovel**	Akermans 10MB
	Trucks	Volvo
	Loaders	Komatsu
CRUSHING PLANT	**Primary**	Parker 36 × 24 Jaw
	Secondary	Pegson 1200 Autocone
		Nordberg 4.25′ Cone
	Tertiary	(2)Goodwin Barsby 30 × 6 Jaw

BULLIMORES SAND AND GRAVEL LTD
Head Office *South Witham, Grantham, Lincs NG33 5QE*
Phone *057283 393/627*
Personnel *C Bullimore, Managing Director*
N Bullimore, Production Director
Quarries *Clipsham (Big Pits), Collyweston*

BULLIMORES SAND AND GRAVEL LTD

CLIPSHAM QUARRY (BIG PITS QUARRY)   135

Clipsham, Stamford, Lincs PE9 1PY
Phone 0780 81634
OS Map 130   **OS Grid Reference** SK968145
Personnel N Bullimore, Production Director and Unit Manager

PRODUCT DATA	**Rock type**	Jurassic Oolitic Limestone
	Colour	Creamy brown
	Grain	Fine
	Products	Dressed or Lump Stone
		Crushed Stone
		Agricultural Lime
DRILLING DATA		Hole Diameter 105mm
		Bench Height 10-12m
	Drills	Holman Voltrak
COMPRESSORS		CompAir RO37HP
LOAD AND HAUL	**Trucks**	None
	Loaders	(2)Zettelmeyer ZL1801
CRUSHING PLANT	**Primary**	Parker Jaw
	Secondary	Parker Jaw
	Tertiary	Parker Jaw

NO FURTHER INFORMATION PROVIDED

BULLIMORES SAND AND GRAVEL LTD

COLLYWESTON QUARRY   136

Collyweston, Stamford, Lincs PE9 3PY
Phone 0780 83353
OS Map 130   **OS Grid Reference** TF001015
Personnel T Todkill, Unit Manager

PRODUCT DATA	**Rock type**	Jurassic Oolitic Limestone
	Colour	Creamy grey
	Grain	Fine
	Product	Crushed Stone

DRILLING DATA		Hole Diameter 105mm
		Bench Height 15m
	Drills	Holman Voltrak

| COMPRESSORS | | CompAir RO37HP |

| SECONDARY BREAKING | | Hydraulic Hammer - Furukawa on Komatsu PC210 |

LOAD AND HAUL	**Face shovel**	Akermans H14B
	Trucks	(3)Nordvick
	Loaders	(2)Komatsu WA450

CRUSHING PLANT	**Primary**	Pegson (Mobile)
	Secondary	Hazemag 1013K Impactor
	Tertiary	None

BURLINGTON SLATE LTD
Head Office *Cavendish House, Kirkby-in-Furness, Cumbria LA17 7UN*
Phone *022989 661/666* **Fax** *022989 466* **Telex** *65157*
Personnel *D C Wallace, Managing Director*
      H Ogden, Director
      G M Brownlee, Director
      T Southward, Director
      M Dickinson, General Manager
Quarries *Baycliff, Brandy Crag, Brathay, Broughton Moor, Burlington, Bursting Stone, Elterwater, Kirkby, Moss Rigg, Skiddaw, Spout Cragg*

BURLINGTON SLATE LTD
BAYCLIFF QUARRY   **137**
Ulverston, Cumbria LA12 9QE
Phone 022989 661
OS Map 96 **OS Grid Reference** SD288780

QUARRY ON STAND

| ROCK DATA | **Rock type** | Ordovician Borrowdale Slate |

BURLINGTON SLATE LTD
BRANDY CRAG QUARRY   **138**
Coniston, Cumbria LA21 8DJ
Phone 022989 661
OS Map 96 **OS Grid Reference** SD284985
Personnel M Dickinson, General Manager

PRODUCT DATA	**Rock type**	Ordovician Borrowdale Slate
	Colour	Silver grey
	Grain	Fine
	Product	Dressed or Lump Stone

| DRILLING DATA | | Hand-held equipment |

BURLINGTON SLATE LTD
BRATHAY QUARRY
Ambleside, Cumbria LA22 0HN
Phone 022989 661
OS Map 90 **OS Grid Reference** NY357017

139

QUARRY ON STAND

ROCK DATA **Rock type** Silurian Brathay Slate

BURLINGTON SLATE LTD
BROUGHTON MOOR QUARRY
Coniston, Lancs LA21 8DJ
Phone 022989 661
OS Map 96 **OS Grid Reference** SD255946
Personnel M Dickinson, General Manager

140

PRODUCT DATA	**Rock type**	Ordovician Borrowdale Slate
	Colour	Olive green
	Grain	Fine
	Product	Dressed or Lump Stone

DRILLING DATA Hand-held equipment

BURLINGTON SLATE LTD
BURLINGTON QUARRY
Kirkby-in-Furness, Coniston, Cumbria LA20 2DE
Phone 022989 661
OS Map 96 **OS Grid Reference** SD319998
Personnel M Dickinson, General Manager

141

PRODUCT DATA	**Rock type**	Ordovician Borrowdale Slate
	Colour	Bluish grey
	Grain	Fine
	Product	Dressed or Lump Stone

DRILLING DATA Hand-held equipment

BURLINGTON SLATE LTD
BURSTING STONE QUARRY
Coniston, Cumbria LA21 8HD
Phone 022989 661
OS Map 96 **OS Grid Reference** SD278974
Personnel M Dickinson, General Manager

142

PRODUCT DATA	**Rock type**	Ordovician Borrowdale Slate
	Colour	Light olive green
	Grain	Fine
	Product	Dressed or Lump Stone

DRILLING DATA Hand-held equipment

BURLINGTON SLATE LTD
ELTERWATER QUARRY

143

Little Langdale, Ambleside, Cumbria LA22 8DJ
Phone 022989 661
OS Map 90   **OS Grid Reference** NY325048
Personnel M Dickinson, General Manager

PRODUCT DATA	**Rock type**	Ordovician Borrowdale Slate
	Colour	Light green
	Grain	Fine
	Product	Dressed or Lump Stone

DRILLING DATA		Hand-held equipment

BURLINGTON SLATE LTD
KIRKBY QUARRY

144

Broughton-in-Furness, Cumbria LA20 6DE
Phone 022989 661
OS Map 96   **OS Grid Reference** SD245836
Personnel M Dickinson, General Manager

PRODUCT DATA	**Rock type**	Silurian Brathay Slate
	Colour	Bluish grey
	Grain	Fine
	Product	Dressed or Lump Stone

DRILLING DATA		Hand-held equipment

BURLINGTON SLATE LTD
MOSS RIGG QUARRY

145

Skelwith Bridge, Ambleside, Cumbria LA22 9HT
Phone 022989 661
OS Map 90   **OS Grid Reference** NY314025
Personnel M Dickinson, General Manager

PRODUCT DATA	**Rock type**	Ordovician Borrowdale Slate
	Colour	Olive green
	Grain	Fine
	Product	Dressed or Lump Stone

DRILLING DATA		Hand-held equipment

BURLINGTON SLATE LTD
SKIDDAW QUARRY

146

Baycliff, Ulverston, Cumbria LA12 0QA
Phone 022989 661
OS Map 96   **OS Grid Reference** SD288723

QUARRY ON STAND

ROCK DATA	**Rock type**	Ordovician Borrowdale Slate
	Colour	Light buff, dark cream
	Grain	Fine

BURLINGTON SLATE LTD

SPOUT CRAGG QUARRY

147

Little Langdale, Ambleside, Cumbria LA22 8DJ
Phone 022989 661
OS Map 90 **OS Grid Reference** NY308052
Personnel M Dickinson, General Manager

PRODUCT DATA	**Rock type**	Ordovician Borrowdale Slate
	Colour	Greyish green
	Grain	Fine
	Product	Dressed or Lump Stone
DRILLING DATA		Hand-held equipment

NEIL BUTCHER
Head Office *Park Nook Quarry, Bannister Lane, Skelbrooke, Doncaster, S Yorks DN6 8LT*
Phone *0302 727924*
Personnel *N Butcher, Director*
Quarries *Park Nook*

NEIL BUTCHER

PARK NOOK QUARRY

148

Bannister Lane, Skelbrooke, Doncaster, S Yorks DN6 8LT
Phone 0302 727924
OS Map 111 **OS Grid Reference** SE515126
Personnel N Butcher, Director and Unit Manager

PRODUCT DATA	**Rock type**	Permian Magnesian Limestone
	Colour	Dark buff
	Grain	Fine
	Product	Dressed or Lump Stone

BUTTERLEY AGGREGATES LTD (RMC GROUP plc)
Head Office *17–21 West Parade, Lincoln, Lincs LN1 1NP*
Phone *0522 523391* **Fax** *0522 511230*
Personnel *A D Lamond, Managing Director*
J B Cooper, Divisional Director
D R Swinson, Divisional Manager
A Singleton, Area Manager
Quarries *Greetwell, Kirton Lindsey*

BUTTERLEY AGGREGATES LTD (RMC GROUP plc)

GREETWELL QUARRY

149

Fiskerton Road, Lincoln, Lincs, LN3 4NH
Phone 0522 523391
OS Map 121 **OS Grid Reference** TF004723
Personnel T Atkin, Unit Manager

PRODUCT DATA	**Rock type**	Jurassic Inferior Oolitic Limestone
	Colour	White
	Grain	Fine
	Product	Crushed Stone

DRILLING DATA		Hole Diameter 105mm
		Bench Height 5–6m
	Drills	Holman Voltrak

COMPRESSORS	CompAir RO60–250

SECONDARY BREAKING		Dropball on Ruston-Bucyrus 22RB
Load and Haul		
	Face shovel	Liebherr 952
	Loaders	Volvo L160
	Trucks	(2)Heathfield 25t

CRUSHING PLANT	No information provided

BUTTERLEY AGGREGATES LTD (RMC GROUP plc)

KIRTON LINDSEY QUARRY　　　　　150

Gainsthorpe Road, Kirton Lindsey, Gainsborough, Lincs, DN21 4JH
Phone 0522 523391
OS Map 112　**OS Grid Reference** SK928975
Personnel T Atkin, Unit Manager

PRODUCT DATA	**Rock type**	Jurassic Inferior Oolitic Limestone
	Colour	Cream
	Grain	Fine
	Product	Crushed Stone

DRILLING DATA		Hole Diameter 105mm
		Bench Height 10m
	Drills	Atlas Copco ROC404-04

COMPRESSORS	Compair 650

SECONDARY BREAKING		Dropball on Ruston-Bucyrus 22RB
Load and Haul		
	Loaders	Volvo
	Trucks	(4)Heathfield 25t

CRUSHING PLANT	No information provided

CASTLE CEMENT (KETTON) LTD
Head Office *Priestgate, Peterborough, Cambs PE1 1DF*
Phone *0733 310505*
Personnel *L V Hewitt, Managing Director*
Quarries *Ketton (Grange Top), Ribblesdale Works*

CASTLE CEMENT (KETTON) LTD

KETTON QUARRY (GRANGE TOP QUARRY)　　151

Stamford, Lincs PE9 3SX
Phone 0780 720501
OS Map 141　**OS Grid Reference** SK976057
Personnel K Hudson, Unit Manager

PRODUCT DATA	**Rock type**	Jurassic Inferior Oolitic Limestone
	Colour	Cream to brownish buff
	Grain	Fine
	Product	Industrial Limestone

entry continues overleaf

DRILLING DATA		Hole Diameter 130mm
		Bench Height 15m
	Drills	Hausherr HBM120
LOAD AND HAUL	**Face shovel**	Komatsu
	Trucks	(5)Aveling Barford RD50
	Loaders	Caterpillar 950
		Caterpillar 966C
		Caterpillar 980C
CRUSHING PLANT	**Primary**	Krupp Hammermill
		Sheepbridge 62 × 72 Jaw
	Secondary	None
	Tertiary	None

CASTLE CEMENT (KETTON) LTD

RIBBLESDALE WORKS 152

Clitheroe, Lancs BB7 4QA
Phone 0200 22401 **Fax** 0200 28969
OS Map 103 **OS Grid Reference** SD760440
Personnel A MacKechnie, Unit Manager

PRODUCT DATA	**Rock type**	Carboniferous Limestone
	Colour	Brown to buff
	Grain	Fine
	Product	Industrial Limestone
DRILLING DATA	**Drills**	(2)Halco MPD60-25T
		Halco 650H
RIPPING	**Ripper**	Caterpillar
LOAD AND HAUL	**Ripper**	Caterpillar
	Face shovel	(3)Ruston-Bucyrus 110-RB
		(2)O&K RH30C
		Caterpillar
	Trucks	(14)Terex 3307
	Loaders	(2)Caterpillar
		Terex
CRUSHING PLANT	**Primary**	Sheepbridge 72 × 60 Jaw
		Sheepbridge 60 × 48 Jaw
	Secondary	GEC Reversible Impactor
	Tertiary	None

CASTLE HILL QUARRY COMPANY LTD
Head Office *Castle Hill Quarry, Cannington, Bridgwater, Somerset TA5 2QD*
Phone *0278 652280*
Personnel *W G King, Owner*
Quarries *Castle Hill*

CASTLE HILL QUARRY COMPANY LTD

CASTLE HILL QUARRY

153

Cannington, Bridgwater, Somerset TA5 2QD
Phone 0278 652280
OS Map 182 **OS Grid Reference** ST252405
Personnel S Ford, Unit Manager

PRODUCT DATA	**Rock type**	Carboniferous Limestone
	Colour	Not commercially significant
	Product	Crushed Stone
DRILLING DATA		Hole Diameter 105mm
		Bench Height 12m
		Contractor Ritchies Equipment
SECONDARY BREAKING		Hydraulic Hammer - Furukawa on Komatsu
LOAD AND HAUL	**Trucks**	(3)Foden Leyland
	Loaders	Bray 500
		Bray 1100
CRUSHING PLANT	**Primary**	BJD Impactor

NO FURTHER INFORMATION PROVIDED

CAST VALE QUARRIES (HARGREAVES QUARRIES LTD)
Head Office *Vale Road Quarry, Mansfield Woodhouse, Mansfield, Notts NG19 8DP*
Phone *0623 24570*
Personnel *M Wallen, General Manager*
Quarries *Vale Road*

CAST VALE QUARRIES (HARGREAVES QUARRIES LTD)

VALE ROAD QUARRY

154

Mansfield Woodhouse, Mansfield, Notts NG19 8DP
Phone 0623 24570
OS Map 120 **OS Grid Reference** SK530650
Personnel M Wallen, General Manager

PRODUCT DATA	**Rock type**	Permian Magnesian Limestone
	Colour	Not commercially significant
	Grain	Fine
	Products	Dressed or Lump Stone
		Crushed Stone
		Asphalt or Tarmacadam
DRILLING DATA		Hole Diameter 110mm
		Bench Height 15 & 30m
	Drills	Halco 410C
COMPRESSORS		CompAir 650HE
SECONDARY BREAKING		Hydraulic Hammer - Montabert
LOAD AND HAUL	**Trucks**	Terex R35T
	Loaders	Ford A62
		Ford A64

entry continues overleaf

71

| CRUSHING PLANT | **Primary** | Hazemag SAP5 Impactor |
| | **Secondary** | Hazemag APK40 Impactor |

CAT CASTLE QUARRIES LTD
Head Office *Cat Castle Quarry, Deepdale, Barnard Castle, Durham DL12 9DQ*
Phone *0833 50681*
Personnel *D C Green, Managing Director*
         P D M Snaith, Director
Quarries *Cat Castle*

CAT CASTLE QUARRIES LTD
CAT CASTLE QUARRY
155

Deepdale, Barnard Castle, Durham DL12 9DQ
Phone 0833 50681
OS Map 92   **OS Grid Reference** NZ010164
Personnel P D M Snaith, Director and Unit Manager

PRODUCT DATA	**Rock type**	Carboniferous Sandstone
	Colour	White to streaky brown
	Grain	Coarse
	Product	Dressed or Lump Stone

| DRILLING DATA | | Hand-held equipment |

CENTRAL REGION COUNCIL
Head Office *Roads & Transport Dept, Viewforth, Stirling, Central FK8 2ET*
Phone *0786 73111* **Fax** *0786 50802*
Personnel *G I Crindle, Technical Director*
Quarries *Northfield*

CENTRAL REGION COUNCIL
NORTHFIELD QUARRY
156

Nr Denny, Central FK6 6RB
Phone 0324 823777
OS Map 65   **OS Grid Reference** NS800854
Personnel T Wallis, Unit Manager

PRODUCT DATA	**Rock type**	Quartzitic Dolerite Sill
	Colour	Dark grey
	Grain	Medium
	Products	Crushed Stone
		Asphalt or Tarmacadam

DRILLING DATA		Hole Diameter 100mm
		Bench Height 15-20m
		Contractor Drilling & Blasting

| SECONDARY BREAKING | | Dropball on Ruston-Bucyrus 22-RB |

| LOAD AND HAUL | **Trucks** | Aveling Barford RD17 |
| | **Loaders** | Zettelmeyer |

CRUSHING PLANT	**Primary**	Parker 50×36 Jaw (Single Toggle)
	Secondary	Hewitt Robins 4' Cone
	Tertiary	Allis Chalmers 5.5' Cone

D E CHANCE LTD
Head Office *Bableigh Wood Quarry, 2 Acre Road, Horns Cross, Bideford, Devon EX39 5EB*
Phone *02375 279*
Personnel *D E Chance, Owner*
Quarries *Bableigh Wood*

D E CHANCE LTD

BABLEIGH WOOD QUARRY   **157**

Bableigh Barton, Buckland Brewer, Bideford, Devon EX39 5EB
Phone 02375 279
OS Map 190   **OS Grid Reference** SS393200
Personnel D E Chance, Owner and Unit Manager

PRODUCT DATA | **Rock type** | Devonian Sandstone
| **Colour** | Dark brown
| **Grain** | Medium
| **Products** | Dressed or Lump Stone
| | Crushed Stone

RIPPABLE STRATA

DRILLING DATA   Contractor Celtic Rock Services

CHARTER CONSOLIDATED plc
Corporate Head Office *40 Holborn Viaduct, London EC1P 1A*
Phone *071 353 1545* **Fax** *071 583 2950* **Telex** *929582*
Personnel *C H Parker, Director - New Business and Planning*
Companies *Hargreaves Quarries Ltd, Penryn Granite Ltd* — see separate entry

---

CHIC-GRIT LTD
Head Office *Unit 12, Industrial Estate, Norwich Road, Watton, Norfolk IP25 6DR*
Phone *0953 881642*
Personnel *D J Morfoot, Managing Director*
Quarries *Mount Pleasant*

CHIC-GRIT LTD

MOUNT PLEASANT QUARRY   **158**

Rocklands, Norfolk NR17 1XG
Phone 0953 881642
OS Map 144   **OS Grid Reference** TL990970
Personnel D J Morfoot, Managing Director and Unit Manager

PRODUCT DATA | **Rock type** | Cretaceous Upper Chalk (Flint)
| **Colour** | Blue
| **Grain** | Round
| **Product** | Agricultural Lime

DRILLING DATA   Hand-held equipment

CRUSHING PLANT | **Primary** | Jaw

CLARK CONTRACTORS
Head Office *Grove Farm, Welders Lane, Chalfont St Peter, Bucks SL9 8TU*
Phone *0247 3200*
Personnel *T Clarke, Director*
        B Clarke, Director
Quarries *Flaunden*

CLARK CONTRACTORS

FLAUNDEN QUARRY   **159**

Flaunden Hill, Flaunden, Hemel Hempstead, Herts, HP3 0PS
Phone 0442 824245
OS Map 166   **OS Grid Reference** TL015015
Personnel B Clarke, Director and Unit Manager

PRODUCT DATA	**Rock type**	Cretaceous Upper Chalk
	Colour	White
	Grain	Fine
	Product	Agricultural Lime
DRILLING DATA		No drilling
LOAD AND HAUL	**Face shovel**	Massey Ferguson
	Loaders	Volvo 1641
		Kramer Allard 512SE
	Trucks	None
CRUSHING PLANT		None

CLIPSHAM QUARRY COMPANY
Head Office *Clipsham Hall, Clipsham, Leics LE15 7QS*
Phone *0780 410204*
Personnel *Sir D Davenport, Chairman*
Landowner of Clipsham Quarry — see separate entry

CLOBURN QUARRY COMPANY LTD
Head Office *Pettinain Quarry, Pettinain, Lanark, Strathclyde ML11 8SR*
Phone *0555 4898*
Personnel *R W N Durward, Managing Director*
Quarries *Pettinain*

CLOBURN QUARRY COMPANY LTD

PETTINAIN QUARRY   **160**

Pettinain, Lanark, Strathclyde ML11 8SR
Phone 0555 4898
OS Map 72   **OS Grid Reference** NS943411
Personnel R W N Durward, Managing Director and Unit Manager

PRODUCT DATA	**Rock type**	Microgranite
	Colour	Bright red
	Grain	Fine
	Product	Crushed Stone
DRILLING DATA		Hole Diameter 105mm
		Bench Height 18-21m
		Contractor Ritchies Equipment

SECONDARY BREAKING		Dropball
LOAD AND HAUL	**Face shovel**	Akermans 25C
	Trucks	None
	Loaders	Komatsu WA470
		Komatsu WA500
CRUSHING PLANT	**Primary**	Kue Ken 120 42×32 Jaw (Double Toggle)
	Secondary	Kue Ken 105 42×10 Jaw
	Tertiary	Kobelco 45″ Autofine

D R COLLYER
Head Office *Ridgeway Quarry, Crich Lane, Ridgeway, Belper, Derbys DE4 5BQ*
Phone *077 3852400*
Personnel *D R Collyer, Owner*

D R COLLYER

RIDGEWAY QUARRY   161

Crich Lane, Ridgeway, Belper, Derbys DE4 5BQ
Phone 077 3852400
OS Map 119   **OS Grid Reference** SK357513
Personnel D R Collyer, Owner and Unit Manager

PRODUCT DATA	**Rock type**	Carboniferous Sandstone
	Colour	Fawn, pink
	Grain	Coarse
	Product	Dressed or Lump Stone
DRILLING DATA		Hand-held equipment

CORNISH ROADSTONES (BRIGHOUSE BRICK, TILE AND STONE CO LTD)
Head Office *Tower Farm, Hollins Lane, Steeton, Keighley, W Yorks BD20 6LY*
Phone *0535 654864*
Personnel *K Atkinson, Managing Director*
         L Evans, Director
Quarries *Trevassack*

CORNISH ROADSTONES (BRIGHOUSE BRICK, TILE AND STONE CO LTD)

TREVASSACK QUARRY   162

Goonhilly, Mawgan, Cornwall TR12 6LH
Phone 032622 522
OS Map 202   **OS Grid Reference** SW703222
Personnel B Semmens, Unit Manager

PRODUCT DATA	**Rock type**	Serpentine
	Colour	Green, red, yellow, white
	Grain	Medium to coarse
	Product	Crushed Stone
DRILLING DATA		Hole Diameter 110mm
		Bench Height 12m
		Contractor Saxton Co (Deep Drillers)

entry continues overleaf

COMPRESSORS		CompAir 650H
SECONDARY BREAKING		Hydraulic Hammer
LOAD AND HAUL	**Trucks**	(2)Terex 25t
	Loaders	Caterpillar 966D
		Liebherr 405
CRUSHING PLANT	**Primary**	Goodwin Barsby Jaw
	Secondary	Kue Ken Impactor

CORNISH RUSTIC STONE LTD
Head Office *Bowithic Quarry, Trewarmet, Tintagel, Cornwall PL34 0AN*
Phone *0836 350082*
Personnel *C Hawkey, Managing Director*
Quarries *Bowithic*

CORNISH RUSTIC STONE LTD
BOWITHIC QUARRY   163
Trewarmet, Tintagel, Cornwall PL34 0AN
Phone 0836 350082
OS Map 200   **OS Grid Reference** SX069861
Personnel

PRODUCT DATA	**Rock type**	Upper Devonian Slate
	Colour	Rustic
	Grain	Fine
	Product	Dressed or Lump Stone
DRILLING DATA		Hand-held equipment

COTSWOLD WALLING AND MASONRY
Head Office *The Old School, Naunton, Cheltenham, Glos GL54 3BA*
Phone *04515 775*
Personnel *J C Ward, Managing Director*
       J Wiltshire, Director
       T Beeston, Operations Manager
Quarries *Brockhill, Cotswold Hill*

COTSWOLD WALLING AND MASONRY
BROCKHILL QUARRY   164
Naunton, Cheltenham, Glos GL54 3BA
Phone 04515 775
OS Map 163   **OS Grid Reference** SP125253
Personnel T Beeston, Operations Manager

PRODUCT DATA	**Rock type**	Jurassic Limestone
	Colour	Brownish yellow
	Grain	Fine
	Product	Dressed or Lump Stone

NO FURTHER INFORMATION PROVIDED

COTSWOLD WALLING AND MASONRY

COTSWOLD HILL QUARRY

165

Ford, Temple Guiting, Cheltenham, Glos GL54 5RU
Phone 038673 566
OS Map 163   **OS Grid Reference** SP086294
Personnel T Beeston, Operations Manager

PRODUCT DATA	**Rock type**	Jurassic Inferior Oolitic Limestone
	Colour	Creamy yellow, yellowish orange
	Grain	Fine
	Products	Dressed or Lump Stone
		Crushed Stone

RIPPABLE STRATA

LOAD AND HAUL	**Face shovel**	Akermans 16C
	Trucks	Ford
		Aveling Barford
	Loaders	Hannomag 70E

| CRUSHING PLANT | | No information provided |

W S CROSSLEY (YORK STONE) LTD

Head Office *Norcliffe Lane, Southowram, Halifax, W Yorks HX3 8PL*
Phone *0422 44263*
Personnel *W S Crossley, Owner*
Quarries *Thumpas*

W S CROSSLEY (YORK STONE) LTD

THUMPAS QUARRY

166

Pinnar Lane, Southowram, Halifax, W Yorks HX3 9QT
Phone 0422 44263
OS Map 104   **OS Grid Reference** SE107238
Personnel W S Crossley, Owner and Unit Manager

PRODUCT DATA	**Rock type**	Carboniferous Sandstone (Flagstone)
	Colour	Light buff to fawn
	Grain	Fine
	Product	Dressed or Lump Stone

| DRILLING DATA | | Hand-held equipment |

CULTS LIME LTD

Head Office *Skelpie Mine, Limehills, Pitlessie, Cupar, Fife KY7 7TF*
Phone *0334 52548*
Personnel *J D J Cochrane, Owner*
Quarries *Skelpie*

CULTS LIME LTD

SKELPIE MINE

167

Limehills, Pitlessie, Cupar, Fife KY7 7TF
Phone 0334 52548
OS Map 59   **OS Grid Reference** NO335095
Personnel J D J Cochrane, Owner and Unit Manager   *entry continues overleaf*

| PRODUCT DATA | **Rock type** | Carboniferous Limestone |
| | **Product** | Agricultural Lime |

| DRILLING DATA | | Hand-held equipment |

CUMBRIA STONE QUARRIES LTD
Head Office *Silver Street, Crosby Ravensworth, Penrith, Cumbria CA10 3JA*
Phone *09315 227* **Fax** *09315 367*
Personnel *D Rice, Managing Director*
J G Myers, Director
Mrs D T Jackson, Director
J Wignall, Quarry Manager
Quarries *Bankend, Birkhams, Broad Oak, Lazonby Fell, Pickerings, Rooks, Salterwath, Scratchmill Scar*

CUMBRIA STONE QUARRIES LTD

BANKEND QUARRY   168
Sandwith St Bees, Whitehaven, Cumbria CA28 9UG
Phone 09315 227
OS Map 89   **OS Grid Reference** NX992127
Personnel J Wignall, Quarry Manager

PRODUCT DATA	**Rock type**	Triassic St Bees Sandstone
	Colour	Dark red
	Grain	Fine
	Product	Dressed or Lump Stone

| DRILLING DATA | | Hand-held equipment |

CUMBRIA STONE QUARRIES LTD

BIRKHAMS QUARRY   169
Sandwith St Bees, Whitehaven, Cumbria CA28 9UU
Phone 09315 227
OS Map 89   **OS Grid Reference** NX955154
Personnel J Wignall, Quarry Manager

PRODUCT DATA	**Rock type**	Triassic St Bees Sandstone
	Colour	Dull red
	Grain	Fine
	Product	Dressed or Lump Stone

| DRILLING DATA | | Hand-held equipment |

CUMBRIA STONE QUARRIES LTD

BROAD OAK QUARRY   170
Waberthwaite, Millom, Cumbria LA19 5YJ
Phone 09315 227
OS Map 96   **OS Grid Reference** SD115946
Personnel J Wignall, Quarry Manager

PRODUCT DATA	**Rock type**	Biotite Granodiorite
	Colour	Speckled light bluish grey
	Grain	Fine
	Product	Dressed or Lump Stone

| DRILLING DATA | | Hand-held equipment |

CUMBRIA STONE QUARRIES LTD
LAZONBY FELL QUARRY
171

Lazonby, Plumpton, Penrith, Cumbria CA10 1BY
Phone 09315 227
OS Map 90 **OS Grid Reference** NY517380
Personnel J Wignall, Quarry Manager

PRODUCT DATA	**Rock type**	Permian Penrith Sandstone
	Colour	Light red, dark pink
	Grain	Fine to medium
	Product	Dressed or Lump Stone

DRILLING DATA   Hand-held equipment

CUMBRIA STONE QUARRIES LTD
PICKERINGS QUARRY
172

Scar, Orton, Penrith, Cumbria CA10 9LL
Phone 09315 227
OS Map 91 **OS Grid Reference** NY642155
Personnel J Wignall, Quarry Manager

PRODUCT DATA	**Rock type**	Carboniferous Great Scar Limestone
	Colour	Light grey
	Product	Dressed or Lump Stone

DRILLING DATA   Hand-held equipment

CUMBRIA STONE QUARRIES LTD
ROOKS QUARRY
173

Orton, Penrith, Cumbria CA10 1RL
Phone 09315 227
OS Map 91 **OS Grid Reference** NY598092
Personnel J Wignall, Quarry Manager

PRODUCT DATA	**Rock type**	Carboniferous Limestone
	Colour	Light grey
	Product	Dressed or Lump Stone

DRILLING DATA   Hand-held equipment

CUMBRIA STONE QUARRIES LTD
SALTERWATH QUARRY
174

Crosby, Ravensworth Fell, Penrith, Cumbria CA10 3QY
Phone 09315 227
OS Map 91 **OS Grid Reference** NZ599029
Personnel J Wignall, Quarry Manager

PRODUCT DATA	**Rock type**	Carboniferous Limestone
	Colour	Light bluish grey
	Product	Dressed or Lump Stone

DRILLING DATA   Hand-held equipment

CUMBRIA STONE QUARRIES LTD

SCRATCHMILL SCAR QUARRY

175

Plumpton, Penrith, Cumbria CA11 9PF
Phone 09315 227
OS Map 90 **OS Grid Reference** NY490371
Personnel J Wignall, Quarry Manager

PRODUCT DATA	**Rock type**	Permian Sandstone
	Colour	Light grey
	Grain	Fine
	Product	Dressed or Lump Stone

| DRILLING DATA | | Hand-held equipment |

J CURTIS AND SONS LTD
Head Office Thrupp Lane, Radley, Abingdon, Oxon OX14 3NQ
Phone 0235 24545
Personnel J Curtis, Managing Director
G Barrett, Director
Quarries Town

J CURTIS AND SONS LTD

TOWN QUARRY

176

Charlbury, Oxon OX7 3PR
Phone 0608 81373
OS Map 164 **OS Grid Reference** SP365198
Personnel K Fawdrey, Unit Manager

PRODUCT DATA	**Rock type**	Jurassic Great Oolitic Limestone
	Colour	Light grey
	Grain	Fine
	Product	Crushed Stone

RIPPABLE STRATA

| DRILLING DATA | | Contractor Railside Engineering |

| SECONDARY BREAKING | | Dropball on Ruston-Bucyrus 22-RB |

LOAD AND HAUL	**Face shovel**	Akermans H9
		Akermams H14
	Trucks	Volvo
	Loaders	Caterpillar
		Volvo

| CRUSHING PLANT | **Primary** | Goodwin Barsby 42 × 36 Jaw |
| | **Secondary** | Goodwin Barsby 30 × 18 Jaw |

DALGETY AGRICULTURE LTD
Head Office Parkmore Quarry, Parkmore, Dufftown, Keith, Grampian AB5 4DL
Phone 0340 20200
Personnel C Aitken, Managing Director
Quarries Parkmore

DALGETY AGRICULTURE LTD

PARKMORE QUARRY

177

Parkmore, Dufftown, Keith, Grampian AB5 4DL
Phone 0340 20200
OS Map 28   **OS Grid Reference** NJ332408
Personnel N Phillips, Unit Manager

PRODUCT DATA	**Rock type**	Pre-Cambrian Dalradian Limestone
	Colour	Not commercially significant
	Products	Crushed Stone
		Asphalt or Tarmacadam

DRILLING DATA		Hole Diameter 110mm
		Bench Height 15m
		Contractor Drilling & Blasting

| SECONDARY BREAKING | | Dropball on Ruston-Bucyrus 22-RB |

CRUSHING PLANT	**Primary**	Pegson 30 × 42 Jaw
	Secondary	Pegson 4' Gyrasphere
	Tertiary	(3)Barmac Duopactor

DARRINGTON QUARRIES LTD
Head Office *Darrington Quarry, Darrington Leys, Cridling Stubbs, Knottingley, N Yorks WF11 0AH*
Phone *0977 82368/82369*
Personnel *A Downham, Managing Director*
B Grist, General Manager
Quarries *Darrington, Foxcliffe, Jackdaw Crag, Long Lane, Newthorpe, Skelbrooke*

DARRINGTON QUARRIES LTD

DARRINGTON QUARRY

178

Darrington Leys, Cridling Stubbs, Knottingley, N Yorks WF11 0AH
Phone 0977 82368
OS Map 105   **OS Grid Reference** SE507212
Personnel B Grist, General Manager

PRODUCT DATA	**Rock type**	Permian Upper Magnesian Limestone
	Products	Dressed or Lump Stone
		Crushed Stone

DRILLING DATA		Hole Diameter 90mm
		Bench Height 3-5m
		Contractor occasionally

DARRINGTON QUARRIES LTD

FOXCLIFFE QUARRY

179

Brotherton, N Yorks WF11 9EG
Phone 0977 83507
OS Map 105   **OS Grid Reference** SE482257
Personnel B Grist, General Manager

PRODUCT DATA	**Rock type**	Permian Upper Magnesian Limestone
	Products	Dressed or Lump Stone
		Crushed Stone

entry continues overleaf

RIPPABLE STRATA

LOAD AND HAUL	**Loaders**	Komatsu WA650
CRUSHING PLANT	**Primary**	Wakefield 56 Impactor
	Secondary	Bolliden Allis Cone
		Nordberg 4' Cone

DARRINGTON QUARRIES LTD

JACKDAW CRAG QUARRY

180

Stutton, Tadcaster, N Yorks LS24 9BG
Phone 0937 832131
OS Map 105   **OS Grid Reference** SE465415
Personnel B Grist, General Manager

PRODUCT DATA	**Rock type**	Permian Lower Magnesian Limestone
	Colour	Dark cream
	Grain	Fine
	Products	Dressed or Lump Stone
		Crushed Stone
DRILLING DATA		Bench Height 3m
		Hand-held equipment
		Contractor R&B Rock Drillers
SECONDARY BREAKING		Dropball on Ruston-Bucyrus 22-RB

DARRINGTON QUARRIES LTD

LONG LANE QUARRY

181

Kirk Smeaton, Pontefract, W Yorks WF8 3JX
Phone 0302 700329
OS Map 105   **OS Grid Reference** SE515150
Personnel B Grist, General Manager

PRODUCT DATA	**Rock type**	Permian Lower Magnesian Limestone
	Colour	Buff
	Products	Dressed or Lump Stone
		Crushed Stone
DRILLING DATA		Bench Height 2–3m
		Hand-held equipment

DARRINGTON QUARRIES LTD

NEWTHORPE QUARRY

182

South Milford, Leeds, W Yorks LS25 5LX
Phone 0977 682512
OS Map 105   **OS Grid Reference** SD460320
Personnel B Grist, General Manager

PRODUCT DATA	**Rock type**	Permian Magnesian Limestone
	Product	Dressed or Lump Stone

RIPPABLE STRATA

DARRINGTON QUARRIES LTD
SKELBROOKE QUARRY
183

Doncaster Lane, Skelbrooke, Doncaster, S Yorks DN6 8LY
Phone 0977 82368
OS Map 111 **OS Grid Reference** SE515114
Personnel B Grist, General Manager

PRODUCT DATA **Rock type** Permian Magnesian Limestone
Colour White to cream
Product Dressed or Lump Stone

RIPPABLE STRATA

N DAVID

Head Office 134 Victoria Rise, London SW4 0NW
Phone 081 720 8764
Personnel N David, Owner
Quarries Callow

N DAVID
CALLOW QUARRY
184

Buckholt, Monmouth, Gwent NP5 3RL
Phone 081 720 8764
OS Map 162 **OS Grid Reference** SO508168

QUARRY ON STAND

ROCK DATA **Rock type** Devonian Sandstone
Colour Brown
Grain Fine

R DAVIDSON

Head Office Beacon Lodge Quarry, Long Lane, Southowram, Halifax, W Yorks HX3 9UD
Phone 0422 63674
Personnel R Davidson, Owner
Quarries Beacon Lodge

R DAVIDSON
BEACON LODGE QUARRY
185

Long Lane, Southowram, Halifax, W Yorks HX3 9UD
Phone 0422 63674
OS Map 104 **OS Grid Reference** SD108235
Personnel R Davidson, Owner and Unit Manager

PRODUCT DATA **Rock type** Carboniferous Sandstone
Colour Creamy yellow
Grain Fine
Product Dressed or Lump Stone

DRILLING DATA Hand-held equipment

DAVIDSON AND MURISON LTD
Head Office *The Garage, Maryculter, Aberdeen, Grampian AB1 0BT*
Phone *0224 732391*
Personnel *W Davidson, Managing Director and Unit Manager*
Quarries *North Mains, Savoch*

DAVIDSON AND MURISON LTD

NORTH MAINS QUARRY

186

Findon, Aberdeen, Grampian AB1 4SJ
Phone 0224 732391
OS Map 38 **OS Grid Reference** NO937983
Personnel W Davidson, Unit Manager

PRODUCT DATA	**Rock type**	Gabbro and Schist
	Colour	Dark grey to black. Schist - blue
	Grain	Coarse
	Product	Crushed Stone
DRILLING DATA		Hole Diameter 105mm
		Bench Height 22m
		Contractor Drilling & Blasting
SECONDARY BREAKING		Dropball
LOAD AND HAUL	**Trucks**	None
	Loaders	(2)Hannomag 55
		(3)Hannomag 66
CRUSHING PLANT	**Primary**	Goodwin Barsby 36 × 24 Jaw
	Secondary	Nordberg 4.5' Cone

DAVIDSON AND MURISON LTD

SAVOCH QUARRY

187

Longside, Peterhead, Grampian AB4 7YR
Phone 0224 732391
OS Map 30 **OS Grid Reference** NK054471
Personnel W Davidson, Unit Manager

PRODUCT DATA	**Rock type**	Granite
	Colour	Not commercially significant
	Grain	Coarse to very coarse
	Product	Crushed Stone
DRILLING DATA		Hole Diameter 105mm
		Bench Height 20-22m
		Contractor Drilling & Blasting
SECONDARY BREAKING		Dropball
LOAD AND HAUL	**Trucks**	None
	Loaders	Hannomag 55
		(3)Hannomag 66
CRUSHING PLANT	**Primary**	Parker 36 × 24 Jaw
	Secondary	Hazemag APK40 Impactor

MANSELL DAVIES AND SON LTD
Head Office Station Yard, Crymmych, Dyfed SA35 0BZ
Phone 023973 631
Personnel M Davies, Managing Director
Quarries Glogue

MANSELL DAVIES AND SON LTD

GLOGUE QUARRY 188

Glogue, Dyfed SA36 0EE
Phone 023973 631
OS Map 145 **OS Grid Reference** SN222328
Personnel M Davies, Managing Director and Unit Manager

PRODUCT DATA	**Rock type**	Silurian Slate
	Colour	Deep blue
	Grain	Fine
	Product	Dressed or Lump Stone

| DRILLING DATA | | Hand-held equipment |

DEVON COUNTY COUNCIL
Head Office Civic Centre, North Walk, Barnstable, Devon EX31 1ED
Phone 0271 47067 **Fax** 0271 47067
Personnel R Hawkins, Technical Director
Quarries Barton Wood

DEVON COUNTY COUNCIL

BARTON WOOD QUARRY 189

Brayford, Barnstaple, Devon EX32 7QB
Phone 0598 8272
OS Map 180 **OS Grid Reference** SS688344
Personnel E W Cornish, Unit Manager

PRODUCT DATA	**Rock type**	Devonian Sandstone
	Colour	Medium to dark grey
	Grain	Medium
	Product	Crushed Stone

DRILLING DATA		Hole Diameter 89mm
		Bench Height 11–20m
	Drills	Holman Voltrak

LOAD AND HAUL	**Face shovel**	Ruston-Bucyrus
	Trucks	Aveling Barford
		Euclid
	Loaders	JCB 430

CRUSHING PLANT	**Primary**	Goodwin Barsby 30 × 18 Jaw (Single Toggle)
	Secondary	Babbitless 2.5′ Gyratory
	Tertiary	Pegson 3′ Gyrasphere

DEVON QUARRY AND TARPAVING COMPANY LTD
Head Office *Blackaller Quarry, Drewsteignton, Exeter, Devon EX6 6RA*
Phone *0647 21226*
Personnel *Mrs E T Hodge, Company Secretary*
Quarries *Blackaller*

DEVON QUARRY AND TARPAVING COMPANY LTD

BLACKALLER QUARRY

Drewsteignton, Exeter, Devon EX6 6RA
Phone 0647 21226
OS Map 191   **OS Grid Reference** SX730912
Personnel R Chudley, Unit Manager

PRODUCT DATA	**Rock type**	Carboniferous Limestone
	Colour	Grey
	Products	Crushed Stone
		Asphalt or Tarmacadam
DRILLING DATA		Hole Diameter 105mm
		Bench Height 12m
	Drills	Holman Voltrak
COMPRESSORS		CompAir 650HE
SECONDARY BREAKING		Hydraulic Hammer
LOAD AND HAUL	**Trucks**	Terex
	Loaders	JCB
CRUSHING PLANT	**Primary**	Broadbent 50 × 40 Jaw
	Secondary	None
	Tertiary	None

190

DRINKWATER SABEY LTD (ATTWOODS plc)
Head Office *The Pickeridge, Stoke Common Road, Fulmer, Bucks SL3 6HA*
Phone *02816 2700*   **Fax** *02816 2464*
Personnel *A Pontin, Managing Director*
        T J Penfold, Managing Director
Quarries *Barnsdale Bar, Barnshaw*

DRINKWATER SABEY LTD (ATTWOODS plc)

BARNSDALE BAR QUARRY

Kirk Smeaton, Pontefract, W Yorks WF8 3JX
Phone 0302 700980
OS Map 105   **OS Grid Reference** SE512142
Personnel P Wood, Unit Manager

PRODUCT DATA	**Rock type**	Permian Limestone
	Colour	Grey
	Products	Dressed or Lump Stone
		Crushed Stone
DRILLING DATA		Hole Diameter 110mm
		Bench Height 6-12m
		Contractor R&B Rock Drillers

191

LOAD AND HAUL   **Trucks**   Caterpillar
                       Loaders   Caterpillar

NO FURTHER INFORMATION PROVIDED

DRINKWATER SABEY LTD (ATTWOODS plc)
BARNSHAW QUARRY   **192**
Fell Lane, Oakworth, Keighley, W Yorks BD22 6BZ
Phone 0302 700980
OS Map 104   **OS Grid Reference** SE038388

QUARRY ON STAND

ROCK DATA   **Rock type**   Carboniferous Sandstone

DUMFRIES AND GALLOWAY REGION COUNCIL
Head Office *Council Offices, Dumfries, Dumfries & Galloway DG2 2DD*
Phone *0387 53141*
Personnel *H D B Murray, Director of Roads*
Quarries *Boreland Fell*

DUMFRIES AND GALLOWAY REGION COUNCIL
BORELAND FELL QUARRY   **193**
Spittal, Wigtown, Dumfries, Dumfries & Galloway DG8 0BP
Phone 0387 53141
OS Map 83   **OS Grid Reference** NX350590

QUARRY ON STAND

ROCK DATA   **Rock type**   Silurian Greywacke

DUNHOUSE QUARRY COMPANY LTD
Head Office *7 South Church Road, Bishop Auckland, Durham DL14 7LB*
Phone *0388 602322*   **Fax** *0388 603832*
Personnel *J B Blake, General Manager*
Quarries *Corncockle, Corsehill, Dunhouse*

DUNHOUSE QUARRY COMPANY LTD
CORNCOCKLE QUARRY   **194**
Templand, Lochmaben, Lockerbie, Dumfries & Galloway DG11 1TE
Phone 0388 602322
OS Map 78   **OS Grid Reference** NY083865
Personnel S Braithwaite, Unit Manager

PRODUCT DATA   **Rock type**   Silurian Sandstone
                   Colour   Red
                   Grain   Medium
                   Product   Dressed or Lump Stone

DRILLING DATA   Hand-held equipment

DUNHOUSE QUARRY COMPANY LTD

CORSEHILL QUARRY

195

Eaglesfield Road, Annan, Dumfries & Galloway DG12 5LN
Phone 0388 602322
OS Map 75 **OS Grid Reference** NY707703
Personnel S Braithwaite, Unit Manager

PRODUCT DATA	**Rock type**	Silurian Sandstone
	Colour	Red
	Grain	Fine
	Product	Dressed or Lump Stone

| DRILLING DATA | | Hand-held equipment |

DUNHOUSE QUARRY COMPANY LTD

DUNHOUSE QUARRY

196

Cleatlam, Darlington, Durham DL2 3QU
Phone 0388 60749
OS Map 92 **OS Grid Reference** NZ141183
Personnel D Dartmell, Unit Manager

PRODUCT DATA	**Rock type**	Carboniferous Sandstone
	Colour	Creamy brown
	Grain	Medium
	Product	Dressed or Lump Stone

| DRILLING DATA | | Hand-held equipment |

DUNLOSSIT TRUSTEES LTD
Head Office *Ballygrant Quarry, Ballygrant, Isle of Islay, Strathclyde PA45 7QL*
Phone *049684 232/652*
Personnel *Major McTroy-Forbes, Managing Director*
Quarries *Ballygrant*

DUNLOSSIT TRUSTEES LTD

BALLYGRANT QUARRY

197

Ballygrant, Isle of Islay, Strathclyde PA45 7QL
Phone 049684 232/652
OS Map 60 **OS Grid Reference** NR397660
Personnel Major McTroy-Forbes, Managing Director and Unit Manager

PRODUCT DATA	**Rock type**	Pre-Cambrian Dalradian Limestone
	Colour	Not commercially significant
	Product	Crushed Stone

DRILLING DATA		Hole Diameter 105mm
		Bench Height 18m
		Contractor Ritchies Equipment

| SECONDARY BREAKING | | Dropball |

| CRUSHING PLANT | | No information provided |

A M & K O EARL
Head Office *Sycamore Quarry, Windmill Road, Kerridge, Macclesfield, Cheshire SK10 5AZ*
Phone *0625 72125*
Personnel *Miss A M Earl, Director*
        K O Earl, Director
Quarries *Sycamore, Wimberry Moss*

A M & K O EARL

SYCAMORE QUARRY   198

Windmill Road, Kerridge, Macclesfield, Cheshire SK10 5AZ
Phone 0625 72125
OS Map 118   **OS Grid Reference** SJ940765
Personnel K O Earl, Director and Unit Manager

PRODUCT DATA	**Rock type**	Carboniferous Milnrow Sandstone
	Colour	Fawn, grey
	Grain	Fine
	Product	Dressed or Lump Stone

RIPPABLE STRATA

A M & K O EARL

WIMBERRY MOSS QUARRY   199

Smith Street, Rainow, Macclesfield, Cheshire SK10 5AZ
Phone 0625 72125
OS Map 118   **OS Grid Reference** SJ953738
Personnel K O Earl, Director and Unit Manager

PRODUCT DATA	**Rock type**	Carboniferous Millstone Grit
	Colour	Pink to lilac
	Grain	Fine
	Product	Dressed or Lump Stone

RIPPABLE STRATA

ECC CALCIUM CARBONATES LTD (ENGLISH CHINA CLAYS LTD)
Head Office *Wilton Road, Quidhampton, Salisbury, Wilts SP2 9AD*
Phone *0722 743444* **Fax** *0722 744900* **Telex** *477054*
Personnel *G R W Lovering, Chairman*
        R M L Hexter, Director and General Manager
Quarries *East Grinstead, Queensgate, Quidhampton*

ECC CALCIUM CARBONATES LTD (ENGLISH CHINA CLAYS LTD)

EAST GRINSTEAD QUARRY   200

Salisbury, Wilts SP2 9AD
Phone 0722 743444
OS Map 184   **OS Grid Reference** SU226270
Personnel D Phillips, Unit Manager

PRODUCT DATA	**Rock type**	Cretaceous Upper Chalk
	Colour	White
	Grain	Fine
	Product	Industrial Limestone

RIPPABLE STRATA

ECC CALCIUM CARBONATES LTD (ENGLISH CHINA CLAYS LTD)

QUEENSGATE QUARRY   **201**

Westwood, Beverley, Humbers HU17 8RQ
Phone 0482 881234
OS Map 107   **OS Grid Reference** TA021382
Personnel C Bayley, Unit Manager

PRODUCT DATA		
	Rock type	Cretaceous Flamborough Chalk
	Colour	White
	Grain	Fine
	Product	Industrial Limestone

RIPPABLE STRATA

ECC CALCIUM CARBONATES LTD (ENGLISH CHINA CLAYS LTD)

QUIDHAMPTON QUARRY   **202**

Fugglestone St Peter, Quidhampton, Salisbury, Wilts SP2 9AD
Phone 0722 74344
OS Map 184   **OS Grid Reference** SU114313
Personnel D Phillips, Unit Manager

PRODUCT DATA		
	Rock type	Cretaceous Upper Chalk
	Colour	Creamy white
	Grain	Fine
	Product	Industrial Limestone

RIPPABLE STRATA

ECC QUARRIES LTD (ENGLISH CHINA CLAYS LTD)
Head Office *Greystones, Huntcote Road, Croft, Leics LE9 6GS*
Phone *0455 285200*   **Fax** *0455 283837*
Personnel *Dr J W Inglis, Managing Director*
H Robinson, Regional Director - Midlands and North
C Denman, Regional Director - South and South West

ECC QUARRIES LTD (ENGLISH CHINA CLAYS LTD) Midlands and North Area
Area Office *Croft Quarry, Croftworks, Croft, Leics LE9 6GS*
Phone *0455 282601*
Personnel *T Cann, General Manager*
Quarries *Back Lane, Croft, Holme Park, Lea, Lillishall, Shavers End*

ECC QUARRIES LTD (ENGLISH CHINA CLAYS LTD) Midlands and North Area

BACK LANE QUARRY   **203**

Carnforth, Lancs LA6 1BP
Phone 0524 733512
OS Map 97   **OS Grid Reference** SD508690
Personnel A Hunter, Unit Manager

PRODUCT DATA		
	Rock type	Carboniferous Limestone
	Colour	Light grey
	Products	Crushed Stone
		Industrial Limestone
		Agricultural Lime

DRILLING DATA		Hole Diameter 115–130mm
		Bench Height 25m
	Drills	Halco 410C

COMPRESSORS	Ingersoll-Rand

SECONDARY BREAKING	Dropball on Ruston-Bucyrus 22-RB

LOAD AND HAUL	**Trucks**	Terex
	Loaders	Caterpillar 988B

CRUSHING PLANT	**Primary**	Pegson 42 × 48 Jaw (Single Toggle)
	Secondary	Mansfield No5 Hammermill
	Tertiary	Christy & Norris 24 × 16 Pulverizer

ECC QUARRIES LTD (ENGLISH CHINA CLAYS LTD) Midlands and North Area

CROFT QUARRY

204

Croftworks, Croft, Leics LE9 6GS
Phone 0455 282601
OS Map 140   **OS Grid Reference** SP513963
Personnel A Robinson, Unit Manager

PRODUCT DATA	**Rock type**	Diorite
	Colour	Medium grey to pink
	Grain	Coarse
	Products	Crushed Stone
		Asphalt or Tarmacadam
		Pre-cast Concrete Products

DRILLING DATA		Hole Diameter 105–125mm
		Bench Height 18m
	Drills	(3)Ingersoll-Rand CM351
		Holman Voltrak

COMPRESSORS	(3)Atlas Copco XRH350
	Ingersoll-Rand DXL600P

SECONDARY BREAKING	(2)Dropball on Ruston-Bucyrus 38-RB

LOAD AND HAUL	**Face shovel**	Demag H185
		Liebherr 982
		Caterpillar 988B
	Trucks	Aveling Barford RD40
		(2)Caterpillar 777
		(2)Caterpillar 773B
		Aveling Barford RD40
	Loaders	Komatsu
		Caterpillar

CRUSHING PLANT	**Primary**	Allis Chalmers Superior 54 × 74 Gyratory
		Allis Chalmers R150–120 Jaw
	Secondary	(2)Nordberg 5.5′ Cone
	Tertiary	Pegson 489F Cone
		Parker 1350 Cone
		Nordberg 4.75′ Cone

ECC QUARRIES LTD (ENGLISH CHINA CLAYS LTD) Midlands and North Area

HOLME PARK QUARRY

205

Burton, Carnforth, Lancs LA6 1NZ
Phone 0524 781441
OS Map 97   **OS Grid Reference** SD536788
Personnel A Hunter, Unit Manager

PRODUCT DATA	**Rock type**	Carboniferous Great Scar Limestone
	Colour	Light grey
	Products	Crushed Stone
		Industrial Limestone
		Agricultural Lime
DRILLING DATA		Hole Diameter 110mm
		Bench Height 12.5–18m
	Drills	Ingersoll-Rand Resko
SECONDARY BREAKING		Dropball on Ruston-Bucyrus 22-RB
LOAD AND HAUL	**Trucks**	Aveling Barford RD40
		Terex
	Loaders	Caterpillar
CRUSHING PLANT	**Primary**	Pegson 42 × 48 Jaw (Single Toggle)
	Secondary	(3)Mansfield No 4 Hammermill
	Tertiary	Lanway No 4 Hammermill
		Mansfield No 4 Hammermill

ECC QUARRIES LTD (ENGLISH CHINA CLAYS LTD) Midlands and North Area

LEA QUARRY

206

Much Wenlock
Phone 0952 727324
OS Map 134   **OS Grid Reference** SO593984
Personnel K Bradley, Unit Manager

PRODUCT DATA	**Rock type**	Silurian Wenlock Limestone
	Colour	Mottled grey
	Products	Crushed Stone
		Industrial Limestone
		Agricultural Lime
DRILLING DATA		Hole Diameter 110–115mm
		Bench Height 20m
	Drills	Halco 410C
		Halco 650H
COMPRESSORS		(2)Atlas Copco XA350
SECONDARY BREAKING		Hydraulic Hammer - Krupp
LOAD AND HAUL	**Trucks**	(3)Aveling Barford RD40
		Terex R35T
	Loaders	Caterpillar 966C
		Caterpillar 966E

CRUSHING PLANT	**Primary**	Goodwin Barsby 40 × 24 Jaw (Single Toggle)
		(2)Pegson 42 × 48 Jaw
	Secondary	Lightning Hammermill
		Sheepbridge Hammermill
	Tertiary	(3)Sheepbridge Hammermill

ECC QUARRIES LTD (ENGLISH CHINA CLAYS LTD) Midlands and North Area

LILLISHALL QUARRY　　　　207

Much Wenlock, Shrops TF13 6DG
Phone 0445 282601
OS Map 138　**OS Grid Reference** SO575967

QUARRY ON STAND

| ROCK DATA | **Rock type** | Silurian Limestone |

ECC QUARRIES LTD (ENGLISH CHINA CLAYS LTD) Midlands and North Area

SHAVERS END QUARRY　　　　208

Stourport-on-Severn, Hereford & Worcs DY13 0UT
Phone 0299 896473
OS Map 150　**OS Grid Reference** SO769678
Personnel D Newman, Unit Manager

PRODUCT DATA	**Rock type**	Silurian Limestone
	Colour	Medium bluish grey
	Product	Crushed Stone
DRILLING DATA		Hole Diameter 105mm
		Bench Height 10m
	Drills	Halco 410C
LOAD AND HAUL	**Face shovel**	Zettelmeyer
	Trucks	Terex
	Loaders	Komatsu WA350

ECC QUARRIES LTD (ENGLISH CHINA CLAYS LTD) Southern Area
Area Office *Rockbeare Hill Quarry, Exeter, Devon EX5 2HB*
Phone 0404 822494
Personnel *J A West, General Manager*
Quarries *Callow Rock, Colemans, Stoneycombe*

ECC QUARRIES LTD (ENGLISH CHINA CLAYS LTD) Southern Area

CALLOW ROCK QUARRY　　　　209

Shipman Gorge, Cheddar, Somerset BS27 3DQ
Phone 0934 742621
OS Map 182　**OS Grid Reference** ST445558
Personnel R Sheffield, Unit Manager

PRODUCT DATA	**Rock type**	Carboniferous Limestone
	Colour	Medium to light grey
	Products	Crushed Stone
		Pre-cast Concrete Products
		Industrial Limestone
		Agricultural Lime

entry continues overleaf

DRILLING DATA		Hole Diameter 135mm
		Bench Height 20m
	Drills	Halco 650H

COMPRESSORS	CompAir 750HE

SECONDARY BREAKING	Dropball on Ruston-Bucyrus 30-RB

LOAD AND HAUL	**Face shovel**	Caterpillar 245
	Trucks	Caterpillar
	Loaders	Caterpillar 966C
		Caterpillar 980C
		Caterpillar 988B

CRUSHING PLANT	**Primary**	Pegson 48 × 42 Jaw
		Pegson 60 × 48 Jaw (Mobile)
	Secondary	Sheepbridge Impactor
		Nordberg 5.5' Cone
	Tertiary	(3)Symons Cone
		Nordberg 4.25' Cone
		Lanway No3 Impactor

ECC QUARRIES LTD (ENGLISH CHINA CLAYS LTD) Southern Area

COLEMANS QUARRY 210

Frome, Somerset BA11 4PX
Phone 0404 822494
OS Map 183 **OS Grid Reference** ST726451

QUARRY ON STAND

ROCK DATA	**Rock type**	Carboniferous Limestone
	Colour	Light buff to grey
	Grain	Medium

ECC QUARRIES LTD (ENGLISH CHINA CLAYS LTD) Southern Area

STONEYCOMBE QUARRY 211

Newton Abbot, Devon TQ12 5LL
Phone 0804 72193
OS Map 202 **OS Grid Reference** SX862672
Personnel E R Tomkyn, Unit Manager

PRODUCT DATA	**Rock type**	Devonian Limestone
	Colour	Light pinkish grey
	Products	Crushed Stone
		Asphalt or Tarmacadam
		Pre-cast Concrete Products
		Agricultural Lime

DRILLING DATA		Hole Diameter 108mm
		Bench Height 10m
	Drills	Halco SPD43–22

COMPRESSORS	CompAir 750HE

SECONDARY BREAKING	Dropball on Ruston-Bucyrus 22-RB

LOAD AND HAUL	**Trucks**	Aveling Barford 28
		(4)Aveling Barford RD40
		(2)Heathfield
	Loaders	Caterpillar 966C
		Caterpillar 988B
CRUSHING PLANT	**Primary**	Hazemag AP5 Impactor
	Secondary	Hazemag APK40 Impactor
		Sheepbridge
	Tertiary	Parker Kubitizer

ECC QUARRIES LTD (ENGLISH CHINA CLAYS LTD) South West Area
Area Office *Moorcroft Quarry, Billacombe, Plymouth, Devon PL9 8AJ*
Phone *0752 402661*
Personnel *M Booth, General Manager*
Quarries *Greystone, Luxulyan, Moorcroft, New England, Venn, Westleigh*

ECC QUARRIES LTD (ENGLISH CHINA CLAYS LTD) South West Area

GREYSTONE QUARRY

212

Lawhitton, Launceston, Cornwall PL15 9PS
Phone 0566 2392
OS Map 201 **OS Grid Reference** SX363805
Personnel D Payne, Unit Manager

PRODUCT DATA	**Rock type**	Dolerite
	Colour	Medium greenish grey
	Grain	Medium
	Products	Crushed Stone
		Asphalt or Tarmacadam
DRILLING DATA		Hole Diameter 90mm
		Bench Height 10m
	Drills	Ingersoll-Rand CM351
COMPRESSORS		Ingersoll-Rand DXL600P
SECONDARY BREAKING		Dropball on Ruston-Bucyrus 30-RB
LOAD AND HAUL	**Face shovel**	Ruston-Bucyrus
	Trucks	Aveling Barford
	Loaders	Broyt
CRUSHING PLANT		No information provided

ECC QUARRIES LTD (ENGLISH CHINA CLAYS LTD) South West Area

LUXULYAN QUARRY

213

Luxulyan, St Austell, Cornwall PL30 5DX
Phone 0726 850482
OS Map 200 **OS Grid Reference** SX053585
Personnel J P Symmonds, Unit Manager

PRODUCT DATA	**Rock type**	St Austell Granite
	Colour	Light greyish buff
	Grain	Coarse
	Products	Crushed Stone
		Asphalt or Tarmacadam

entry continues overleaf

DRILLING DATA		Hole Diameter 76mm
		Bench Height 10m
	Drills	Ingersoll-Rand LM401C

LOAD AND HAUL	**Trucks**	(2)Heathfield
	Loaders	Caterpillar 966E
		Caterpillar 966C
		Broyt

CRUSHING PLANT	**Primary**	Goodwin Barsby 36 × 24 Jaw
		Broadbent 48 × 42 Jaw
	Secondary	Broadbent 36 × 6 Granulator
	Tertiary	Allis Chalmers 1336 Gyratory
		Nordberg 3′ Cone (Shorthead)

ECC QUARRIES LTD (ENGLISH CHINA CLAYS LTD) South West Area

MOORCROFT QUARRY 214

Billacombe, Plymouth, Devon PL9 8AJ
Phone 0752 402661
OS Map 202   **OS Grid Reference** SX525540
Personnel J Evenden, Unit Manager

PRODUCT DATA	**Rock type**	Devonian Limestone
	Colour	Grey
	Products	Crushed Stone
		Asphalt or Tarmacadam
		Pre-cast Concrete Products
		Industrial Limestone
		Agricultural Lime

DRILLING DATA		Hole Diameter 75–102mm
		Bench Height 6–11m
	Drills	(2)Bohler DTC122H

SECONDARY BREAKING		Hydraulic Hammer
		Dropball on Ruston-Bucyrus 22-RB

LOAD AND HAUL	**Trucks**	(5)Aveling Barford RD40
	Loaders	(2)Caterpillar 988B

CRUSHING PLANT	**Primary**	Pegson 60 × 48 Jaw
	Secondary	Christy & Norris
	Tertiary	Hazemag

ECC QUARRIES LTD (ENGLISH CHINA CLAYS LTD) South West Area

NEW ENGLAND QUARRY 215

Yealmpton, Plymouth, Devon PL7 5BA
Phone 0752 880322
OS Map 202   **OS Grid Reference** SX597545
Personnel S Elson, Unit Manager

PRODUCT DATA	**Rock type**	Dolerite
	Colour	Greenish grey
	Grain	Medium
	Product	Crushed Stone

DRILLING DATA		Hole Diameter 105mm
		Bench Height 12m
	Drills	Ingersoll-Rand CM350

COMPRESSORS	Ingersoll-Rand

SECONDARY BREAKING	Dropball

LOAD AND HAUL	**Trucks**	(3)Aveling Barford RD40
	Loaders	Caterpillar 966C

CRUSHING PLANT	**Primary**	Nordberg V10.08 Gyratory
		Nordberg 42×36 Jaw
	Secondary	Allis Chalmers Superior 11×36 Cone
	Tertiary	(2)Nordberg 3' Cone (Shorthead)

ECC QUARRIES LTD (ENGLISH CHINA CLAYS LTD) South West Area

VENN QUARRY

216

Barnstaple, Devon EX32 9NU
Phone 0271 830831
OS Map 180 **OS Grid Reference** SS582306
Personnel J Fowler, Unit Manager

PRODUCT DATA	**Rock type**	Carboniferous Sandstone
	Colour	Medium to dark grey
	Grain	Medium
	Products	Crushed Stone
		Asphalt or Tarmacadam

DRILLING DATA		Hole Diameter 115-135mm
		Bench Height 15-20m
	Drills	Halco 450H

COMPRESSORS	CompAir 700HE

SECONDARY BREAKING	Dropball on Ruston-Bucyrus 22-RB

LOAD AND HAUL	**Trucks**	Heathfield
	Loaders	Caterpillar 966C

CRUSHING PLANT	**Primary**	Goodwin Barsby 24×36 Jaw (Single Toggle)
	Secondary	Goodwin Barsby 3' Cone
		(2)Symons Cone (Standard)

ECC QUARRIES LTD (ENGLISH CHINA CLAYS LTD) South West Area

WESTLEIGH QUARRY

217

Burlescombe, Tiverton, Devon EX16 7JB
Phone 0823 672353
OS Map 181 **OS Grid Reference** ST064174
Personnel J Pitt, Unit Manager

PRODUCT DATA	**Rock type**	Carboniferous Limestone
	Colour	Medium purplish grey
	Products	Crushed Stone
		Asphalt or Tarmacadam

entry continues overleaf

DRILLING DATA		Hole Diameter 105–125mm
		Bench Height 10m
	Drills	Bohler DTC122 DHD
		Halco 410C
		Atlas Copco ROC604
COMPRESSORS		Atlas Copco XR350
SECONDARY BREAKING		Dropball on Ruston-Bucyrus 22-RB
LOAD AND HAUL	**Trucks**	Aveling Barford RD40
	Loaders	Caterpillar 966C
		Caterpillar 968C
		Caterpillar 980B
CRUSHING PLANT	**Primary**	Pegson 50 × 60 Jaw (Single Toggle)
	Secondary	Hazemag Impactor
	Tertiary	Hazemag Impactor

EDENHALL CONCRETE PRODUCTS
Head Office *Old Mill Quarry, Beith, Strathclyde KA15 1HY*
Phone *05055 2721*
Personnel *K Shaw, Director*

EDENHALL CONCRETE PRODUCTS
OLD MILL QUARRY   218
Beith, Strathclyde KA15 1HY
Phone 05055 2721
OS Map 63   **OS Grid Reference** NS392524
Personnel J Wilson, Unit Manager

PRODUCT DATA	**Rock type**	Carboniferous Dockra Limestone
	Product	Dressed or Lump Stone
DRILLING DATA		Hand-held equipment

R W T EDWORTHY (HEATHFIELD MINERALS) LTD
Head Office *Hayne Quarry, Johnsland, Bow, Crediton, Devon EX17 7HG*
Phone *03633 283*
Personnel *D Edworthy, Owner*
Quarries *Hayne*

R W T EDWORTHY (HEATHFIELD MINERALS) LTD
HAYNE QUARRY   219
Johnsland, Bow, Crediton, Devon EX17 7HG
Phone 03633 283
OS Map 191   **OS Grid Reference** SS715032
Personnel D Edworthy, Owner and Unit Manager

PRODUCT DATA	**Rock type**	Carboniferous Sandstone
	Colour	Buff to brown
	Grain	Medium
	Product	Dressed or Lump Stone

DRILLING DATA   Hand-held equipment
                                  Contractor Celtic Rock Services

---

ELGIN PRECAST CONCRETE LTD (ROBERTSONS CONTRACTING)
Head Office Edgar Road, Elgin, Grampian IV30 3YQ
Phone 0343 549358/549634
Personnel W G Robertson, Managing Director
Quarries Gedloch

ELGIN PRECAST CONCRETE LTD (ROBERTSONS CONTRACTING)

GEDLOCH QUARRY

220

Longmorn, Elgin, Grampian IV30 3SH
Phone 0343 533283
OS Map 28   **OS Grid Reference** NJ234552
Personnel G Welsh, Unit Manager

PRODUCT DATA	**Rock type**	Pre-Cambrian Moine Schist
	Colour	Not commercially significant
	Product	Crushed Stone

DRILLING DATA   Hole Diameter 95mm
                                  Bench Height 12m
                                  Contractor Rockblast

SECONDARY BREAKING   Dropball

| LOAD AND HAUL | **Trucks** | None |
| | **Loaders** | JCB |

CRUSHING PLANT	**Primary**	Baxter 50 × 40 Jaw (Single Toggle)
	Secondary	Kue Ken 36 × 12 Jaw
	Tertiary	Symons 3′ Cone (Shorthead)

R J ENGLAND
Head Office 22–25 Brympton Way, Lynx W Trading Estate, Yeovil, Somerset BA20 2HP
Phone 0935 78173
Personnel R J England, Owner
Quarries Ham Hill

R J ENGLAND

HAM HILL QUARRY

221

Ham Hill, Stoke sub Hamdon, Yeovil, Somerset TA14 6RW
Phone 0935 78173
OS Map 183   **OS Grid Reference** ST477173
Personnel R J England, Owner and Unit Manager

PRODUCT DATA	**Rock type**	Jurassic Oolitic Limestone
	Colour	Honey brown, greyish brown
	Grain	Shelly
	Product	Dressed or Lump Stone

RIPPABLE STRATA

ENGLISH CHINA CLAYS GROUP plc
Corporate Head Office John Keay House, St Austell, Cornwall, PL25 4DJ
Phone 0726 74482 **Fax** 0726 623019 **Telex** 45526
Personnel Lord Chilver, Chairman
D A Langford, Company Secretary
Companies English China Clays Ltd

ENGLISH CHINA CLAYS LTD (ENGLISH CHINA CLAYS GROUP plc)
Head Office Northernhay House East, Northernhay Place, Exeter, Devon EX4 3QP
Phone 0392 52231 **Fax** 0392 412132 **Telex** 42838
Personnel T W Stobart, Managing Director
Companies ECC Calcium Carbonates Ltd, ECC Quarries Ltd, Grinshill Stone Quarries Ltd — see separate entry

ESKETT QUARRIES LTD (EVERED HOLDINGS plc)
Head Office Stoneleigh, Park End Road, Workington, Cumbria CA14 4DG
Phone 0900 62271
Personnel P Hughes, Director
Quarries Avon Dassett, Blencowe, Cool Scar, Eskett, Ghyll Scaur, Highmoor, Hutchbank, Whitworth

---

ESKETT QUARRIES LTD (EVERED HOLDINGS plc)

AVON DASSETT QUARRY
222

Avon Dassett, Warmington, Warwicks CV32 6AS
Phone 0900 62271
OS Map 151 **OS Grid Reference** SP416508
Personnel P Lee, Unit Manager

PRODUCT DATA	**Rock type**	Jurassic Limestone
	Colour	Not commercially significant
	Product	Crushed Stone

RIPPABLE STRATA

| LOAD AND HAUL | **Face shovel** | Poclain |
| | **Trucks** | (2)Caterpillar |

ESKETT QUARRIES LTD (EVERED HOLDINGS plc)

BLENCOWE QUARRY
223

Blencowe, Penrith, Cumbria CA11 0DE
Phone 0900 62271
OS Map 90 **OS Grid Reference** NY460298

QUARRY ON STAND

ROCK DATA	**Rock type**	Carboniferous Limestone
	Colour	Brownish grey
	Grain	Fine

ESKETT QUARRIES LTD (EVERED HOLDINGS plc)

COOL SCAR QUARRY
224

Kilnsey, Nr Skipton, N Yorks BD23 5PS
Phone 0756 752497
OS Map 98 **OS Grid Reference** SD968677
Personnel A K Mallinson, Unit Manager

PRODUCT DATA	**Rock type**	Carboniferous Great Scar Limestone
	Colour	Light creamy grey
	Products	Crushed Stone
		Industrial Limestone
		Agricultural Lime
DRILLING DATA		Hole Diameter 105mm
		Bench Height 25m
	Drills	Tamrock DHA1000
SECONDARY BREAKING		Dropball on Ruston-Bucyrus 22-RB
LOAD AND HAUL	**Trucks**	Volvo
	Loaders	Volvo
CRUSHING PLANT	**Primary**	BJD 45 Impactor
	Secondary	BJD 33 Impactor

ESKETT QUARRIES LTD (EVERED HOLDINGS plc)

ESKETT QUARRY 225

Winder Frizington, Cumbria CA26 3UN
Phone 0946 861366/861561
OS Map 89 **OS Grid Reference** NY054170
Personnel G Skelham, Unit Manager

PRODUCT DATA	**Rock type**	Carboniferous Limestone
	Colour	Light pinkish grey
	Grain	Medium
	Products	Crushed Stone
		Asphalt or Tarmacadam
		Industrial Limestone
		Agricultural Lime
DRILLING DATA		Hole Diameter 110mm
		Bench Height 20m
		Contractor Rocklift
	Drills	Tamrock DHA800
SECONDARY BREAKING		Dropball on Ruston-Bucyrus 22-RB
LOAD AND HAUL	**Trucks**	Volvo
	Loaders	Volvo
CRUSHING PLANT	**Primary**	Sheepbridge Impactor
	Secondary	Mansfield No 3 Hammermill
		Mansfield No 4 Hammermill
	Tertiary	Allis Chalmers Hydrocone 200

ESKETT QUARRIES LTD (EVERED HOLDINGS plc)

GHYLL SCAUR QUARRY 226

Millom, Cumbria LA18 5HB
Phone 065 74222
OS Map 96 **OS Grid Reference** SD171828
Personnel I Charwood, Unit Manager

entry continues overleaf

PRODUCT DATA	**Rock type**	Ordovician Borrowdale Andesitic Tuff
	Colour	Dark green
	Grain	Fine
	Product	Crushed Stone

DRILLING DATA		Hole Diameter 110mm
		Bench Height 20m
		Contractor Rocklift

| SECONDARY BREAKING | | Hydraulic Hammer - Montabert on JCB |

| LOAD AND HAUL | **Face shovel** | Caterpillar 235C |
| | **Trucks** | Volvo BM A25 |

CRUSHING PLANT	**Primary**	Kue Ken Jaw
	Secondary	Kue Ken Jaw
	Tertiary	Allis Chalmers 3' Gyratory
	Quaternary	Lokomo 30" Cone
		Barmac Duopactor

ESKETT QUARRIES LTD (EVERED HOLDINGS plc)

HIGHMOOR QUARRY

227

Doctor Lane, Southead, Oldham, Lancs OL4 3RY
Phone 0457 875157
OS Map 109 **OS Grid Reference** SD973067
Personnel G Loynds, Unit Manager

PRODUCT DATA	**Rock type**	Carboniferous Lower Coal Measures Sandstone
	Colour	Not Commercially Significant
	Grain	Fine
	Product	Crushed Stone

RIPPABLE STRATA

| LOAD AND HAUL | **Trucks** | Volvo |
| | **Loaders** | Poclain 350 |

NO FURTHER INFORMATION PROVIDED

ESKETT QUARRIES LTD (EVERED HOLDINGS plc)

HUTCHBANK QUARRY

228

Grawe Road, Haslingden, Rossendale, Lancs BB4 5PB
Phone 0706 224338
OS Map 103 **OS Grid Reference** SD777231
Personnel S Simpson, Unit Manager

PRODUCT DATA	**Rock type**	Carboniferous Sandstone (Haslingden Flags)
	Colour	Grey
	Grain	Medium
	Product	Crushed Stone

DRILLING DATA		Hole Diameter 105mm
		Bench Height 18m
		Contractor Ritchies Equipment

SECONDARY BREAKING		Dropball on Ruston-Bucyrus 22-RB
LOAD AND HAUL	**Trucks**	Volvo
	Loaders	Volvo
CRUSHING PLANT	**Primary**	Kue Ken 120S Jaw
	Secondary	Kue Ken 4.25′ Cone

ESKETT QUARRIES LTD (EVERED HOLDINGS plc)

WHITWORTH QUARRY

229

Tong End Lane, Whitworth, Nr Rochdale, Lancs OL12 8BE
Phone 0706 853295
OS Map 109   **OS Grid Reference** SD875185
Personnel C Topham, Unit Manager

PRODUCT DATA	**Rock type**	Carboniferous Sandstone (Haslingden Flags)
	Colour	Grey
	Grain	Fine
	Product	Crushed Stone
DRILLING DATA		Hole Diameter 105mm
		Bench Height 30m
	Drills	Holman Voltrak
SECONDARY BREAKING		Dropball on Ruston-Bucyrus 22-RB
LOAD AND HAUL	**Face shovel**	Volvo 4500
	Trucks	Volvo
	Loaders	Komatsu WA500
		Volvo BM L160
		Volvo 1240
CRUSHING PLANT	**Primary**	Parker ST50 × 36 Jaw (Single Toggle)
	Secondary	Parker 105
	Tertiary	SBM Impactor

EVERED HOLDINGS plc

Corporate Head Office *6th Floor, Radcliffe House, Blenheim Court, Lode Lane, Solihull, W Midlands B91 2AA*
Phone *021 711 1717*
Personnel *R Kettle, Chief Executive*
K Harris, Financial Director
Companies *Eskett Quarries Ltd, Evered Quarry Products (England) Ltd, Ogden Roadstone Ltd, Peterborough Quarries Ltd, Tractor Shovels Tawse Ltd*

EVERED QUARRY PRODUCTS (ENGLAND) LTD (EVERED HOLDINGS plc)

Head Office *Evered House, Burnwick Terrace, Newport Road, Stafford, Staffs ST16 1BB, W Midlands B91 2AA*
Phone *0785 57500*
Personnel *G Plant, Managing Director*
S Gibbs, General Manager
Quarries *Heights*

entry continues overleaf

EVERED QUARRY PRODUCTS (ENGLAND) LTD (EVERED HOLDINGS plc)

HEIGHTS QUARRY

230

Westgate, Weardale, Bishop Auckland, Durham DL13 1PE
Phone 0388 517381
OS Map 92   **OS Grid Reference** NY925389
Personnel F Dooris, Unit Manager

PRODUCT DATA	**Rock type**	Carboniferous Limestone
	Products	Crushed Stone
		Asphalt or Tarmacadam
DRILLING DATA		Hole Diameter 110mm
		Bench Height 18-20m
	Drills	Halco 410C
COMPRESSORS		CompAir 650HE
SECONDARY BREAKING		Dropball on Ruston-Bucyrus 22-RB
LOAD AND HAUL	**Trucks**	DJB 25
	Loaders	Caterpillar 950
		Caterpillar 980C
CRUSHING PLANT	**Primary**	Kue Ken 42×30 Jaw
	Secondary	Mansfield No 4 Hammermill
		Kue Ken Hammermill
	Tertiary	Mansfield No 3 Hammermill

FARMINGTON STONE LTD
Head Office *37 Roslyn Road, Redland, Bristol, Avon BS6 6NJ*
Phone *0272 247336*
Personnel *J D Ferris, Director*
Quarries *Farmington*

FARMINGTON STONE LTD

FARMINGTON QUARRY

231

Northleach, Cheltenham, Glos GL54 3NG
Phone 04516 0280
OS Map 163   **OS Grid Reference** SP125143
Personnel D Selby, Unit Manager

PRODUCT DATA	**Rock type**	Jurassic Limestone
	Colour	White to grey
	Grain	Fine
	Product	Dressed or Lump Stone
DRILLING DATA		Hole Diameter 38mm
		Hand-held equipment
COMPRESSORS		Atlas Copco
LOAD AND HAUL	**Loaders**	Massey Ferguson - Climax

G FARRAR (QUARRIES) LTD
Head Office *Bradford Street, Keighley, W Yorks BD21 3EB*
Phone *0535 602344*
Personnel *R A Farrar, Owner*
Quarries *Northowram Hills*

G FARRAR (QUARRIES) LTD
NORTHOWRAM HILLS QUARRY
232
Northowram, Halifax, W Yorks HX3 5LL
Phone 0535 602344
OS Map 104 **OS Grid Reference** SE109272
Personnel R A Farrar, Owner and Unit Manager

PRODUCT DATA	**Rock type**	Carboniferous Sandstone
	Colour	Buff to grey
	Grain	Fine
	Product	Dressed or Lump Stone

DRILLING DATA		Hand-held equipment

FENSTONE QUARRIES LTD
Head Office *Market Bridge, Bielby, York, W Yorks YO4 4JR*
Phone 0759 318666
Personnel K Turton, Director
Quarries Settrington, Stonegrave

FENSTONE QUARRIES LTD
SETTRINGTON QUARRY
233
Settrington, N Yorks YO17 8NP
Phone 0759 318666
OS Map 100 **OS Grid Reference** SE830710
Personnel K Turton, Director and Unit Manager

PRODUCT DATA	**Rock type**	Jurassic Limestone
	Product	Crushed Stone

DRILLING DATA		Hole Diameter 108mm
		Bench Height 17m
		Contractor R&B Rock Drillers

CRUSHING PLANT		No information provided

FENSTONE QUARRIES LTD
STONEGRAVE QUARRY
234
Stonegrave, York, N Yorks YO6 4LJ
Phone 0759 318666
OS Map 100 **OS Grid Reference** SE659779
Personnel K Turton, Director and Unit Manager

PRODUCT DATA	**Rock type**	Jurassic Limestone
	Product	Crushed Stone

entry continues overleaf

DRILLING DATA	Hole Diameter 108mm
	Bench Height 17m
	Contractor R&B Rock Drillers
CRUSHING PLANT	No information provided

FFESTINIOG SLATE QUARRY LTD
Head Office *Blaenau Ffestiniog, Gwynedd LL41 3NB*
Phone *0766 830204/830664/830218* **Fax** *0766 831105*
Personnel *O G Williams, Managing Director*
Quarries *Cwt-y-Bugail, Gloddfa Ganol, Nantlle, Oakeley*

FFESTINIOG SLATE QUARRY LTD
CWT-Y-BUGAIL QUARRY 235
Manod, Blaenau Ffestiniog, Gwynedd LL41 3NB
Phone 0766 830204
OS Map 115 **OS Grid Reference** SH690470
Personnel G Edwards, Unit Manager

PRODUCT DATA	**Rock type**	Ordovician Slate
	Colour	Bluish grey
	Grain	Fine
	Product	Dressed or Lump Stone
DRILLING DATA		Hand-held equipment

FFESTINIOG SLATE QUARRY LTD
GLODDFA GANOL QUARRY 236
Blaenau Ffestiniog, Gwynedd LL41 3NB
Phone 0766 830204
OS Map 115 **OS Grid Reference** SH690470
Personnel W Roberts, Unit Manager

PRODUCT DATA	**Rock type**	Ordovician Slate
	Colour	Bluish grey
	Grain	Fine
	Product	Dressed or Lump Stone
DRILLING DATA		Hand-held equipment

FFESTINIOG SLATE QUARRY LTD
NANTLLE QUARRY 237
Penygroes, Caernafon, Gwynedd LL54 6AE
Phone 0766 830204
OS Map 115 **OS Grid Reference** SH467531
Personnel P Hughes, Unit Manager

PRODUCT DATA	**Rock type**	Cambrian Slate
	Colour	Heather blue
	Grain	Fine
	Product	Dressed or Lump Stone
DRILLING DATA		Hand-held equipment

FFESTINIOG SLATE QUARRY LTD

OAKELEY QUARRY

238

Blaenau Ffestiniog, Gwynedd LL41 3NB
Phone 0766 830204
OS Map 115   **OS Grid Reference** SH690470
Personnel G Roberts, Unit Manager

PRODUCT DATA	**Rock type**	Ordovician Slate
	Colour	Bluish grey
	Grain	Fine
	Product	Dressed or Lump Stone

DRILLING DATA		Hand-held equipment

FILKINS QUARRIES LTD
Head Office Brook House, Cricklade, Wilts SN6 6DD
Phone 0793 750150
Personnel S Aitken, Managing Director
Quarries Filkins

FILKINS QUARRIES LTD

FILKINS QUARRY

239

Filkins, Burford, Oxford, Oxon OX9 1BN
Phone 0793 750150
OS Map 163   **OS Grid Reference** SP231058
Personnel M Swinfors, Unit Manager

PRODUCT DATA	**Rock type**	Jurassic Great Oolitic Limestone
	Colour	Honey brown
	Grain	Fine, fossiliferous
	Product	Dressed or Lump Stone

RIPPABLE STRATA

LOAD AND HAUL	**Face shovel**	Hymac
	Trucks	None
	Loaders	JCB

CRUSHING PLANT		No information provided

W E FORD AND SON (QUARRIES)
Head Office Fords Quarry, Bacup Road, Burnley, Lancs BB11 3RL
Phone 0282 56815
Personnel W E Ford, Owner
Quarries Fords

W E FORD AND SON (QUARRIES)

FORDS QUARRY

240

Bacup Road, Burnley, Lancs BB11 3RL
Phone 0282 56815
OS Map 103   **OS Grid Reference** SD839320
Personnel W E Ford, Owner and Unit Manager

entry continues overleaf

PRODUCT DATA	**Rock type**	Carboniferous Sandstone (Gritstone)
	Colour	Greyish buff
	Grain	Fine
	Product	Dressed or Lump Stone

| DRILLING DATA | | Hand-held equipment |

FOREST OF DEAN STONE FIRMS LTD
Head Office Bixland Stone Works, Parkend, Lydney, Glos GL15 4JS
Phone 0594 562304 **Fax** 0594 564184
Personnel P Scott-Russell, Managing Director
Quarries Bixhead

FOREST OF DEAN STONE FIRMS LTD
BIXHEAD QUARRY   241
Barnhill, Coleford, Glos GL16 7EB
Phone 0594 562304
OS Map 162 **OS Grid Reference** SO598107
Personnel L J Hinton, Unit Manager

PRODUCT DATA	**Rock type**	Carboniferous Sandstone
	Colour	Greyish blue, greyish green
	Grain	Fine to medium
	Product	Dressed or Lump Stone

| DRILLING DATA | | Hand-held equipment |

FOSTER YEOMAN LTD – see under YEOMAN

GALA QUARRY COMPANY
Head Office Abbey Place, Melrose, Borders TD6 9LQ
Phone 089682 2595
Personnel R Landakhers, Director
Quarries Blynlee

GALA QUARRY COMPANY
BLYNLEE QUARRY   242
Galashiels, Borders TD1 1TQ
Phone 0896 55788
OS Map 73 **OS Grid Reference** NT490362
Personnel R Landakhers, Director and Unit Manager

PRODUCT DATA	**Rock type**	Silurian Sandstone
	Products	Dressed or Lump Stone
		Crushed Stone

| DRILLING DATA | | Contractor occasionally |

NO FURTHER INFORMATION PROVIDED

J & A GARDNER AND COMPANY LTD
Head Office *288 Clyde Street, Glasgow, Strathclyde G1 4JS*
Phone *041 221 7845*
Personnel *E G Struthers, Managing Director*
Quarries *Bonawe*

J & A GARDNER AND COMPANY LTD

BONAWE QUARRY

Connel, Strathclyde PA37 1RL
Phone 0631 75275
OS Map 50   **OS Grid Reference** NN015335
Personnel C Struthers, Operations Manager
W Galloway, Unit Manager

243

PRODUCT DATA	**Rock type**	Etive Granite
	Colour	Light pinkish grey
	Grain	Coarse
	Products	Crushed Stone
		Asphalt or Tarmacadam
		Ready-mixed Concrete
		Pre-cast Concrete Products

DRILLING DATA		Hole Diameter 105mm
		Bench Height 15–25m
		Contractor Albion Drilling Services

| SECONDARY BREAKING | | Dropball |

| LOAD AND HAUL | **Trucks** | None |
| | **Loaders** | (2)Caterpillar |

CRUSHING PLANT	**Primary**	AB Swing Hammer
	Secondary	Nordberg 4' Cone
	Tertiary	Nordberg 3' Gyradisc
		Barmac Duopactor

D & J GARRICK QUARRIES LTD
Head Office *Vatseter Quarry, Gott, Shetland ZE2 9SB*
Phone *059584 279*
Personnel *R Garrick, Managing Director*
Quarries *Vatseter*

D & J GARRICK QUARRIES LTD

VATSETER QUARRY

Gott, Shetland ZE2 9SB
Phone 059 584279
OS Map 2   **OS Grid Reference** HU549898
Personnel R Garrick, Managing Director and Unit Manager

244

PRODUCT DATA	**Rock type**	Basalt
	Colour	Greyish black
	Grain	Fine
	Product	Crushed Stone

NO FURTHER INFORMATION PROVIDED

GASKELL BROS (W M & C) LTD
Head Office Bryn Road, Ashton-in-Makerfield, Wigan, Lancs WN4 8AH
Phone 0942 725722 **Fax** 0942 271189
Personnel W M Gaskell, Director
Quarries Borrow Pit

GASKELL BROS (W M & C) LTD

BORROW PIT QUARRY   **245**

Winwick Lane, Croft, Warrington, Cheshire WA3 7EW
Phone 0942 725722
OS Map 109 **OS Grid Reference** SJ635934
Personnel W M Gaskell, Director and Unit Manager

PRODUCT DATA	**Rock type**	Triassic Sandstone
	Product	Crushed Stone
RIPPABLE STRATA		

LOAD AND HAUL	**Face shovel**	Komatsu PC300
	Loaders	Komatsu WA240
		(3)Komatsu WA350

NO FURTHER INFORMATION PROVIDED

DAVID GEDDES (QUARRIES) LTD
Head Office Swirlburn, Colliston-by-Arbroath, Tayside DD11 3SH
Phone 024189 266 **Fax** 024189 445
Personnel N Geddes, Director
Quarries Ardownie

DAVID GEDDES (QUARRIES) LTD

ARDOWNIE QUARRY   **246**

Monifieth, Tayside DD5 4HW
Phone 0382 533050
OS Map 54 **OS Grid Reference** NO488340
Personnel R Andrews, Unit Manager

PRODUCT DATA	**Rock type**	Dolerite
	Colour	Greyish black
	Grain	Medium
	Product	Crushed Stone

DRILLING DATA		Hole Diameter 110mm
		Bench Height 30m
		Contractor Ritchies Equipment

SECONDARY BREAKING		Dropball on Ransomes & Rapier NCK-Rapier 305B

LOAD AND HAUL	**Face shovel**	Akermans
	Trucks	(2)Heathfield
	Loaders	Volvo

CRUSHING PLANT	**Primary**	Kue Ken 42×36 Jaw
	Secondary	Kue Ken 1300CT
	Tertiary	Pegson 900 Autocone
		Barmac Duopactor

T W GIBBS
Head Office *Horn Park Quarry, Broadwindsor, Beaminster, Dorset DT8 3PT*
Phone *0308 68419*
Personnel *T W Gibbs, Owner*
Quarries *Horn Park*

T W GIBBS
HORN PARK QUARRY
247
Broadwindsor, Beaminster, Dorset DT8 3PT
Phone 0308 68419
OS Map 193   **OS Grid Reference** ST457022
Personnel T W Gibbs, Owner and Unit Manager

PRODUCT DATA	**Rock type**	Jurassic Oolitic Limestone
	Colour	White
	Grain	Fine
	Product	Dressed or Lump Stone

RIPPABLE STRATA

DENNIS GILLSON AND SON (HAWORTH) LTD
Head Office *Naylor Hill Quarries, Blackmore Road, Keighley, W Yorks BD22 9SW*
Phone *0535 43317*
Personnel *F Gillson, Managing Director*
Quarries *Naylor Hill*

DENNIS GILLSON AND SON (HAWORTH) LTD
NAYLOR HILL QUARRIES
248
Blackmore Road, Keighley, W Yorks BD22 9SW
Phone 0535 43317
OS Map 104   **OS Grid Reference** SE040365
Personnel F Gillson, Managing Director and Unit Manager

PRODUCT DATA	**Rock type**	Carboniferous Sandstone
	Colour	Buff to brown
	Grain	Fine
	Product	Dressed or Lump Stone
DRILLING DATA		Hand-held equipment

F GILMAN LTD
Head Office *Snowdrop Lane, Haverfordwest, Dyfed SA61 1ET*
Phone *0437 765226* **Fax** *0437 760810*
Personnel *J Ingle, General Manager*
Quarries *Alltgoch, Bolton Hill, Cerrigyrwyn, Coygen, Middle Mill*

F GILMAN LTD
ALLTGOCH QUARRY
249
Cwrt-newydd, Llanybythen, Dyfed SA40 9YL
Phone 0570 46337
OS Map 146   **OS Grid Reference** SN491485
Personnel E G Williams, Unit Manager

entry continues overleaf

PRODUCT DATA	**Rock type**	Ordovician Sandstone
	Colour	Not Commercially Significant
	Grain	Medium
	Products	Crushed Stone
		Asphalt or Tarmacadam
		Ready-mixed Concrete

DRILLING DATA		Hole Diameter 110mm
		Bench Height 20m
	Drills	Holman Universal

| COMPRESSORS | | CompAir 37HP |

| SECONDARY BREAKING | | Dropball on Ruston-Bucyrus 30-RB |

| CRUSHING PLANT | | No information provided |

F GILMAN LTD

BOLTON HILL QUARRY

250

Quarry Tiers Cross, Haverfordwest, Dyfed SA62 3ER
Phone 0437 890560
OS Map 158   **OS Grid Reference** SM918114
Personnel R Payne, Unit Manager

PRODUCT DATA	**Rock type**	Diorite
	Colour	Speckled black and white
	Grain	Coarse
	Products	Crushed Stone
		Asphalt or Tarmacadam
		Ready-mixed Concrete

DRILLING DATA		Hole Diameter 110mm
		Bench Height 20m
	Drills	Holman Voltrak

| COMPRESSORS | | CompAir 650HE |

| SECONDARY BREAKING | | Dropball on Ruston-Bucyrus 22-RB |

| CRUSHING PLANT | | No information provided |

F GILMAN LTD

CERRIGYRWYN QUARRY

251

Llangynog, Carmarthen, Dyfed SA32 5HU
Phone 0437 890511
OS Map 159   **OS Grid Reference** SN337161
Personnel R Payne, Unit Manager

PRODUCT DATA	**Rock type**	Dolerite
	Colour	Black
	Grain	Medium
	Product	Crushed Stone

DRILLING DATA		Hole Diameter 110mm
		Bench Height 12.5m
	Drills	Halco 400C

SECONDARY BREAKING		Dropball
LOAD AND HAUL	**Trucks**	(2)Heathfield
		Aveling Barford RD30
	Loaders	JCB
CRUSHING PLANT		No information provided

F GILMAN LTD

COYGEN QUARRY

252

Laugharne, Carmarthen, Dyfed SA33 4RR
Phone 0994 21621
OS Map 158   **OS Grid Reference** SN285092
Personnel E Istead, Unit Manager

PRODUCT DATA	**Rock type**	Dolerite
	Colour	Greyish black
	Grain	Medium
	Products	Crushed Stone
		Asphalt or Tarmacadam
DRILLING DATA		Hole Diameter 105mm
		Bench Height 25m
	Drills	Holman Voltrak
COMPRESSORS		CompAir 650HE
SECONDARY BREAKING		Dropball on Ruston-Bucyrus 22-RB
CRUSHING PLANT	**Primary**	Pegson Jaw
	Secondary	BJD 44 Rotary
	Tertiary	(2)Lanway Hammermill

F GILMAN LTD

MIDDLE MILL QUARRY

253

Solva, Haverfordwest, Dyfed SA62 4LD
Phone 0437 5226
OS Map 158   **OS Grid Reference** SM804259

QUARRY ON STAND

ROCK DATA	**Rock type**	Granite

GLENDINNING GROUP

Head Office *Glentor, Balland Lane, Ashburton, Newton Abbot, Devon TQ13 7LF*
Phone *0364 52601*
Personnel *I Ashford, Managing Director*
           I Glendinning, Operations Director
Quarries *Linhay Hill*

entry continues overleaf

GLENDINNING GROUP

LINHAY HILL QUARRY

254

Balland Lane, Ashburton, Newton Abbott, Devon TQ13 7LF
Phone 0364 52601
OS Map 202 **OS Grid Reference** SX710770
Personnel I Glendinning, Operations Director and Unit Manager

PRODUCT DATA	**Rock type**	Devonian Limestone
	Colour	Dark grey with red and white veins
	Products	Crushed Stone
		Asphalt or Tarmacadam
		Ready-mixed Concrete
		Agricultural Lime
DRILLING DATA		Hole Diameter 105mm
		Bench Height 12m
	Drills	(2)Halco 410
SECONDARY BREAKING		Dropball on Ruston-Bucyrus 22-RB
LOAD AND HAUL	**Face shovel**	O&K RH30
	Trucks	Aveling Barford RD50
		Aveling Barford RD70
	Loaders	Michigan 475C
		Caterpillar 966C
		Caterpillar 988B
CRUSHING PLANT	**Primary**	Pegson 60 × 48 Jaw
	Secondary	Mansfield No 5 Hammermill

GLOUCESTER SAND AND GRAVEL COMPANY LTD
Head Office Overbury Estate Office, Overbury, Tewkesbury, Glos GL20 7NS
Phone 038689 217
Personnel I Ralph, General Manager
Quarries Syreford

GLOUCESTER SAND AND GRAVEL COMPANY LTD

SYREFORD QUARRY

255

Andoversford, Cheltenham, Glos GL54 5SJ
Phone 0242 674744
OS Map 163 **OS Grid Reference** SP025227
Personnel - Not yet in production

PRODUCT DATA	**Rock type**	Carboniferous Limestone

GOETRE LTD
Head Office Station Yard, Abermule, Montgomery, Powys SY15 6NL
Phone 068686 667
Personnel G Bayliss, General Manager
Quarries Penstrowed

GOETRE LTD
PENSTROWED QUARRY
256

Abermule, Montgomery, Powys SY15 6NL
Phone 068686 667
OS Map 136   **OS Grid Reference** SJ161949
Personnel G Bayliss, General Manager and Unit Manager

| PRODUCT DATA | **Rock type** | Silurian Sandstone |
| | **Product** | Crushed Stone |

| DRILLING DATA | | No drilling |

GRAMPIAN REGION COUNCIL
Head Office *Director of Works, Woodhill House, Ashgrove Road West, Aberdeen, Grampian AB9 2LU*
Phone *0224 682222* **Fax** *0224 697445*
Personnel *G Kirkbride, Director of Works*
Quarries *Balmedie, Bluehill, Craiglash, Pitcaple, Western New Forres*

GRAMPIAN REGION COUNCIL
BALMEDIE QUARRY
257

Balmedie, Aberdeen, Grampian AB4 0YT
Phone 0358 42203
OS Map 38   **OS Grid Reference** NJ944181
Personnel J Lamb, Unit Manager

PRODUCT DATA	**Rock type**	Gabbro
	Colour	Dark grey
	Grain	Coarse
	Product	Crushed Stone

DRILLING DATA		Hole Diameter 105mm
		Bench Height 12m
		Contractor Drilling & Blasting

| SECONDARY BREAKING | | Dropball on Ruston-Bucyrus 22-RB |

LOAD AND HAUL	**Trucks**	Aveling Barford RD35
	Loaders	Caterpillar
		Aveling Barford

CRUSHING PLANT	**Primary**	Goodwin Barsby 36 × 38 Jaw (Double Toggle)
	Secondary	Nordberg 4.25′ Cone
	Tertiary	Nordberg 3′ Cone (Shorthead)
		Kennedy 25″ Cone

GRAMPIAN REGION COUNCIL
BLUEHILL QUARRY
258

Dufftown, Keith, Grampian AB5 4BW
Phone 03405 376
OS Map 37   **OS Grid Reference** NJ296438
Personnel J Lamb, Unit Manager

PRODUCT DATA	**Rock type**	Pre-Cambrian Moine Schist
	Colour	Not commercially significant
	Product	Crushed Stone

entry continues overleaf

DRILLING DATA		Hole Diameter 105mm Bench Height 12m Contractor Drilling & Blasting
SECONDARY BREAKING		Dropball on Ruston-Bucyrus 22-RB
LOAD AND HAUL	**Trucks** **Loaders**	(2)Aveling Barford Caterpillar Aveling Barford
CRUSHING PLANT	**Primary** **Secondary** **Tertiary**	Marsden 36×24 Jaw (Double Toggle) Kue Ken 24×10 Jaw (Double Toggle) (2)Kue Ken 28″ Cone

GRAMPIAN REGION COUNCIL

CRAIGLASH QUARRY

259

Glassel, Banchory, Grampian AB3 4EA
Phone 033982 230
OS Map 37 **OS Grid Reference** NO622987
Personnel J Lamb, Unit Manager

PRODUCT DATA	**Rock type** **Colour** **Product**	Pre-Cambrian Dalradian Limestone Not commercially significant Crushed Stone
DRILLING DATA		Hole Diameter 105mm Bench Height 12m Contractor Drilling & Blasting
SECONDARY BREAKING		Dropball on Ruston-Bucyrus 22-RB
LOAD AND HAUL	**Trucks** **Loaders**	(2)Aveling Barford Caterpillar Aveling Barford
CRUSHING PLANT	**Primary** **Secondary** **Tertiary**	Goodwin Barsby 36×38 Jaw (Double Toggle) Nordberg 4.25′ Cone (2)Nordberg 3′ Cone

GRAMPIAN REGION COUNCIL

PITCAPLE QUARRY

260

Pitcaple, Inverurie, Grampian AB5 9PB
Phone 04676 218
OS Map 38
Personnel J Lamb, Unit Manager

PRODUCT DATA	**Rock type** **Colour** **Grain** **Product**	Gabbro Dark grey Coarse Crushed Stone
DRILLING DATA		Hole Diameter 105mm Bench Height 12m Contractor Drilling & Blasting

SECONDARY BREAKING		Dropball on Ruston-Bucyrus 22-RB
LOAD AND HAUL	**Trucks**	(2)Aveling Barford
	Loaders	Caterpillar
		Aveling Barford
CRUSHING PLANT	**Primary**	Baxter 36 × 24 Jaw (Double Toggle)
	Secondary	Nordberg 4′ Cone
	Tertiary	(2)Kue Ken 28″ Cone

GRAMPIAN REGION COUNCIL

WESTERN NEW FORRES

261

Quarry, Forres, Grampian IV36 0RQ
Phone 0224 682222
OS Map 27 **OS Grid Reference** NJ063579
Personnel J Lamb, Unit Manager

PRODUCT DATA	**Rock type**	Pre-Cambrian Moine Schist
	Colour	Not commercially significant
	Product	Crushed Stone
DRILLING DATA		Hole Diameter 105mm
		Bench Height 12m
		Contractor Drilling & Blasting
SECONDARY BREAKING		Dropball on Ruston-Bucyrus 22-RB
LOAD AND HAUL	**Trucks**	(2)Aveling Barford
	Loaders	Caterpillar
		Aveling Barford
CRUSHING PLANT	**Primary**	Kue Ken 36 × 24 Jaw (Double Toggle)
	Secondary	Kue Ken 24 × 12 Jaw
	Tertiary	(2)Kennedy 19″ Cone

GREAVES WELSH SLATE CO LTD
Head Office *Llechwedd Slate Mines, Blaenau Ffestiniog, Gwynedd LL41 3NB*
Phone 0766 830522 **Fax** *0766 831064*
Personnel J W Greaves, Chairman
R H Davies, Managing Director
Quarries Braich Ddu, Diphwys Casson, Llechwedd Slate Mine, Maenofferen

GREAVES WELSH SLATE CO LTD

BRAICH DDU QUARRY

262

Blaenau Ffestiniog, Gwynedd LL41 3NB
Phone 0766 830522
OS Map 115 **OS Grid Reference** SH715464
Personnel F J Jones, Unit Manager

PRODUCT DATA	**Rock type**	Ordovician Slate
	Colour	Bluish grey
	Grain	Fine
	Product	Dressed or Lump Stone
DRILLING DATA		Hand-held equipment

GREAVES WELSH SLATE CO LTD

DIPHWYS CASSON QUARRY

263

Blaenau Ffestiniog, Gwynedd LL41 3NB
Phone 0766 830522
OS Map 115   **OS Grid Reference** SH701470
Personnel F J Jones, Unit Manager

PRODUCT DATA	**Rock type**	Ordovician Slate
	Colour	Bluish grey
	Grain	Fine
	Product	Dressed or Lump Stone

| DRILLING DATA | | Hand-held equipment |

GREAVES WELSH SLATE CO LTD

LLECHEDD SLATE MINE

264

Blaenau Ffestiniog, Gwynedd LL41 3NB
Phone 0766 830522
OS Map 115   **OS Grid Reference** SH705470
Personnel F J Jones, Unit Manager

PRODUCT DATA	**Rock type**	Ordovician Slate
	Colour	Bluish grey
	Grain	Fine
	Product	Dressed or Lump Stone

| DRILLING DATA | | Hand-held equipment |

GREAVES WELSH SLATE CO LTD

MAENOFFEREN QUARRY

265

Blaenau Ffestiniog, Gwynedd LL41 3NB
Phone 0766 830522
OS Map 115   **OS Grid Reference** SH712468
Personnel F J Jones, Unit Manager

PRODUCT DATA	**Rock type**	Ordovician Slate
	Colour	Bluish grey
	Grain	Fine
	Product	Dressed or Lump Stone

| DRILLING DATA | | Hand-held equipment |

D GREENWOOD

Head Office *Soil Hill Quarry, Coal Lane, Causeway Foot, Halifax, W Yorks HX2 9PG*
Phone *0274 832384*
Personnel *D Greenwood, Owner*
Quarries *Soil Hill*

D GREENWOOD

SOIL HILL QUARRY

266

Coal Lane, Causeway Foot, Halifax, W Yorks HX2 9PG
Phone 0274 832384
OS Map 104   **OS Grid Reference** SE079314
Personnel D Greenwood, Owner and Unit Manager

PRODUCT DATA	**Rock type**	Carboniferous Sandstone
	Colour	Light buff
	Grain	Fine
	Product	Dressed or Lump Stone
DRILLING DATA		Hand-held equipment

GREGORY QUARRIES LTD
Head Office *184 Nottingham Road, Mansfield, Notts NG18 5AP*
Phone *0623 23092*
Personnel *E E Abbott, Director*
Quarries *Ancaster, Glebe, Mansfield*

GREGORY QUARRIES LTD

ANCASTER QUARRY 267

Ancaster, Grantham, Lincs NG32 3QE
Phone 0623 23092
OS Map 130   **OS Grid Reference** SK991407
Personnel P King, Unit Manager

PRODUCT DATA	**Rock type**	Jurassic Limestone
	Colour	Creamy white
	Grain	Fine
	Product	Dressed or Lump Stone
DRILLING DATA		Hand-held equipment

GREGORY QUARRIES LTD

GLEBE QUARRY 268

Ancaster, Grantham, Lincs NG32 3RH
Phone 0623 23092
OS Map 130   **OS Grid Reference** SK990410
Personnel P King, Unit Manager

PRODUCT DATA	**Rock type**	Jurassic Limestone
	Colour	Hard white
	Grain	Fine
	Product	Dressed or Lump Stone
DRILLING DATA		Hand-held equipment

GREGORY QUARRIES LTD

MANSFIELD QUARRY 269

Mansfield, Notts NG18 5AP Road, Mansfield, Notts NG18 5AP
Phone 0623 23092
OS Map 120   **OS Grid Reference** SK538599
Personnel P King, Unit Manager

PRODUCT DATA	**Rock type**	Permian Magnesian Limestone
	Colour	Buff with green to grey bands
	Grain	Fine
	Product	Dressed or Lump Stone
DRILLING DATA		Hand-held equipment

GRINSHILL STONE QUARRIES LTD (ECC QUARRIES LTD)
Head Office *Grinshill Quarry, Clive, Shrewsbury, Shrops SY4 5PU*
Phone *093928 523*
Personnel *J O'Hare, Managing Director*
Quarries *Grinshill, Webscott*

GRINSHILL STONE QUARRIES LTD (ECC QUARRIES LTD)

GRINSHILL QUARRY

270

Clive, Shrewsbury, Shrops SY4 5PU
Phone 093928 523
OS Map 126 **OS Grid Reference** SJ528242
Personnel J O'Hare, Managing Director and Unit Manager

PRODUCT DATA		
	Rock type	Triassic Sandstone
	Colour	Creamy buff
	Grain	Fine
	Product	Dressed or Lump Stone

DRILLING DATA	Hand-held equipment

GRINSHILL STONE QUARRIES LTD (ECC QUARRIES LTD)

WEBSCOTT QUARRY

271

Myddle, Shrewsbury, Shrops SY4 3QU
Phone 093928 523
OS Map 126 **OS Grid Reference** SJ474226
Personnel J O'Hare, Managing Director and Unit Manager

PRODUCT DATA		
	Rock type	Triassic Sandstone
	Colour	Dark red
	Grain	Fine
	Product	Dressed or Lump Stone

DRILLING DATA	Hand-held equipment

GRYPHON QUARRIES LTD
Head Office *Old Mill Works, Pontllanfraith, Gwent NP2 2AH*
Phone *0495 227331*
Personnel *A Gilson, Managing Director*
Quarries *Trefil*

GRYPHON QUARRIES LTD

TREFIL QUARRY COMPLEX

272

Trefil, Tredegar, Gwent NP2 4HG
Phone 0633 843497
OS Map 34 **OS Grid Reference** SO126128
Personnel K Williams, Unit Manager

PRODUCT DATA		
	Rock type	Carboniferous Limestone
	Colour	Dark grey
	Grain	Medium
	Products	Crushed Stone
		Industrial Limestone

DRILLING DATA		Hole Diameter 125mm
		Bench Height 33m
		Contractor A Jones Rock Drillers
SECONDARY BREAKING		Hydraulic Hammer on Cat 215
		Dropball
LOAD AND HAUL	**Face shovel**	Caterpillar 215
	Trucks	Caterpillar 769C
		Volvo BM L90
	Loaders	Caterpillar 966C
CRUSHING PLANT	**Primary**	Kue Ken Jaw
	Secondary	Kue Ken Impactor
	Tertiary	Hazemag AP4

NO FURTHER INFORMATION PROVIDED

GWYNDY QUARRIES LTD

Head Office *Gwyndy Quarry, Llandrygarn, Llanerchymedd, Anglesey, Gwynedd LL71 7AS*
Phone *0407 720236*
Personnel *W C Hutchins, Director*
   W L Hutchins, Director
   J W Hutchins, Director
   K W Attwell, Director
Quarries *Gwyndy*

GWYNDY QUARRIES LTD

GWYNDY QUARRY   273

Llandrygarn, Llanerchymedd, Anglesey, Gwynedd LL71 7AS
Phone 0407 720236
OS Map 114   **OS Grid Reference** SH396795
Personnel W L Hutchins, Director and Unit Manager

PRODUCT DATA	**Rock type**	Granite
	Colour	Light grey
	Grain	Coarse
	Products	Crushed Stone
		Pre-cast Concrete Products
DRILLING DATA		Hole Diameter 105mm
		Bench Height 16m
		Contractor K&H Rockdrillers
	Drills	Ingersoll-Rand Rig
COMPRESSORS		Atlas Copco
SECONDARY BREAKING		Dropball
LOAD AND HAUL	**Face shovel**	(2)Broyt
	Trucks	Terex R25
		Euclid
	Loaders	JCB, Ford
		Zettelmeyer ZL2002

entry continues overleaf

CRUSHING PLANT	**Primary**	Kue Ken 42 × 27 Jaw (Double Toggle)
	Secondary	Kue Ken 24 × 10 Jaw (Single Toggle)
	Tertiary	Nordberg
		Pegson

ALEXANDER HALL AND SON (BUILDERS) LTD
Head Office *Granitehill, Northfield, Aberdeen, Grampian AB9 2AW*
Phone *0224 693155*
Personnel *J Birnie, Managing Director*
Quarries *Greenbrey*

ALEXANDER HALL AND SON (BUILDERS) LTD
GREENBREY QUARRY
274

Hopeman, Grampian IV30 4RW
Phone 0224 693155
OS Map 28   **OS Grid Reference** NJ137692

QUARRY ON STAND

ROCK DATA	**Rock type**	Permian Sandstone
	Colour	Fawn
	Grain	Fine

JOHN HANCOCK AND SONS (BATH) LTD
Head Office *1A Sommer Dale Ave, Odd Down, Bath, Avon BA2 2PG*
Phone *0225 833428*
Personnel *J Hancock, Owner*
Quarries *Upper Lawn*

JOHN HANCOCK AND SONS (BATH) LTD
UPPER LAWN QUARRY
275

Combe Down, Bath, Avon BA2 7BA
Phone 0225 833337
OS Map 172   **OS Grid Reference** ST766624
Personnel J Hancock, Owner and Unit Manager

PRODUCT DATA	**Rock type**	Jurassic Limestone
	Colour	Cream
	Grain	Medium
	Product	Dressed or Lump Stone

| DRILLING DATA | | No drilling |

| LOAD AND HAUL | **Loaders** | (2)Hannomag |

HANSON plc
Corporate Head Office *1 Grosvenor Place, London SW1X 7JH*
Phone *071 245 1245* **Fax** *071 235 3455* **Telex** *917698*
Personnel *Lord Hanson, Chairman*
A G L Alexander, Chief Operating Officer (UK)
Companies *ARC Ltd* — see separate entry

HARDROCK LTD
Head Office *Stoney Brow Quarry, Roby Mill, Upholland, Wigan, Lancs WN8 0QE*
Phone *0695 622950*
Personnel *P Ratcliffe, Director*
Quarries *Stoney Brow*

HARDROCK LTD

STONEY BROW QUARRY 276

Roby Mill, Upholland, Wigan, Lancs WN8 0QE
Phone 0695 622950
OS Map 108 **OS Grid Reference** SD515048
Personnel P Ratcliffe, Director and Unit Manager

PRODUCT DATA	**Rock type**	Carboniferous Sandstone
	Product	Dressed or Lump Stone

DRILLING DATA		Hand-held equipment

HARGREAVES QUARRIES LTD (CHARTER CONSOLIDATED plc)
Head Office *PO Box 1, Crossgate Lane, Pickering, N Yorks YO18 7ER*
Phone *0751 72231* **Fax** *0751 76380*
Personnel *P Fuchs, Managing Director*
  G Overfield, Production Manager
Companies *Cast Vale Quarries* — see separate entry
Quarries *Black, Force Garth, Hartley, High Force, Kilmond Wood, Newbridge, Spaunton*

HARGREAVES QUARRIES LTD (CHARTER CONSOLIDATED plc)

BLACK QUARRY 277

Leyburn, N Yorks DL8 5LA
Phone 0969 23197
OS Map 99 **OS Grid Reference** SE098914
Personnel J Parker, Unit Manager

PRODUCT DATA	**Rock type**	Carboniferous Limestone
	Colour	Dark grey
	Grain	Medium
	Products	Dressed or Lump Stone
		Crushed Stone
		Asphalt or Tarmacadam
DRILLING DATA		Hole Diameter 105mm
		Bench Height 17m
	Drills	Ingersoll-Rand CM300
COMPRESSORS		Ingersoll-Rand
LOAD AND HAUL	**Trucks**	None
	Loaders	Caterpillar
		Volvo
CRUSHING PLANT	**Primary**	Sheepbridge Jaw
	Secondary	Allis Chalmers
	Tertiary	Allis Chalmers

HARGREAVES QUARRIES LTD (CHARTER CONSOLIDATED plc)

FORCE GARTH QUARRY

278

Middleton in Teesdale, Durham DL12 0XJ
Phone 0833 22255
OS Map 92   **OS Grid Reference** NY874282
Personnel J Deacon, Unit Manager

PRODUCT DATA	**Rock type**	Dolerite Whin Sill
	Colour	Very dark grey
	Grain	Medium
	Products	Crushed Stone
		Asphalt or Tarmacadam
DRILLING DATA		Hole Diameter 110mm
		Bench Height 20m
	Drills	Halco 410C
COMPRESSORS		CompAir 700HE
LOAD AND HAUL	**Trucks**	(2)Heathfield
	Loaders	(2)Broyt
CRUSHING PLANT	**Primary**	Kue Ken 150 Jaw
	Secondary	Allis Chalmers 1645 Gyratory
	Tertiary	Allis Chalmers 3' Cone
	Quaternary	Pegson 900 Autocone
		Allis Chalmers 736

HARGREAVES QUARRIES LTD (CHARTER CONSOLIDATED plc)

HARTLEY QUARRY

279

Kirkby Stephen, Cumbria CA17 4JJ
Phone 0930 71740
OS Map 91   **OS Grid Reference** NY792084
Personnel G Sowerby, Unit Manager

PRODUCT DATA	**Rock type**	Carboniferous Great Scar Limestone
	Colour	Light grey
	Products	Crushed Stone
		Industrial Limestone
		Burnt Limestone
		Agricultural Lime
DRILLING DATA		Hole Diameter 75mm
		Bench Height 15–30m
	Drills	Ingersoll-Rand CM300
COMPRESSORS		Ingersoll-Rand
SECONDARY BREAKING		Dropball on Ruston-Bucyrus 30-RB
LOAD AND HAUL	**Face shovel**	Volvo
	Trucks	(2)Terex R25
	Loaders	Caterpillar 950
		Caterpillar 966C
CRUSHING PLANT	**Primary**	Kue Ken 42×36 Jaw (Double Toggle)
	Secondary	Mansfield No 3 Hammermill

HARGREAVES QUARRIES LTD (CHARTER CONSOLIDATED plc)

HIGH FORCE QUARRY

280

Middleton-in-Teeside, Barnard Castle, Durham DL12 0EP
Phone 0833 22255
OS Map 92   **OS Grid Reference** NY873282
Personnel J Deacon, Unit Manager

PRODUCT DATA	**Rock type**	Carboniferous Limestone
	Colour	Very dark grey
	Products	Crushed Stone
		Asphalt or Tarmacadam
DRILLING DATA		Hole Diameter 105mm
		Bench Height 15–30m
	Drills	Halco 410C
SECONDARY BREAKING		Dropball on Ruston-Bucyrus 22-RB
LOAD AND HAUL	**Trucks**	Heathfield
	Loaders	Volvo
CRUSHING PLANT	**Primary**	Kue Ken 150 Jaw
	Secondary	Allis Chalmers
	Tertiary	Pegson 900 Autocone
		Allis Chalmers 736

HARGREAVES QUARRIES LTD (CHARTER CONSOLIDATED plc)

KILMOND WOOD QUARRY

281

Bowes, Barnard Castle, Durham DL12 9SW
Phone 0833 72231
OS Map 92   **OS Grid Reference** NZ023135

QUARRY ON STAND

ROCK DATA	**Rock type**	Carboniferous Limestone
	Colour	Dark grey

HARGREAVES QUARRIES LTD (CHARTER CONSOLIDATED plc)

NEWBRIDGE QUARRY

282

Pickering, N Yorks YO18 7ER
Phone 0751 72257
OS Map 100   **OS Grid Reference** SE799860
Personnel R Pettitt, Unit Manager

PRODUCT DATA	**Rock type**	Jurassic Corallion Limestone
	Colour	Light creamy grey
	Grain	Fine
	Products	Crushed Stone
		Asphalt or Tarmacadam
DRILLING DATA		Hole Diameter 105mm
		Bench Height 17m
	Drills	Ingersoll-Rand CM300

entry continues overleaf

COMPRESSORS		Ingersoll-Rand
LOAD AND HAUL	**Trucks**	None
	Loaders	Caterpillar
		Volvo
CRUSHING PLANT	**Primary**	Sheepbridge Jaw
	Secondary	Mansfield No 1 Hammermill
		Mansfield No 3 Hammermill

HARGREAVES QUARRIES LTD (CHARTER CONSOLIDATED plc)

SPAUNTON QUARRY   283

Kirkbymoorside, N Yorks YO6 8JL
Phone 0751 31244
OS Map 100   **OS Grid Reference** SE721865
Personnel P Woodward, Unit Manager

PRODUCT DATA	**Rock type**	Jurassic Corallion Limestone
	Colour	Light grey
	Grain	Fine
	Products	Crushed Stone
		Asphalt or Tarmacadam
DRILLING DATA		Hole Diameter 105mm
		Bench Height 17m
	Drills	Halco 410C
SECONDARY BREAKING		Dropball on Ruston-Bucyrus 22-RB
LOAD AND HAUL	**Trucks**	(2)Heathfield
	Loaders	Komatsu WA350
CRUSHING PLANT	**Primary**	Sheepbridge Jaw
	Secondary	Mansfield No 1 Hammermill
	Tertiary	Allis Chalmers Hammermill

C F HARRIS LTD
Head Office *3 High Street, South Milford, Leeds, N Yorks LS25 5AA*
Phone *0977 682337*
Personnel *C F Harris, Owner*
Quarries *Sherburn*

C F HARRIS LTD

SHERBURN QUARRY   284

Copley Lane, Sherburn in Elmet, Leeds, N Yorks LS25 3ED
Phone 0977 682337
OS Map 105   **OS Grid Reference** SE482348
Personnel C F Harris, Owner and Unit Manager

PRODUCT DATA	**Rock type**	Permian Magnesian Limestone
	Colour	Cream
	Grain	Fine
	Product	Dressed or Lump Stone
DRILLING DATA		Hand-held equipment

HARRIS QUARRIES (STONECRAFT YORKSHIRE)
Head Office *61 Lockwood Avenue, South Anston, Sheffield, S Yorks S31 7GQ*
Phone *0742 872478*
Personnel *B P Harris, Director*
Quarries *Blackmoor*

HARRIS QUARRIES (STONECRAFT YORKSHIRE)

BLACKMOOR QUARRY

285

Ulley, Sheffield, S Yorks S31 0YH
Phone 0742 872478
OS Map 120 **OS Grid Reference** SK461874
Personnel B P Harris, Director and Unit Manager

PRODUCT DATA	**Rock type**	Carboniferous Sandstone
	Colour	Mauve to deep red, blue
	Grain	Fine
	Product	Dressed or Lump Stone

DRILLING DATA		Hand-held equipment

W J HAYSOM AND SON
Head Office *St Aldhelm's Quarry, Worth Matravers, Swanage, Dorset BH19 3HL*
Phone *092943 217*
Personnel *W J Haysom, Managing Director*
Quarries *St Aldhelms, Southard*

W J HAYSOM AND SON

ST ALDHELM'S QUARRY

286

Worth Matravers, Swanage, Dorset BH19 3HL
Phone 092943 217
OS Map 195 **OS Grid Reference** SY964761
Personnel W J Haysom, Managing Director and Unit Manager

PRODUCT DATA	**Rock type**	Jurassic Limestone
	Colour	Greyish white
	Grain	Fine
	Product	Dressed or Lump Stone

RIPPABLE STRATA

LOAD AND HAUL	**Loaders**	Caterpillar

W J HAYSOM AND SON

SOUTHARD QUARRY

287

Townsend, Swanage, Dorset BH19 3DY
Phone 092943 217
OS Map 195 **OS Grid Reference** SY965810
Personnel W J Haysom, Managing Director and Unit Manager

PRODUCT DATA	**Rock type**	Jurassic Limestone
	Colour	Greyish white
	Grain	Fine
	Product	Dressed or Lump Stone

entry continues overleaf

RIPPABLE STRATA

LOAD AND HAUL   **Loaders**   Caterpillar

HEARSON QUARRY (SWIMBRIDGE)
Head Office *Hearson Quarry, Swimbridge, Barnstable, Devon EX32 0QH*
Phone *0271 830055*
Personnel *D C Griffiths, Managing Director*
Quarries *Hearson*

HEARSON QUARRY (SWIMBRIDGE)

HEARSON QUARRY   **288**

Swimbridge, Barnstable, Devon EX32 0QH
Phone 0271 830055
OS Map 180   **OS Grid Reference** SS605293
Personnel D C Griffiths, Managing Director and Unit Manager
PRODUCT DATA   **Rock type**   Devonian Sandstone
                      Colour   Blue with orange tint
                      Grain   Medium
                      Product   Dressed or Lump Stone

RIPPABLE STRATA
                              Ripper Caterpillar D4H

HIGHLAND LIME COMPANY
Head Office *Torlundy Quarry, Torlundy, Fort William, Highland PH33 6SW*
Phone *0397 2227*
Personnel *D Howie, Managing Director*
Quarries *Torlundy*

HIGHLAND LIME COMPANY

TORLUNDY QUARRY   **289**

Torlundy, Fort William, Highland PH33 6SW
Phone 0397 2227
OS Map 41   **OS Grid Reference** NM180777
Personnel R Thom, Unit Manager

PRODUCT DATA   **Rock type**   Pre-Cambrian Dalradian Limestone
                      Colour   Not commercially significant
                      Products   Crushed Stone
                                      Agricultural Lime

DRILLING DATA   Hole Diameter 75mm
                               Bench Height 10m
                      Drills   Halco 17HR Wagon Drill

COMPRESSORS   CompAir 37HP

SECONDARY BREAKING   Dropball on Ruston-Bucyrus 22-RB

LOAD AND HAUL   **Face shovel**   JCB
                 Loaders   Caterpillar 936
                                Caterpillar 963

CRUSHING PLANT	**Primary**	Kue Ken 36×24 Jaw (Double Toggle)
	Secondary	Kue Ken 24×15 Jaw (Double Toggle)
	Tertiary	Barmac 843 Duopactor

HIGHLAND REGION COUNCIL
Head Office *Regional Buildings, Glenurquhart Road, Inverness, Highland IV3 5NX*
Phone *0463 234121* **Fax** *0463 223201*
Personnel *G K N MacFarlane, Technical Director*
Quarries *Sconser*

HIGHLAND REGION COUNCIL
SCONSER QUARRY   290
Sconser, Isle of Skye, Highland IV49 9BA
Phone 0478 52202
OS Map 32   **OS Grid Reference** NG515318
Personnel W Morrison, Unit Manager

PRODUCT DATA	**Rock type**	Devonian Sandstone
	Products	Crushed Stone
		Asphalt or Tarmacadam
DRILLING DATA		Hole Diameter 105-115mm
		Bench Height 17-18m
	Drills	Atlas Copco ROC304-04
LOAD AND HAUL	**Loaders**	Hannomag 70E
CRUSHING PLANT	**Primary**	Blake 36×24 Jaw
	Secondary	Lokomo Cone
	Tertiary	Barmac Duopactor

SETH HILL AND SON LTD
Head Office *The Quarries, Bonvilston, Cardiff, S Glam CF5 6TQ*
Phone *04468 207*
Personnel *S Hill, Owner*
Quarries *Pantyffynnon*

SETH HILL AND SON LTD
PANTYFFYNNON QUARRY   291
Bonvilston, Cardiff, S Glam CF5 6TQ
Phone 04468 207
OS Map 170   **OS Grid Reference** ST047740
Personnel S Hill, Owner and Unit Manager

PRODUCT DATA	**Rock type**	Carboniferous Limestone
	Colour	Grey
	Grain	Medium
	Product	Crushed Stone
DRILLING DATA		Hole Diameter 105mm
		Bench Height 12m
	Drills	Halco 410C

entry continues overleaf

COMPRESSORS		CompAir 650HE
SECONDARY BREAKING		Dropball on Ruston-Bucyrus 19-RB
LOAD AND HAUL	**Face shovel**	O&K
	Trucks	(4)Foden 14t
	Loaders	(2)Bray
CRUSHING PLANT	**Primary**	Parker Jaw
	Secondary	Sheepbridge

HILLHOUSE QUARRY COMPANY
Head Office *Hillhouse Quarry, Troon, Strathclyde KA10 7HX*
Phone *0292 313311* **Fax** *0292 314640*
Personnel *S B Vernon, Director*
Quarries *Hillhouse*

HILLHOUSE QUARRY COMPANY

HILLHOUSE QUARRY

292

Troon, Strathclyde KA10 7HX
Phone 0292 313311
OS Map 70 **OS Grid Reference** NS345342
Personnel J Cairns, Unit Manager

PRODUCT DATA	**Rock type**	Dolerite Sill (Whinstone)
	Colour	Very dark grey
	Grain	Medium
	Products	Crushed Stone
		Asphalt or Tarmacadam
		Ready-mixed Concrete
DRILLING DATA		Hole Diameter 110mm
		Bench Height 18m
	Drills	Ingersoll-Rand CM350
COMPRESSORS		Ingersoll-Rand
SECONDARY BREAKING		Dropball on Ruston-Bucyrus 22-RB
LOAD AND HAUL	**Loaders**	Caterpillar
CRUSHING PLANT	**Primary**	Allis Chalmers Cone
	Secondary	Allis Chalmers Cone
	Tertiary	Allis Chalmers Cone

R HINCHCLIFFE AND SON LTD (MARSHALS HALIFAX plc)
Head Office *Appleton Quarry, Shepley, Huddersfield, W Yorks HD8 8BB*
Phone *0484 606390*
Personnel *R Hinchcliffe,*
Quarries *Appleton*

R HINCHCLIFFE AND SON LTD (MARSHALS HALIFAX plc)

APPLETON QUARRY

293

Shepley, Huddersfield, W Yorks HD8 8BB
Phone 0484 606390
OS Map 110   **OS Grid Reference** SE193087
Personnel R Hinchcliffe, Director and Unit Manager

PRODUCT DATA	**Rock type**	Carboniferous Sandstone
	Colour	Blue
	Grain	Fine
	Product	Dressed or Lump Stone

| DRILLING DATA | | Hand-held equipment |

HORN CRAG FARM LTD
Head Office Horn Crag Quarry, Barclay Bank Chambers, Kirkgate, Silsden, Keighley, W Yorks BD20 0AJ
Phone 0535 55442
Personnel N D Wilkinson, Managing Director
         G E Sedgwick, Director
Quarries Horn Crag

HORN CRAG FARM LTD

HORN CRAG QUARRY

294

Fishbeck Lane, Silsden, Keighley, W Yorks BD20 0NP
Phone 0535 55442
OS Map 104   **OS Grid Reference** SE053480
Personnel G E Sedgwick, Director and Unit Manager

PRODUCT DATA	**Rock type**	Carboniferous Millstone Grit
	Colour	Brown
	Grain	Fine
	Product	Dressed or Lump Stone

| DRILLING DATA | | Hand-held equipment |

HORTON QUARRIES LTD
Head Office Edgehill, Banbury, Oxon OX15 6HS
Phone 029 587238
Personnel A W Stanley, Managing Director
Quarries Horton

HORTON QUARRIES LTD

HORTON QUARRY

295

Ratley, Warmington, Warwicks OX95 6DI
Phone 029 587238
OS Map 151   **OS Grid Reference** SP375471
Personnel B J Hayward, Unit Manager

PRODUCT DATA	**Rock type**	Jurassic Limestone
	Colour	Greenish blue, brown
	Grain	Fine
	Product	Dressed or Lump Stone

| DRILLING DATA | | Hand-held equipment |

ROGER HUGHES AND COMPANY

Head Office *Plas Gwilym Quarry, Llysfaen Road, Old Colwyn, Colwyn Bay, Clwyd LL29 9HE*
Phone *0492 515255*
Personnel *R Hughes, Managing Director*
Quarries *Plas Gwilym*

ROGER HUGHES AND COMPANY

PLAS GWILYM QUARRY   296

Llysfaen Road, Old Colwyn, Colwyn Bay, Clwyd LL29 9HE
Phone 0492 515255
OS Map 116 **OS Grid Reference** SH879781
Personnel S Jones, Unit Manager

PRODUCT DATA	**Rock type**	Carboniferous Limestone
	Colour	Pinkish grey
	Product	Dressed or Lump Stone

DRILLING DATA		Hand-held equipment

W H HUMPHREYS

Head Office *Sunnyside, Penygroes, Caernarfon, Gwynedd LL54 6RN*
Phone *0286 880502/881028*
Personnel *W H Humphreys, Owner*
Quarries *Twll Llwyd*

W H HUMPHREYS

TWLL LLWYD QUARRY   297

Tanrallt, Talysarn, Caernafon, Gwynedd LL54 6TA
Phone 0286 880502
OS Map 115 **OS Grid Reference** SH491518
Personnel W H Humphreys, Owner and Unit Manager

PRODUCT DATA	**Rock type**	Pre-Cambrian Slate
	Colour	Greenish grey, red, brown
	Grain	Fine
	Product	Dressed or Lump Stone

DRILLING DATA		Hole Diameter 32mm
		Hand-held equipment

HUNTSMANS QUARRIES LTD (COTSWOLD PLANT COMPANY)

Head Office *The Old School, Naunton, Cheltenham, Glos GL54 3AE*
Phone *0451 5555*
Personnel *J C Milner, Managing Director*
          A Greenwood, Director
Quarries *Hornsleasow, Huntsmans*

HUNTSMANS QUARRIES LTD (COTSWOLD PLANT COMPANY)

HORNSLEASOW QUARRY   298

Snowshill, Broadway, Hereford & Worcs WR12 7JT
Phone 0451 5555
OS Map 150 **OS Grid Reference** SP117370
Personnel A Greenwood, Director and Unit Manager

PRODUCT DATA   **Rock type**   Jurassic Limestone
              Product   Dressed or Lump Stone

NO FURTHER INFORMATION PROVIDED

HUNTSMANS QUARRIES LTD (COTSWOLD PLANT COMPANY)

HUNTSMANS QUARRY   299

Naunton, Cheltenham, Glos GL54 3AE
Phone 04515 628
OS Map 163   **OS Grid Reference** SO125255
Personnel A Greenwood, Director and Unit Manager

PRODUCT DATA	**Rock type**	Jurassic Oolitic Limestone
	Colour	Dark grey, brown, yellow
	Grain	Fine
	Products	Crushed Stone
		Asphalt or Tarmacadam

DRILLING DATA		Hole Diameter 105mm
		Bench Height 10-12m
		Contractor Railside Engineering

SECONDARY BREAKING   Dropball on Ruston-Bucyrus 22-RB

LOAD AND HAUL	**Face shovel**	Akermans H16D
	Trucks	Terex
		Volvo 53-50
	Loaders	Caterpillar 966C
		(2)Hannomag
		Hannomag 60E

CRUSHING PLANT	**Primary**	Impactor
	Secondary	Impactor
	Tertiary	Impactor

HUNTSMANS QUARRIES (PERTON) LTD

Head Office *Perton Quarry, Stoke Edith, Hereford, Hereford & Worcs HR1 4HW*
Phone *0432 79258*
Personnel *J C Milner, Managing Director*
         A Greenwood, Director
Quarries *Perton*

HUNTSMANS QUARRIES (PERTON) LTD

PERTON QUARRY   300

Stoke Edith, Hereford, Hereford & Worcs HR1 4HW
Phone 0432 79258
OS Map 149   **OS Grid Reference** SO595399
Personnel T Brimmel, Unit Manager

PRODUCT DATA	**Rock type**	Silurian Limestone
	Colour	Not commercially significant
	Product	Crushed Stone

DRILLING DATA		Hole Diameter 110mm
		Bench Height 10m
	Drills	Tamrock DHA600

entry continues overleaf

| LOAD AND HAUL | **Trucks** | Caterpillar |
| | **Loaders** | Caterpillar |

| CRUSHING PLANT | | No information provided |

ICI (CHEMICALS AND POLYMERS) LTD
Corporate Head Office *PO Box 14, The Heath, Runcorn, Cheshire SK17 8TH*
Phone *0928 514444* **Telex** *629655*
Personnel *R I Lindsell, Chief Executive*

ICI (CHEMICALS AND POLYMERS) LTD MOND DIVISION
Head Office *PO Box 3, Buxton, Derbys SK17 8TH*
Phone *0298 768444*
Personnel *W H Pilling, General Manager*
Quarries *Tunstead*

ICI (CHEMICALS AND POLYMERS) LTD MOND DIVISION

TUNSTEAD QUARRY

301

Buxton, Derbys SK17 8TH
Phone 0298 768444
OS Map 119 **OS Grid Reference** SK099740
Personnel A R Stevens, Unit Manager

PRODUCT DATA	**Rock type**	Carboniferous Bee Low Limestone
	Colour	Very light grey to grey
	Products	Crushed Stone
		Industrial Limestone
		Burnt Limestone
		Agricultural Lime

DRILLING DATA		Hole Diameter 152mm
		Bench Height 19m
	Drills	Gardner-Denver Drilltec 40K

| SECONDARY BREAKING | | Dropball on Ruston-Bucyrus 22-RB |

LOAD AND HAUL	**Trucks**	(12)Aveling Barford
		Caterpillar
	Loaders	(2)Caterpillar 988B
		Caterpillar 992C
		Demag H121

CRUSHING PLANT	**Primary**	Parker Gyratory (Mobile)
	Secondary	(2)Roller
	Tertiary	(2)Mansfield Hammermill
		Barmac Duopactor

ISLE OF MAN GOVERNMENT DEPT OF HIGHWAYS, PORTS AND PROPERTIES
Head Office *Douglas Borough Council, Town Hall, Douglas, Isle of Man*
Phone *0624 842387*
Personnel *E T Atherton, Technical Director*
Quarries *Poortown, South Barrule*

ISLE OF MAN GOVERNMENT DEPT OF HIGHWAYS, PORTS AND PROPERTIES

POORTOWN QUARRY

302

Peel, Isle of Man
Phone 0624 842387
OS Map 95   **OS Grid Reference** SC246840
Personnel S Taylor, General Manager and Unit Manager

PRODUCT DATA	**Rock type**	Granite
	Product	Crushed Stone
DRILLING DATA		Hole Diameter 105mm
		Bench Height 20m
		Contractor A Jones Rock Drillers
SECONDARY BREAKING		Dropball
LOAD AND HAUL	**Trucks**	Aveling Barford
	Loaders	Caterpillar
CRUSHING PLANT	**Primary**	Goodwin Barsby 42×30 Jaw
	Secondary	Allis Chalmers Cone
		Nordberg Cone

ISLE OF MAN GOVERNMENT DEPT OF HIGHWAYS, PORTS AND PROPERTIES

SOUTH BARRULE QUARRY

303

Peel, Isle of Man
Phone 0624 842387
OS Map 95   **OS Grid Reference** SC246840
Personnel S Taylor, General Manager and Unit Manager

PRODUCT DATA	**Rock type**	Silurian Slate
	Grain	Fine
	Product	Dressed or Lump Stone
DRILLING DATA		Hole Diameter 38mm
		Contractor A Jones Rock Drillers

J A JACKSON (PRESTON) LTD
Head Office *Green Lane, Lightfoot Green, Preston, Lancs PR4 0AP*
Phone *0772 861230*
Personnel *J A Jackson, Managing Director*
Quarries *Ellel, Jamestone*

J A JACKSON (PRESTON) LTD

ELLEL QUARRY

304

Galgate, Lancaster, Lancs LA2 0PY
Phone 0524 751444
OS Map 97   **OS Grid Reference** SD505549
Personnel E McMahon, Unit Manager

PRODUCT DATA	**Rock type**	Carboniferous Millstone Grit
	Colour	Buff with brown streaks
	Grain	Medium
	Products	Dressed or Lump Stone
		Crushed Stone

entry continues overleaf

DRILLING DATA		Hole Diameter 105mm
		Bench Height 5–7m
	Drills	Holman Voltrak

COMPRESSORS		CompAir 650HE

SECONDARY BREAKING		Dropball on Ruston-Bucyrus 22-RB

LOAD AND HAUL	**Trucks**	(2)Terex R18T
	Loaders	Caterpillar 996D

CRUSHING PLANT	**Primary**	Parker
	Secondary	None
	Tertiary	None

J A JACKSON (PRESTON) LTD

JAMESTONE QUARRY
305

Grange Road, Haslingden, Rossendale, Lancs BB4 9LJ
Phone 0613 207215
OS Map 103   **OS Grid Reference** SD759235
Personnel S Grindley, Unit Manager

PRODUCT DATA	**Rock type**	Carboniferous Haslingden Flagstone
	Colour	Light grey, greyish buff
	Grain	Fine
	Product	Dressed or Lump Stone

RIPPABLE STRATA

BILL JARVIS AND SON LTD
Head Office *Barnshill Yard, Otley Road, Baildon, W Yorks BD17 7JF*
Phone *0274 586235*
Personnel *W Jarvis, Owner*
Quarries *Apex, Harden Moor*

BILL JARVIS AND SON LTD

APEX QUARRY
306

Butcher Hill, Horsforth, Leeds, W Yorks LS18 4HV
Phone 0274 586235
OS Map 104   **OS Grid Reference** SE255385
Personnel W Jarvis, Owner and Unit Manager

PRODUCT DATA	**Rock type**	Carboniferous Sandstone
	Colour	Yellowish brown to fawn
	Grain	Medium
	Product	Dressed or Lump Stone

DRILLING DATA		Hand-held equipment

BILL JARVIS AND SON LTD

HARDEN MOOR QUARRY
307

Ryecroft Road, Harden, Bingley, W Yorks BD16 1DH
Phone 0274 586235
OS Map 104   **OS Grid Reference** SE080385
Personnel W Jarvis, Owner and Unit Manager

PRODUCT DATA	**Rock type**	Carboniferous Millstone Grit
	Grain	Medium
	Product	Dressed or Lump Stone

| DRILLING DATA | Hand-held equipment |

DONALD A JOHNSON (NORTH UIST) LTD
Head Office *Cnoc nan Locha, Sollas, North Uist, Western Isles PA82 5BU*
Phone *08766 281*
Personnel *D A Johnson, Managing Director*
Quarries *Crogarry Beag, Steangabhal*

DONALD A JOHNSON (NORTH UIST) LTD

CROGARRY BEAG QUARRY   308
Lochmaddy, North Uist, Western Isles PA82 5AZ
Phone 08763 267
OS Map 22   **OS Grid Reference** NA921678

QUARRY ON STAND

| ROCK DATA | **Rock type** | Basalt |

DONALD A JOHNSON (NORTH UIST) LTD

STEANGABHAL QUARRY   309
Benbecula, North Uist, Western Isles PA88 5GA
Phone 08763 281
OS Map 22   **OS Grid Reference** NA799524
Personnel D A Johnson, Managing Director and Unit Manager

PRODUCT DATA	**Rock type**	Basalt
	Colour	Greyish black
	Grain	Fine
	Product	Crushed Stone

DRILLING DATA	Hole Diameter 105mm
	Bench Height 18m
	Contractor Ritchies Equipment

JOHNSON WELLFIELD QUARRIES LTD
Head Office *Crosland Moor Quarry, Crosland Hill, Huddersfield, W Yorks HD4 7AB*
Phone *0484 652311*
Personnel *J D Myers, Managing Director*
Quarries *Crosland Moor, Honley Wood*

JOHNSON WELLFIELD QUARRIES LTD

CROSLAND MOOR QUARRY   310
Crosland Hill, Huddersfield, W Yorks HD4 7AB
Phone 0484 652311
OS Map 110   **OS Grid Reference** SE116145
Personnel D E Lodge, General Manager and Unit Manager

entry continues overleaf

PRODUCT DATA	**Rock type**	Carboniferous Millstone Grit
	Colour	Light brown
	Grain	Fine
	Product	Dressed or Lump Stone

| DRILLING DATA | | Hand-held equipment |

JOHNSON WELLFIELD QUARRIES LTD

HONLEY WOOD QUARRY

311

Honley, Huddersfield, W Yorks HD7 3DS
Phone 0484 652311
OS Map 110 **OS Grid Reference** SE117118
Personnel D E Lodge, General Manager and Unit Manager

PRODUCT DATA	**Rock type**	Carboniferous Millstone Grit
	Colour	Light brown
	Grain	Fine
	Product	Dressed or Lump Stone

| DRILLING DATA | | Hand-held equipment |

JOHNSTON ROADSTONES LTD
Head Office *Leaton Quarry, Leaton, Wellington, Shrops TF6 5HB*
Phone *0952 86351* **Fax** *0952 86413*
Personnel *A Hughes, Managing Director*
Quarries *Leaton, Leinthall*

JOHNSTON ROADSTONES LTD

LEATON QUARRY

312

Leaton, Wellington, Shrops TF16 5HB
Phone 0952 86351
OS Map 126 **OS Grid Reference** SJ617114
Personnel J M Coates, Unit Manager

PRODUCT DATA	**Rock type**	Olivine Dolerite
	Colour	Dark grey, dark green
	Grain	Fine
	Products	Crushed Stone
		Asphalt or Tarmacadam

DRILLING DATA		Hole Diameter 105mm
		Bench Height 12m
		Contractor A Jones Rock Drillers

| SECONDARY BREAKING | | Hydraulic Hammer |

| LOAD AND HAUL | **Trucks** | (2)Aveling Barford RD30 |
| | **Loaders** | Michigan |

CRUSHING PLANT	**Primary**	Kue Ken 120
	Secondary	(2)Allis Chalmers Cone
	Tertiary	(2)Allis Chalmers Cone

JOHNSTON ROADSTONES LTD
LEINTHALL QUARRY

313

Leinthall Earls, Wigmore, Leominster, Hereford & Worcs HR6 9TR
Phone 0568 86521
OS Map 149 **OS Grid Reference** SO442682
Personnel P James, Unit Manager

PRODUCT DATA	**Rock type**	Silurian Limestone
	Colour	Not commercially significant
	Product	Crushed Stone

DRILLING DATA		Hole Diameter 105mm
		Bench Height 21m
	Drills	Halco 400

| COMPRESSORS | | Ingersoll-Rand 700 |

| SECONDARY BREAKING | | Dropball |

| LOAD AND HAUL | **Trucks** | Heathfield |
| | **Loaders** | Caterpillar 966C |

| CRUSHING PLANT | | No information given |

R A JONAS AND SONS
Head Office *Tredennick Downs Quarry, St Issey, Wadebridge, Cornwall PL27 7QZ*
Phone *0841 540332*
Personnel *R A Jonas, Owner*
Quarries *Tredennick*

R A JONAS AND SONS
TREDENNICK DOWNS QUARRY

314

St Issey, Wadebridge, Cornwall PL27 7QZ
Phone 0841 540332
OS Map 200 **OS Grid Reference** SW925705
Personnel R A Jonas, Owner and Unit Manager

PRODUCT DATA	**Rock type**	Devonian Slate
	Colour	Fawn, red
	Grain	Fine
	Product	Dressed or Lump Stone

| DRILLING DATA | | Hand-held equipment |

D GWYNDOF JONES
Head Office *Vronlog Quarry, Llanllyfni, Caernafon, Gwynedd LL54 6RT*
Phone *0286 880574*
Personnel *D Gwyndof Jones, Owner*
Quarries *Vronlog*

entry continues overleaf

D GWYNDOF JONES

VRONLOG QUARRY

315

Llanllyfni, Caernafon, Gwynedd LL54 6RT
Phone 0286 880574
OS Map 115 **OS Grid Reference** SH487518
Personnel D Gwyndof Jones, Owner and Unit Manager

PRODUCT DATA	**Rock type**	Cambrian Slate
	Colour	Green
	Grain	Fine
	Product	Dressed or Lump Stone

| DRILLING DATA | | Hole Diameter 32–38mm |
| | | Hand-held equipment |

| COMPRESSORS | | CompAir |

D JONES (FFYNON PLANT HIRE COMPANY LTD)

Head Office *Mynydd y Garreg Quarry, Crymmych, Dyfed SA41 3QS*
Phone *0239 73223*
Personnel *D A Jones, Managing Director*
Quarries *Mynydd y Garreg*

D JONES (FFYNON PLANT HIRE COMPANY LTD)

MYNYDD Y GARREG QUARRY

316

Crymmych, Dyfed SA41 3QS
Phone 0239 73223
OS Map 159 **OS Grid Reference** SN434085
Personnel D A Jones, Managing Director and Unit Manager

PRODUCT DATA	**Rock type**	Carboniferous Millstone Grit
	Colour	Not commercially significant
	Grain	Medium
	Product	Crushed Stone

DRILLING DATA		Hole Diameter 105mm
		Bench Height 7–10m
		Contractor Ritchies Equipment

| SECONDARY BREAKING | | Dropball on Ruston-Bucyrus 22-RB |

| LOAD AND HAUL | **Trucks** | Komatsu |
| | **Loaders** | Fiatallis |

| CRUSHING PLANT | **Primary** | Goodwin Barsby Jaw |
| | **Secondary** | Marsden Jaw |

K W & H E KEATES

Head Office *31 Easington Road, Worth Matravers, Swanage, Dorset BH19 3LF*
Phone *092943 207*
Personnel *K W Keates, Director*
Quarries *Keates*

K W & H E KEATES

KEATES QUARRY

317

Corfe Castle, Swanage, Dorset BH19 3LF
Phone 092943 207
OS Map 195 **OS Grid Reference** SY982784
Personnel K W Keates, Director and Unit Manager

PRODUCT DATA	**Rock type**	Jurassic Limestone
	Colour	White, grey, blue
	Grain	Fine
	Product	Dressed or Lump Stone

RIPPABLE STRATA

| LOAD AND HAUL | **Loaders** | Caterpillar |

KEY QUARRIES LTD
Head Office Quickburn Quarry, Salters Gate, Tow Law, Bishop Auckland, Durham DL13 4JN
Phone 0388 730500
Personnel P Linacre, Director
Quarries Quickburn

KEY QUARRIES LTD

QUICKBURN QUARRY

318

Salters Gate, Tow Law, Bishop Auckland, Durham DL13 4JN
Phone 0388 730500
OS Map 88 **OS Grid Reference** NZ080428
Personnel K Gibson, Unit Manager

PRODUCT DATA	**Rock type**	Carboniferous Millstone Grit
	Grain	Medium
	Products	Dressed or Lump Stone
		Crushed Stone

RIPPABLE STRATA

| LOAD AND HAUL | **Face shovel** | Komatsu PC85 |

KINTYRE FARMERS
Head Office Glengyle, Glebe Street, Campbeltown, Strathclyde PA28 6LS
Phone 0586 52602
Personnel Vacant, Chief Executive
Quarries Calliburn

KINTYRE FARMERS

CALLIBURN QUARRY

319

Gobsgrennan Road, Campbeltown, Strathclyde PA28 6NX
Phone 0586 53742
OS Map 68 **OS Grid Reference** NR726221
Personnel K Reed, Unit Manager

PRODUCT DATA	**Rock type**	Pre-Cambrian Dalradian Limestone
	Products	Crushed Stone
		Agricultural Lime

entry continues overleaf

DRILLING DATA		Hole Diameter 105mm Bench Height 14–20m Contractor Albion Drillers
SECONDARY BREAKING		Dropball
LOAD AND HAUL	**Trucks**	None
	Loaders	Aveling Barford Fiatallis
CRUSHING PLANT	**Primary**	Baxter RB36 × 24 Jaw Goodwin Barsby 36 × 24 Jaw (Mobile)
	Secondary	Kue Ken 36 × 9 Jaw
	Tertiary	(2)Pulverizers

KIRKSTONE GREEN SLATE QUARRIES LTD
Head Office *Skelwith Bridge, Ambleside, Cumbria LA22 9NN*
Phone *05394 33296* **Fax** *05394 34006* **Telex** *65106*
Personnel *H Fecitt, Chairman*
          N Fecitt, Managing Director
Quarries *Pets*

KIRKSTONE GREEN SLATE QUARRIES LTD
PETS QUARRY   **320**
Skelwith Bridge, Ambleside, Cumbria LA22 9NN
Phone 05394 33296
OS Map 90 **OS Grid Reference** NY325050
Personnel N Fecitt, Managing Director and Unit Manager

PRODUCT DATA	**Rock type**	Ordovician Borrowdale Slate
	Colour	Light sea green
	Grain	Fine
	Product	Dressed or Lump Stone
DRILLING DATA		Hole Diameter 35-45mm
	Drills	Holman Wagon Drill

C KNIVETON LTD
Head Office *Turkeyland Quarry, Ballasalla, Isle of Man*
Phone *0624 823594*
Personnel *C Kniveton, Managing Director*
Quarries *Turkeyland*

C KNIVETON LTD
TURKEYLAND QUARRY   **321**
Ballasalla, Isle of Man
Phone 0624 823594
OS Map 95 **OS Grid Reference** SC282700
Personnel C Kniveton, Managing Director and Unit Manager

PRODUCT DATA	**Rock type**	Pre-Cambrian Dalradian Limestone
	Colour	Not commercially significant
	Products	Crushed Stone Ready-mixed Concrete Agricultural Lime

DRILLING DATA		Hole Diameter 105mm
		Bench Height 17–20m
		Contractor A Jones Rock Drillers
SECONDARY BREAKING		Dropball
LOAD AND HAUL	**Trucks**	Aveling Barford RD30
	Loaders	Caterpillar 966C
CRUSHING PLANT	**Primary**	Parker 36 × 24 Jaw
	Secondary	(2)Marsden Granulators

LADYCROSS STONE COMPANY
Head Office 72 Churchill Close, Shotley Bridge, Consett, Durham DH8 0EU
Phone 043473 302
Personnel C Jewitt, Director
Quarries Ladycross

LADYCROSS STONE COMPANY

LADYCROSS QUARRY   322
Slaley, Hexham, Northumb NE47 0BY
Phone 043473 302
OS Map 87   **OS Grid Reference** NY952550
Personnel C Jewitt, Director and Unit Manager

PRODUCT DATA	**Rock type**	Carboniferous Millstone Grit
	Colour	Buff, grey
	Grain	Fine
	Product	Dressed or Lump Stone
DRILLING DATA		Hand-held equipment

JOHN LAING (CIVIL) LTD
Corporate Head Office *The Marlowes, Hemel Hempstead, Herts HP2 4TP*
Phone *0442 65566*
Personnel *J J Hall, Director*

LAING STONEMASONRY (JOHN LAING (CIVIL) LTD)
Regional Head Office Dalston Road, Carlisle, Cumbria CA2 5NR
Phone 0228 21401
Personnel B Stonehouse, Regional Director
Quarries Shawk

LAING STONEMASONRY (JOHN LAING (CIVIL) LTD)

SHAWK QUARRY   323
Thursby, Cumbria CA5 6PS
Phone 0228 21401
OS Map 85   **OS Grid Reference** NY344484
Personnel A Sharpe, Unit Manager

PRODUCT DATA	**Rock type**	Triassic Gritstone
	Colour	Dull red
	Grain	Fine
	Product	Dressed or Lump Stone
DRILLING DATA		Hand-held equipment

LANDELLE LTD
Head Office *Kirby Clough Farm, Kettleshulme, Stockport, Greater Manchester SK12 5AY*
Phone *0663 46871*
Personnel *J L Potts, Managing Director*
Quarries *Hayfield*

LANDELLE LTD

HAYFIELD QUARRY

324

New Mills Road, Birchvale, Stockport, Greater Manchester SK12 5BT
Phone 0663 46871
OS Map 110 **OS Grid Reference** SK030868
Personnel B Oxford, Unit Manager

PRODUCT DATA	**Rock type**	Bunter Sandstone
	Grain	Medium
	Product	Dressed or Lump Stone

| DRILLING DATA | | Contractor occasionally |

LANDERS QUARRIES
Head Office *Landers Quarry, Kingston Road, Langton Matravers, Swanage, Dorset BH19 3JP*
Phone *0929 43205*
Personnel *C Lander, Director*
Quarries *Landers*

LANDERS QUARRIES

LANDERS QUARRY

325

Kingston Road, Langton Matravers, Swanage, Dorset BH19 3JP
Phone 0929 43205
OS Map 194 **OS Grid Reference** SY978790
Personnel C Lander, Director and Unit Manager

PRODUCT DATA	**Rock type**	Jurassic Limestone
	Colour	Buffish grey, bluish grey
	Grain	Fine
	Product	Dressed or Lump Stone

RIPPABLE STRATA

| LOAD AND HAUL | **Loaders** | Caterpillar |

LAWER BROS
Head Office *Chywoon Quarry, Longdowns, Penryn, Cornwall TR10 9AF*
Phone *0209 860520*
Personnel *C Lawer, Director*
Quarries *Chywoon*

LAWER BROS

CHYWOON QUARRY

326

Longdowns, Penryn, Cornwall TR10 9AF
Phone 0209 860520
OS Map 204 **OS Grid Reference** SW751355
Personnel C Lawer, Director and Unit Manager

PRODUCT DATA	**Rock type**	Carnmenellis Granite
	Colour	Grey
	Grain	Coarse
	Products	Dressed or Lump Stone
		Crushed Stone

DRILLING DATA		Hole Diameter 90mm
		Bench Height 19-20m
	Drills	Holman Tractor Vole Mkl

| COMPRESSORS | | CompAir 650HE |

| SECONDARY BREAKING | | Dropball on Ruston-Bucyrus 30-RB |

LOAD AND HAUL	**Face shovel**	Ruston-Bucyrus
	Trucks	(2)Aveling Barford RD027
	Loaders	(2)Caterpillar 950B

| CRUSHING PLANT | **Primary** | Baxter 36×30 Jaw (Double Toggle) |
| | **Secondary** | Marsden Jaw |

G LAWRIE

Head Office *64 Berelands Road, Prestwick, Strathclyde KA9 1ER*
Phone *0292 79226*
Personnel *G Lawrie, Owner*
Quarries *Hallyards*

G LAWRIE

HALLYARDS QUARRY   **327**

Dundonald, Kilmarnock, Strathclyde KA9 1RD
Phone 0563 850365
OS Map 70   **OS Grid Reference** NS358336
Personnel G Lawrie, Owner and Unit Manager

PRODUCT DATA	**Rock type**	Basalt
	Colour	Greenish black
	Grain	Fine
	Product	Crushed Stone

| DRILLING DATA | | Contractor Ritchies Equipment |

CRUSHING PLANT	**Primary**	Kue Ken Jaw
	Secondary	Kue Ken Jaw
	Tertiary	(3)Cone

NO FURTHER INFORMATION PROVIDED

W E LEACH (SHIPLEY) LTD

Head Office *Westside Mills, Ripley Road, Bradford, W Yorks BD18 1BD*
Phone *0274 727200*
Personnel *W E Leach, Owner*
Quarries *Rawdon*

entry continues overleaf

W E LEACH (SHIPLEY) LTD

RAWDON QUARRY

328

Apperley Lane, Rawdon, Leeds, W Yorks LS19 7EG
Phone 0532 502780
OS Map 104   **OS Grid Reference** SE197392
Personnel W E Leach, Owner and Unit Manager

PRODUCT DATA	**Rock type**	Carboniferous Millstone Grit
	Colour	Fawn to light brown
	Grain	Fine
	Product	Dressed or Lump Stone

| DRILLING DATA | | Hand-held equipment |

LEITH'S TRANSPORT (ABERDEEN) LTD

Head Office Linksfield Road, Aberdeen, Grampian AB2 1RW
Phone 0224 484198
Personnel I Leith, Managing Director
Quarries North Lasts, Torrin

LEITH'S TRANSPORT (ABERDEEN) LTD

NORTH LASTS QUARRY

329

Peterculter, Aberdeen, Grampian AB1 0PE
Phone 0244 484198
OS Map 38   **OS Grid Reference** NJ828041
Personnel S McDonald, Unit Manager

PRODUCT DATA	**Rock type**	Gabbro
	Colour	Dark grey
	Grain	Coarse

Products Crushed Stone

| DRILLING DATA | | Contractor Rockblast |

NO FURTHER INFORMATION PROVIDED

LEITH'S TRANSPORT (ABERDEEN) LTD

TORRIN QUARRY

330

Broadford, Isle of Skye, Highland IV49 9BY
Phone 04712 265
OS Map 32   **OS Grid Reference** NG643235
Personnel C E T Herbert, Unit Manager

PRODUCT DATA	**Rock type**	Ordovician Durness Limestone
	Products	Crushed Stone
		Agricultural Lime

DRILLING DATA		Hole Diameter 105mm
		Bench Height 12m
		Contractor Rockblast

| LOAD AND HAUL | **Trucks** | Foden |
| | **Loaders** | Hannomag |

SPON'S QUARRY GUIDE

Having invested in *Spon's Quarry Guide,* you will have an interest in keeping your data source as up to date as possible. We plan to update the *Guide* and publish new editions regularly. If you register as a purchaser of the first edition by completing this reply-paid card before publication of the second edition we will send you a voucher entitling you to a *10% discount* available through your normal book supplier or direct from us.

When you complete the registration card please let us have your advice and comments on the *Guide* by answering the survey questions. These will be of great value to us in the development of future editions.

Thank you very much in advance.

SPON'S QUARRY GUIDE

1. I heard of the Guide via: ☐ advert or leaflet ☐ trade press
    ☐ word of mouth ☐ book review in...

2. I think the Guide could be improved by including:
    ☐ hard rock importers ☐ quarries in N. Ireland
    ☐ sand and gravel ☐ other mineral workings
    ☐ more plant and equipment listings
    ☐ other information..

3. This copy of Spon's Quarry Guide will probably be referred
    to by ☐ persons as well as myself

4. I am employed in the following industrial/professional category:
    ☐ quarry ownership/management ☐ quarry equipment
    ☐ architecture ☐ building/construction
    ☐ civil engineering ☐ quantity surveying
    ☐ other rock specifier ☐ geology/mineral resources
    ☐ institutional ☐ research
    ☐ other...

5. I would like to order ☐ additional copy/ies of Spon's Quarry Guide
    ☐ Please invoice me @ £85.00 or ☐ our official order will follow

6. ☐ Please register me as a Spon's Quarry Guide: First edition user and
    send me a 10% discount voucher for the next edition.

Name...(Block letters please)
Company/organization name...
Address...
....Postcode................................
Signature..Date..........................

SPON'S QUARRY GUIDE
Building Materials Market Research
FREEPOST (BR871)
BRIGHTON
BN1 1ZW

CRUSHING PLANT	**Primary**	Kue Ken 36 × 24 Jaw
		Marsden 30 × 20 Jaw
	Secondary	Goodwin Barsby 36 × 6 Jaw
		Parker 24 × 6 Jaw
	Tertiary	Goodwin Barsby Ajax Impactor
		Lightning Hammermill

LEWIS LAND SERVICES LTD
Head Office *New Buildings, Newmarket, Stornoway, Isle of Lewis, Western Isles PA86 0ED*
Phone *0851 3232*
Personnel *P Morrison, Director*
Quarries *Lic*

LEWIS LAND SERVICES LTD

LIC QUARRY
331

Newmarket, Stornoway, Isle of Lewis, Western Isles PA86 0ED
Phone 0851 3232
OS Map 8   **OS Grid Reference** NB425355
Personnel C Mackintosh, Unit Manager

PRODUCT DATA	**Rock type**	Gneiss
	Colour	Grey to pink
	Grain	Medium to coarse
	Products	Dressed or Lump Stone
		Crushed Stone
RIPPING	**Ripper**	Komatsu D55
DRILLING DATA		No drilling
RIPPABLE STRATA		

LIME KILN HILL QUARRY COMPANY
Head Office *Lime Kiln Hill Quarry, Mells, Frome, Somerset BA11 3PH*
Phone *0373 812225*
Personnel *R C Pearce, Managing Director*
Quarries *Lime Kiln Hill*

LIME KILN HILL QUARRY COMPANY

LIME KILN HILL QUARRY
332

Mells, Frome, Somerset BA11 3PH
Phone 0373 812225
OS Map 183   **OS Grid Reference** ST545293
Personnel H A Burton, Unit Manager

PRODUCT DATA	**Rock type**	Carboniferous Limestone
	Colour	Bluish grey
	Product	Crushed Stone
DRILLING DATA		Hole Diameter 75mm
		Bench Height 10m
	Drills	Halco 400C
		Holman Voltrak

entry continues overleaf

SECONDARY BREAKING		Dropball on Ruston-Bucyrus 22-RB
CRUSHING PLANT	**Primary**	Pegson
	Secondary	Hazemag APK Impactor
	Tertiary	Mansfield No 2 Hammermill

LINCOLN STONE (1986) LTD
Head Office *32 Lord Street, Gainsborough, Lincs DN21 2DB*
Phone *0427 617464*
Personnel *J Aldershaw, Managing Director*
Quarries *Spittlegate Level*

LINCOLN STONE (1986) LTD
SPITTLEGATE LEVEL QUARRY 333
Grantham, Lincs NG31 7UH
Phone 0476 60777
OS Map 130   **OS Grid Reference** SK915359
Personnel D Gratton, Unit Manager

PRODUCT DATA	**Rock type**	Jurassic Limestone
	Colour	Not commercially significant
	Grain	Fine
	Product	Crushed Stone

RIPPABLE STRATA

G LINDLEY AND SONS
Head Office *Sovereign Quarry, Shepley, Huddersfield, W Yorks HD8 8BH*
Phone *0484 606203*
Personnel *J W Sawyer, Managing Director*
Quarries *Sovereign*

G LINDLEY AND SONS
SOVEREIGN QUARRY 334
Shepley, Huddersfield, W Yorks HD8 8BH
Phone 0484 606203
OS Map 110   **OS Grid Reference** SE194088
Personnel J W Sawyer, Managing Director and Unit Manager

PRODUCT DATA	**Rock type**	Carboniferous Millstone Grit
	Colour	Blue to grey
	Grain	Coarse
	Product	Dressed or Lump Stone
DRILLING DATA		Hand-held equipment

LOCAL STONE COMPANY LTD
Head Office *Bognor Common Quarry, The Grove, Little Bognor, Fittleworth, W Sussex RH20 1JY*
Phone *0798 42888*
Personnel *J E Grinstead, Director*
Quarries *Bognor Common*

LOCAL STONE COMPANY LTD
BOGNOR COMMON QUARRY
335

The Grove, Little Bognor, Fittleworth, W Sussex RH20 1JY
Phone 0798 42888
OS Map 179 **OS Grid Reference** TQ008213
Personnel J E Grinstead, Director and Unit Manager

PRODUCT DATA	**Rock type**	Cretaceous Limestone
	Colour	Green to grey, buff
	Grain	Fine to medium
	Products	Dressed or Lump Stone
		Crushed Stone

RIPPABLE STRATA

NO FURTHER INFORMATION PROVIDED

LONGCLIFFE QUARRIES LTD
Head Office *Brassington, Derbys DE4 4BZ*
Phone *062985 284* **Telex** *377039*
Personnel *D A G Shield, Managing Director*
Quarries *Longcliffe*

LONGCLIFFE QUARRIES LTD
LONGCLIFFE QUARRY
336

Brassington, Derbys DE4 4BZ
Phone 062985 284
OS Map 119 **OS Grid Reference** SK233556
Personnel M R Hart, Unit Manager

| PRODUCT DATA | **Rock type** | Carboniferous Bee Low Limestone |
| | **Product** | Crushed Stone |

DRILLING DATA		Hole Diameter 110mm
		Bench Height 18m
	Drills	Halco 410C

| SECONDARY BREAKING | | Hydraulic Hammer - Montabert |

LOAD AND HAUL	**Trucks**	(2)Terex
		Foden
	Loaders	Caterpillar 966C
		Caterpillar 968

LOTHIAN REGION COUNCIL
Head Office *c/o Department of Highways, 19 Market Street, Edinburgh, Lothian EH1 1BL*
Phone *031 2299292* **Fax** *031 2298600*
Personnel *P J Mason, Technical Director*
Quarries *Hazelbank*

entry continues overleaf

LOTHIAN REGION COUNCIL

HAZELBANK QUARRY

337

Fountainhall, Galashiels, Borders TD1 2SA
Phone 031 2299292
OS Map 66   **OS Grid Reference** NT425504

QUARRY ON STAND

ROCK DATA	**Rock type**	Silurian Greywacke

D & P LOVELL QUARRIES

Head Office *2A Cranbourne Road, Swanage, Dorset, BH19 1EA*
Phone *0929 422657*
Personnel *Miss D Lovell, Director*
   P Lovell, Director
Quarries *Downs*

D & P LOVELL QUARRIES

DOWNS QUARRY

338

Kingston Road, Langton Matravers, Swanage, Dorset BH19 3JP
Phone 092943 255
OS Map 194   **OS Grid Reference** SY972792
Personnel P Lovell, Director and Unit Manager

PRODUCT DATA	**Rock type**	Jurassic Limestone
	Colour	Pale to dark bluish grey, buffish grey
	Grain	Fine
	Product	Dressed or Lump Stone

RIPPABLE STRATA

J LOVIE AND SON LTD

Head Office *Cowbog, New Pitsligo, Fraserburgh, Grampian AB4 4PR*
Phone *07717 272*
Personnel *J Lovie, Director*
Quarries *Cottonhill*

J LOVIE AND SON LTD

COTTONHILL QUARRY

339

Cowbog, New Pitsligo, Fraserburgh, Grampian AB4 4PR
Phone 07717 272
OS Map 29   **OS Grid Reference** NJ724604
Personnel J Lovie, Director and Unit Manager

PRODUCT DATA	**Rock type**	Pre-Cambrian Dalradian Limestone
	Colour	Not commercially significant
	Product	Crushed Stone

DRILLING DATA		Hole Diameter 110mm
		Bench Height 20m
		Contractor Rockblast

| LOAD AND HAUL | **Trucks** | Aveling Barford RD40 |
| | **Loaders** | Caterpillar |

CRUSHING PLANT	**Primary**	Pegson 30 × 42 Jaw
	Secondary	Nordberg 4' Cone (Standard)
	Tertiary	None

R MACASKILL (CONTRACTORS) LTD
Head Office *Ardhasaig Quarry, Harris, Western Isles PA85 3AJ*
Phone *0859 2066*
Personnel *R Macaskill, Director*
Quarries *Ardhasaig*

R MACASKILL (CONTRACTORS) LTD
ARDHASAIG QUARRY   340
Harris, Western Isles PA85 3AJ
Phone 0859 2066
OS Map 14   **OS Grid Reference** NB125026
Personnel R Macaskill, Director and Unit Manager

PRODUCT DATA	**Rock type**	Basalt
	Colour	Greyish black
	Grain	Fine
	Products	Crushed Stone
		Asphalt or Tarmacadam

| DRILLING DATA | | No information provided |

CRUSHING PLANT	**Primary**	Kue Ken 36 × 24 Jaw
	Secondary	Kue Ken 3' Cone
		Kue Ken 28" Cone

MACCLESFIELD STONE QUARRIES LTD
Head Office *Bridge Quarry, Windmill Lane, Kerridge, Macclesfield, Cheshire SK10 5AZ*
Phone *0625 73208*
Personnel *D Tooth, Director*
Quarries *Bridge*

MACCLESFIELD STONE QUARRIES LTD
BRIDGE QUARRY   341
Windmill Lane, Kerridge, Macclesfield, Cheshire SK10 5AZ
Phone 0625 73208
OS Map 118   **OS Grid Reference** SU940762
Personnel D Tooth, Director and Unit Manager

PRODUCT DATA	**Rock type**	Carboniferous Milnrow Sandstone
	Colour	Fawn, grey
	Grain	Fine
	Product	Dressed or Lump Stone

| DRILLING DATA | | Hand-held equipment |

A MACE
Head Office *16 James Street, Oakworth, Keighley, W Yorks BD22 7PE*
Phone *0535 43492*
Personnel *A Mace, Owner*
Quarries *Fly Delph*

A MACE

FLY DELPH QUARRY 342

Warley Moor Reservoir, Oxenhope, W Yorks BD22 0LL
Phone 0535 43492
OS Map 104 **OS Grid Reference** SE034323
Personnel A Mace, Owner and Unit Manager

PRODUCT DATA	**Rock type**	Carboniferous Millstone Grit
	Colour	Fawn, brown, buff
	Grain	Medium
	Product	Dressed or Lump Stone

DRILLING DATA   Hand-held equipment

JOHN MACLENNAN
Head Office *129 Craigston, Castlebay, Isle of Barra, Western Isles PA80 5XS*
Phone *08714 508*
Personnel *J Maclennan, Owner*
Quarries *Cleat and Lower Grean*

JOHN MACLENNAN

CLEAT AND LOWER GREAN QUARRY 343

Castlebay, Isle of Barra, Western Isles PA80 5XS
Phone 08714 508
OS Map 31 **OS Grid Reference** NA665982

QUARRY ON STAND

ROCK DATA   **Rock type**   Basalt

K MACRAE AND COMPANY LTD
Head Office *Borrowston Quarry, Thrumster, Wick, Highland KW1 5TX*
Phone *0955 85383*
Personnel *J Morgan, Managing Director*
Quarries *Borrowston*

K MACRAE AND COMPANY LTD

BORROWSTON QUARRY 344

Thrumster, Wick, Highland KW1 5TX
Phone 0955 85383
OS Map 12 **OS Grid Reference** ND326423
Personnel J Morgan, Managing Director and Unit Manager

PRODUCT DATA	**Rock type**	Devonian Caithness Flagstone
	Products	Crushed Stone
		Asphalt or Tarmacadam
		Ready-mixed Concrete
		Pre-cast Concrete Products

DRILLING DATA		Hole Diameter 110mm
		Bench Height 20m
		Contractor Ritchies Equipment
SECONDARY BREAKING		Hydraulic Hammer
LOAD AND HAUL	**Loaders**	JCB
CRUSHING PLANT	**Primary**	Parker 42 × 20 Jaw
	Secondary	Barmac Duopactor
	Tertiary	None

MANDALE MINING CO LTD
Head Office *Sheldon, Bakewell, Derbys DE4 1QS*
Phone *062981 3676*
Personnel *R M Steele, Director*
Quarries *Once-a-Week*

MANDALE MINING CO LTD
ONCE-A-WEEK QUARRY   345
Sheldon, Bakewell, Derbys DE4 1QS
Phone 062981 3676
OS Map 119 **OS Grid Reference** SJ173687
Personnel R M Steele, Director and Unit Manager

PRODUCT DATA	**Rock type**	Carboniferous Limestone
	Colour	Blue to light grey
	Grain	Fine
	Product	Dressed or Lump Stone
DRILLING DATA		Hand-held equipment

MARLOW STONE (CHILLATON) LTD
Head Office *The Old Chapel, Chillaton, Devon PL16 0HU*
Phone *082286 206*
Personnel *J Marlow, Director*
Quarries *Chillaton*

MARLOW STONE (CHILLATON) LTD
CHILLATON QUARRY   346
The Old Chapel, Chillaton, Devon PL16 0HU
Phone 082286 206
OS Map 201 **OS Grid Reference** SX436817
Personnel J Marlow, Director and Unit Manager

PRODUCT DATA	**Rock type**	Devonian Slate
	Colour	Orangey brown
	Grain	Fine
	Products	Dressed or Lump Stone
		Crushed Stone
DRILLING DATA		Hand-held equipment

MARSHALLS HALIFAX plc
Corporate Head Office *Brier Lodge, Southowram, Halifax, W Yorks HX3 9SY*
Phone *0422 57155* **Fax** *0422 67093*
Personnel *K F Marshall, Chairman*
D Marshall, Managing Director
Companies *R Hinchcliffe and Son Ltd* — see separate entry
Quarries *Clock Face, Cromwell, Fletcher Bank, Glen, Scout Moor*

MARSHALLS HALIFAX plc

CLOCK FACE QUARRY

347

Ripponden, Sowerby Bridge, W Yorks HX6 3BQ
Phone 0422 822520
OS Map 110 **OS Grid Reference** SE047172
Personnel D Cockroft, Unit Manager

PRODUCT DATA	**Rock type**	Carboniferous Millstone Grit
	Colour	Not commercially significant
	Grain	Medium
	Product	Crushed Stone
DRILLING DATA		Hole Diameter 105mm
		Bench Height 15m
	Drills	Halco mast on Cat 2l3
SECONDARY BREAKING		Dropball on Smith
LOAD AND HAUL	**Trucks**	Caterpillar
	Loaders	Caterpillar
CRUSHING PLANT	**Primary**	Baxter RB36 × 24 Jaw (Single Toggle)
	Secondary	None
	Tertiary	None

MARSHALLS HALIFAX plc

CROMWELL QUARRY

348

Southowram, Halifax, W Yorks HX3 9SY
Phone 0422 57155
OS Map 104 **OS Grid Reference** SE124234
Personnel K Mallinson, Unit Manager

PRODUCT DATA	**Rock type**	Carboniferous Sandstone (Flagstone)
	Colour	Buff
	Grain	Fine
	Product	Dressed or Lump Stone
DRILLING DATA		Hand-held equipment

MARSHALLS HALIFAX plc

FLETCHER BANK QUARRY

349

Ramsbottom, Greater Manchester, OL0 0DD
Phone 0706 824911
OS Map 109 **OS Grid Reference** SD803168
Personnel I Kerr, Unit Manager

PRODUCT DATA	**Rock type**	Carboniferous Millstone Grit
	Colour	Not Commercially Significant
	Grain	Medium
	Product	Crushed Stone

DRILLING DATA		Hole Diameter 110mm
		Bench Height 22m
	Drills	Halco MPD50-20

| SECONDARY BREAKING | | Dropball on Ruston-Bucyrus 30-RB |
| | | Dropball on Smith 28 |

LOAD AND HAUL	**Trucks**	(2)Volvo BM A25
		(2)Volvo BM A35
	Loaders	Komatsu WA350

CRUSHING PLANT	**Primary**	Kue Ken Jaw
	Secondary	(2)Nordberg 4.25′ Cone
	Tertiary	Barmac Duopactor

MARSHALLS HALIFAX plc

GLEN QUARRY   350

Ruddlemill Lane, Stainton Maltby, Rotherham, S Yorks S66 7RM
Phone 0709 813151
OS Map 111   **OS Grid Reference** SK546944
Personnel I MacDougal, Unit Manager

PRODUCT DATA	**Rock type**	Permian Magnesian Limestone
	Colour	Not Commercially Significant
	Product	Crushed Stone

DRILLING DATA		Hole Diameter 110mm
		Bench Height 18m
	Drills	Halco 400H

| SECONDARY BREAKING | | Hydraulic Hammer - Montabert |

LOAD AND HAUL	**Trucks**	Caterpillar 769C
	Loaders	(2)Caterpillar 966C
		Caterpillar 980C

CRUSHING PLANT	**Primary**	Baxter ST40×32 Jaw (Single Toggle)
	Secondary	Pegson 36″ Cone (Standard)
	Tertiary	Nordberg Gyradisc

MARSHALLS HALIFAX plc

SCOUT MOOR QUARRY   351

Ramsbottom, Greater Manchester, OL0 0DD
Phone 0706 824911
OS Map 109   **OS Grid Reference** SD799166
Personnel I Kerr, Unit Manager

| PRODUCT DATA | **Rock type** | Carboniferous Sandstone |
| | **Product** | Dressed or Lump Stone |

| DRILLING DATA | | Hand-held equipment |

MARSHALLS STONE COMPANY LTD
Head Office *Spikers Hill Quarry, West Ayton, Scarborough, N Yorks YO13 9LB*
Phone *0723 863175*
Personnel *P Marshall, Managing Director*
T Marshall, Director
Quarries *Spikers Hill*

MARSHALLS STONE COMPANY LTD

SPIKERS HILL QUARRY

352

West Ayton, Scarborough, N Yorks YO13 9LB
Phone 0723 863175
OS Map 101   **OS Grid Reference** SE979866
Personnel T Marshall, Director and Unit Manager

PRODUCT DATA	**Rock type**	Jurassic Corallian Oolitic Limestone
	Grain	Fine
	Products	Crushed Stone
		Asphalt or Tarmacadam
		Agricultural Lime
DRILLING DATA		Hole Diameter 75mm
		Bench Height 12m
		Contractor R & B Rock Drillers
SECONDARY BREAKING		Dropball
LOAD AND HAUL	**Trucks**	None
	Loaders	Terex
		JCB
CRUSHING PLANT	**Primary**	Goodwin Barsby Jaw
	Secondary	None
	Tertiary	None

MARSHINGTON STONE COMPANY LTD
Head Office *Shire Hill Quarry, Sheffield Road, Glossop, Derbys SK13 9PU*
Phone *0663 64211*
Personnel *J A Marshington, Managing Director*
Quarries *Shire Hill*

MARSHINGTON STONE COMPANY LTD

SHIRE HILL QUARRY

353

Sheffield Road, Glossop, Derbys SK13 9PU
Phone 0663 64211
OS Map 110   **OS Grid Reference** SK054945
Personnel J A Marshington, Managing Director and Unit Manager

PRODUCT DATA	**Rock type**	Carboniferous Sandstone
	Colour	Not Commercially Significant
	Grain	Medium
	Product	Crushed Stone
DRILLING DATA		Hole Diameter 105mm
		Bench Height 17m
		Contractor A Sharpe

COMPRESSORS		Broomwade WR600
SECONDARY BREAKING		Dropball
LOAD AND HAUL	**Trucks**	Foden
	Loaders	Caterpillar
CRUSHING PLANT	**Primary**	Goodwin Barsby Jaw (Mobile)
	Secondary	Jaw
	Tertiary	Cones

SIR A McALPINE plc
Corporate Head Office Upton, South Wirral, Merseyside L66 7ND
Phone 051 3394141 **Fax** 051 3394748 **Telex** 627185
Personnel T Scurr, Deputy Chief Executive
Companies Sir A McAlpine Quarry Products Ltd, Sir A McAlpine Quarry Products (Scotland) Ltd,
Sir A McAlpine Slate Products Ltd

SIR A McALPINE QUARRY PRODUCTS LTD (SIR A McALPINE plc)
Head Office Borras Airfield, Holt Road, Wrexham, Clwyd LL13 9SE
Phone 0978 351921 **Fax** 0978 366445 **Telex** 617056
Personnel P Charmbury, Managing Director
K Bowler, Production Director
Quarries Buckton Vale, Dinas, Hendre, Honister, Parish, Torcoedfawr, Wredon

SIR A McALPINE QUARRY PRODUCTS LTD (SIR A McALPINE plc)
BUCKTON VALE QUARRY   **354**
Tameside, Greater Manchester, SK15 3RD
Phone 04575 5328 **Telex** 617056
OS Map 109 **OS Grid Reference** SD990017
Personnel D Wilson, Unit Manager

PRODUCT DATA	**Rock type**	Carboniferous Millstone Grit
	Colour	Not Commercially Significant
	Grain	Medium
	Product	Crushed Stone
DRILLING DATA		Hole Diameter 105mm
		Bench Height 7.5–10m
		Contractor A Jones Rock Drillers
SECONDARY BREAKING		Dropball on Ruston-Bucyrus 22-RB
LOAD AND HAUL	**Trucks**	Caterpillar 769C
		Terex
	Loaders	Poclain
CRUSHING PLANT	**Primary**	Parker 32 × 42 Jaw
	Secondary	Nordberg 4' Cone
	Tertiary	None

SIR A McALPINE QUARRY PRODUCTS LTD (SIR A McALPINE plc)

DINAS QUARRY

355

Llansawel, Llandeilo, Dyfed SA19 7JB
Phone 0269 850341 **Telex** 617056
OS Map 146 **OS Grid Reference** SN627354
Personnel J Davies, Unit Manager

PRODUCT DATA	**Rock type**	Ordovician Sandstone
	Colour	Not Commercially Significant
	Grain	Medium
	Product	Crushed Stone
DRILLING DATA		Hole Diameter 105mm
		Bench Height 10–12.5m
		Contractor A Jones Rock Drillers
SECONDARY BREAKING		Dropball on Ruston-Bucyrus 22-RB
CRUSHING PLANT		No information

SIR A McALPINE QUARRY PRODUCTS LTD (SIR A McALPINE plc)

HENDRE QUARRY

356

Hendre, Mold, Clwyd CH7 5QL
Phone 0352 741200 **Telex** 617056
OS Map 116 **OS Grid Reference** SJ194680
Personnel C J Jones, Unit Manager

PRODUCT DATA	**Rock type**	Carboniferous Limestone
	Colour	Not Commercially Significant
	Grain	Medium
	Products	Crushed Stone
		Asphalt or Tarmacadam
DRILLING DATA		Hole Diameter 105mm
		Bench Height 15m
		Contractor A Jones Rock Drillers
SECONDARY BREAKING		Dropball on Ruston-Bucyrus 22-RB
LOAD AND HAUL	**Face shovel**	Liebherr 974
	Trucks	(2)Aveling Barford RD35
		(2)Aveling Barford RD40
	Loaders	Caterpillar 966C
		Caterpillar 968
CRUSHING PLANT	**Primary**	Hazemag Impactor
	Secondary	None
	Tertiary	None

SIR A McALPINE QUARRY PRODUCTS LTD (SIR A McALPINE plc)

HONISTER QUARRY

357

Borrowdale, Keswick, Cumbria CA12 5XN
Phone 0978 351921 **Telex** 617056
OS Map 90 **OS Grid Reference** NY217137
Personnel B Hughes, Unit Manager

PRODUCT DATA	**Rock type**	Ordovician Borrowdale Group Slate
	Colour	Dark olive green
	Grain	Fine
	Product	Dressed or Lump Stone

DRILLING DATA		Hand-held equipment

SIR A McALPINE QUARRY PRODUCTS LTD (SIR A McALPINE plc)

PARISH QUARRY

358

Via Gellia, Bonsal, Matlock, Derbys DE4 2AJ
Phone 0629 822613 **Telex** 617056
OS Map 120 **OS Grid Reference** SK279573
Personnel A Hindle, Unit Manager

PRODUCT DATA	**Rock type**	Carboniferous Monsal Dale Limestone
	Colour	Not Commercially Significant
	Grain	Medium
	Products	Crushed Stone
		Agricultural Lime

DRILLING DATA		Hole Diameter 105mm
		Bench Height 25m
		Contractor A Jones Rock Drillers

SECONDARY BREAKING		Dropball on Ruston-Bucyrus 22-RB

LOAD AND HAUL	**Face shovel**	Ruston-Bucyrus
	Trucks	(3)Scania
	Loaders	Caterpillar 966C
		Caterpillar 968

CRUSHING PLANT	**Primary**	Goodwin Barsby 36×24 Jaw
	Secondary	Pegson 2030 Impactor
	Tertiary	Christy & Norris 24×24 Swing Hammer

SIR A McALPINE QUARRY PRODUCTS LTD (SIR A McALPINE plc)

TORCOEDFAWR QUARRY

359

Crwbin, Kidwelly, Dyfed SA17 5ED
Phone 0267 86555 **Telex** 617056
OS Map 159 **OS Grid Reference** SN482136
Personnel G Walters, Unit Manager

PRODUCT DATA	**Rock type**	Carboniferous Limestone
	Colour	Not Commercially Significant
	Grain	Medium
	Products	Crushed Stone
		Asphalt or Tarmacadam

DRILLING DATA		Hole Diameter 105mm
		Bench Height 15m
		Contractor A Jones Rock Drillers

SECONDARY BREAKING		Dropball on Ruston-Bucyrus 22-RB

LOAD AND HAUL	**Face shovel**	Liebherr R962
	Trucks	Aveling Barford RD40

entry continues overleaf

CRUSHING PLANT	**Primary**	BJD 55 Rotary
	Secondary	BJD 44 Rotary
	Tertiary	None

SIR A McALPINE QUARRY PRODUCTS LTD (SIR A McALPINE plc)

WREDON QUARRY   360

Caldon Low, Waterhouse, Stoke-on-Trent, Staffs ST10 3HA
Phone 0538 702668 **Telex** 617056
OS Map 119 **OS Grid Reference** SK086473
Personnel D Thomson, Unit Manager

PRODUCT DATA	**Rock type**	Carboniferous Kevin Limestone
	Colour	Not Commercially Significant
	Products	Crushed Stone
		Asphalt or Tarmacadam

DRILLING DATA		Hole Diameter 105mm
		Bench Height 15m
		Contractor A Jones Rock Drillers

| SECONDARY BREAKING | | Dropball on Ruston-Bucyrus 22-RB |

CRUSHING PLANT	**Primary**	Pegson Jaw
	Secondary	Hazemag Impactor
	Tertiary	Parker Impactor

SIR A McALPINE QUARRY PRODUCTS (SCOTLAND) LTD (SIR A McALPINE plc)
Head Office Ashley Drive, Bothwell, Strathclyde G71 8BS
Phone 0698 810404
Personnel J Murtagh, Director and General Manager
Quarries Airdriehill, Bangley, Broadlaw, Underheugh

SIR A McALPINE QUARRY PRODUCTS (SCOTLAND) LTD (SIR A McALPINE plc)

AIRDRIEHILL QUARRY   361

Whiterig, Airdrie, Strathclyde ML6 7SE
Phone 0236 54413
OS Map 64 **OS Grid Reference** NS778670
Personnel J Dishar, Unit Manager

PRODUCT DATA	**Rock type**	Basalt
	Colour	Greyish black
	Grain	Fine
	Product	Crushed Stone

DRILLING DATA		Hole Diameter 110mm
		Bench Height 18m
		Contractor G & N

| SECONDARY BREAKING | | Dropball |

LOAD AND HAUL	**Trucks**	Foden
	Loaders	Caterpillar
		Volvo

CRUSHING PLANT	**Primary**	Goodwin Barsby 42 × 24 Jaw
	Secondary	Parker 30 × 24 Jaw
	Tertiary	Kobelco 3′ Cone

SIR A McALPINE QUARRY PRODUCTS (SCOTLAND) LTD (SIR A McALPINE plc)

BANGLEY QUARRY

362

Haddington, Lothian EH41 3FN
Phone 0620 825811
OS Map 67   **OS Grid Reference** NT487753
Personnel D Grant, Unit Manager

PRODUCT DATA	**Rock type**	Trachyte (Lava)
	Colour	Grey
	Grain	Fine
	Products	Crushed Stone
		Asphalt or Tarmacadam

DRILLING DATA		Hole Diameter 110mm
		Bench Height 18m
		Contractor Rocklift

| SECONDARY BREAKING | | Hydraulic Hammer |

LOAD AND HAUL	**Trucks**	Contractor
	Loaders	Caterpillar
		Zettelmeyer
		Volvo

CRUSHING PLANT	**Primary**	(2)Kue Ken 42 × 32 Jaw
	Secondary	(2)Kue Ken 4′ Cone
	Tertiary	Kue Ken 3′ Cone

SIR A McALPINE QUARRY PRODUCTS (SCOTLAND) LTD (SIR A McALPINE plc)

BROADLAW QUARRY

363

Gorebridge, Lothian EH23 4RQ
Phone 0620 825811
OS Map 99   **OS Grid Reference** NT406100
Personnel D Grant, Unit Manager

PRODUCT DATA	**Rock type**	Basalt
	Colour	Greyish black
	Grain	Fine
	Product	Crushed Stone

DRILLING DATA		Hole Diameter 110mm
		Bench Height 17m
		Contractor Rocklift

| SECONDARY BREAKING | | Hydraulic Hammer |

LOAD AND HAUL	**Trucks**	Contractor
	Loaders	Caterpillar
		Volvo

CRUSHING PLANT	**Primary**	Parker 32 × 42 Jaw
	Secondary	Parker 36 × 10 Jaw
	Tertiary	Nordberg 3′ Cone

SIR A McALPINE QUARRY PRODUCTS (SCOTLAND) LTD (SIR A McALPINE plc)

UNDERHEUGH QUARRY 364

Cloch Road, Lunderstone Bay, Gourock, Strathclyde PA19 1BB
Phone 0474 521195
OS Map 63 **OS Grid Reference** NS203751
Personnel T Cameron, Unit Manager

PRODUCT DATA	**Rock type**	Basalt
	Colour	Black
	Grain	Fine
	Product	Crushed Stone

DRILLING DATA		Hole Diameter 105mm
		Bench Height 17m
		Contractor E Jackson

SECONDARY BREAKING		Dropball

LOAD AND HAUL	**Trucks**	Contractor
	Loaders	Caterpillar
		Volvo

CRUSHING PLANT	**Primary**	Parker 32 × 42 Jaw
	Secondary	Parker 1004 Jaw (Double Toggle)
	Tertiary	Nordberg
		Nordberg 3' Cone

SIR A McALPINE SLATE PRODUCTS LTD (SIR A McALPINE plc)
Head Office *Penrhyn Quarry, Bethesda, Bangor, Gwynedd LL57 4YG*
Phone *0248 600656* **Fax** *0248 601171* **Telex** *61513*
Personnel *G F Drake, Managing Director*
R Owens, Director
I M Abbott, Director
Quarries *Penrhyn*

SIR A McALPINE SLATE PRODUCTS LTD (SIR A McALPINE plc)

PENRHYN QUARRY 365

Bethesda, Bangor, Gwynedd LL57 4YG
Phone 0248 600656 **Telex** 61513
OS Map 115 **OS Grid Reference** SH620650
Personnel A Jones, Pit Manager

PRODUCT DATA	**Rock type**	Cambrian Slate
	Colour	Heather blue, green
	Grain	Fine
	Product	Dressed or Lump Stone

DRILLING DATA		Hole Diameter 108mm
		Bench Height 13-18m
	Drills	(6)Holman Voltrak

COMPRESSORS		(6)CompAir RO75-170

LOAD AND HAUL	**Face shovel**	O&K
	Trucks	Aveling Barford
	Loaders	Caterpillar
		Broyt

McLAREN ROADSTONE LTD (RMC ROADSTONE PRODUCTS LTD)
Head Office *Cragmill Quarry, Belford, Northumb NE70 7EZ*
Phone *06683 866* **Fax** *06683 844*
Personnel *W McLaren, Director*
          J Daglish, General Manager
Quarries *Cragmill, Divethill*

McLAREN ROADSTONE LTD (RMC ROADSTONE PRODUCTS LTD)

CRAGMILL QUARRY
366
Belford, Northumb NE70 7EZ
Phone 0668 213866
OS Map 75   **OS Grid Reference** NU112345
Personnel W Baty, Unit Manager

PRODUCT DATA	**Rock type**	Dolerite Whin Sill
	Colour	Medium grey
	Grain	Coarse
	Products	Crushed Stone
		Asphalt or Tarmacadam
		Ready-mixed Concrete
DRILLING DATA		Hole Diameter 105mm
		Bench Height 11–15m
		Contractor Ritchies Equipment
SECONDARY BREAKING		Dropball on Ruston-Bucyrus 22-RB
LOAD AND HAUL	**Face shovel**	(2)Ruston-Bucyrus 22-RB
	Trucks	Heathfield
		Foden
	Loaders	Komatsu WA300
CRUSHING PLANT	**Primary**	Broadbent 48×36 Jaw (Double Toggle)
	Secondary	Kue Ken 42×14 Jaw
		Nordberg 3' Cone
	Tertiary	Pegson 3' Gyrasphere
		Pegson 2' Gyrasphere
	Quaternary	Barmac Duopactor

McLAREN ROADSTONE LTD (RMC ROADSTONE PRODUCTS LTD)

DIVETHILL QUARRY
367
Capheaton, Newcastle upon Tyne, Northumb NE19 2BG
Phone 0830 30200
OS Map 87   **OS Grid Reference** NY982793
Personnel A Stead, Unit Manager

PRODUCT DATA	**Rock type**	Dolerite Whin Sill
	Colour	Medium grey
	Grain	Coarse
	Products	Crushed Stone
		Ready-mixed Concrete
DRILLING DATA		Hole Diameter 105mm
		Bench Height 12.5m
		Contractor Ritchies Equipment

entry continues overleaf

SECONDARY BREAKING		Dropball on Ruston-Bucyrus 22-RB
LOAD AND HAUL	**Face shovel**	Ruston-Bucyrus
	Trucks	Terex
	Loaders	Caterpillar
		Komatsu
CRUSHING PLANT	**Primary**	Pegson Jaw
	Secondary	Nordberg 4.25′ Cone
	Tertiary	Pegson 3′ Cone
	Quaternary	Barmac Duopactor

P G MEDWELL AND SON

Head Office *Holleywell Quarry, The Quarries, Clipsham, Oakham, Leics LE15 7SQ*
Phone *078081 730*
Personnel *P G Medwell, Owner*
Quarries *Holleywell*

P G MEDWELL AND SON

HOLLEYWELL QUARRY

368

The Quarries, Clipsham, Oakham, Leics LE15 7SQ
Phone 078081 730
OS Map 130   **OS Grid Reference** SK987158
Personnel P G Medwell, Owner and Unit Manager

PRODUCT DATA	**Rock type**	Jurassic Oolitic Limestone
	Colour	Creamy brown
	Grain	Shelly
	Product	Dressed or Lump Stone
DRILLING DATA		Hand-held equipment

B MERRICK

Head Office *Northview, Swallow House Lane, Hayfield, Stockport, Cheshire SK12 5HB*
Phone *0663 43565*
Personnel *B Merrick, Owner*
Quarries *Chinley Moor*

B MERRICK

CHINLEY MOOR QUARRY

369

Chapel Road, Hayfield, Stockport, Greater Manchester, SK12 7EN
Phone 0663 43665
OS Map 109   **OS Grid Reference** SK050853
Personnel B Merrick, Owner and Unit Manager

PRODUCT DATA	**Rock type**	Carboniferous Millstone Grit
	Colour	Buff to light brown
	Grain	Fine to medium
	Product	Dressed or Lump Stone

RIPPABLE STRATA

MILL HILL QUARRY (TAVISTOCK) LTD
Head Office *Longford Quarry, Moorshop, Tavistock, Devon PL19 8NP*
Phone *0822 612483/612786*
Personnel *R J Tate, Director*
Quarries *Longford, Mill Hill*

MILL HILL QUARRY (TAVISTOCK) LTD

LONGFORD QUARRY

370

Moorshop, Tavistock, Devon PL19 8NP
Phone 0822 612786
OS Map 201 **OS Grid Reference** SX330710
Personnel R J Tate, Director and Unit Manager

PRODUCT DATA	**Rock type**	Devonian Slate
	Colour	Brownish blue
	Grain	Fine
	Product	Dressed or Lump Stone

DRILLING DATA		Hand-held equipment

MILL HILL QUARRY (TAVISTOCK) LTD

MILL HILL QUARRY

371

Callington, Tavistock, Devon PL19 8NP
Phone 0822 612786
OS Map 201 **OS Grid Reference** SX319710
Personnel R J Tate, Director and Unit Manager

PRODUCT DATA	**Rock type**	Devonian Slate
	Colour	Brownish blue
	Grain	Fine
	Product	Dressed or Lump Stone

DRILLING DATA		Hand-held equipment

W B MILLER (QUARRY MASTERS) LTD
Head Office *Kennoway, Leven, Fife KY8 5SG*
Phone *0333 350306*
Personnel *W B Miller, Director*
Quarries *Balmullo, Langside*

W B MILLER (QUARRY MASTERS) LTD

BALMULLO QUARRY

372

Quarry Road, Balmullo, St Andrews, Fife KY16 0BH
Phone 0334 870208
OS Map 54 **OS Grid Reference** NO418215
Personnel J W Walker, General Manager

PRODUCT DATA	**Rock type**	Andesitic Devonian Lavas
	Colour	Black
	Grain	Fine
	Product	Crushed Stone

entry continues overleaf

DRILLING DATA		Hole Diameter 90mm
		Bench Height 20m
	Drills	Worthington Rotary Drill

SECONDARY BREAKING		Hydraulic Hammer - Montabert

LOAD AND HAUL	**Trucks**	(2)Cochem
	Loaders	(2)Caterpillar 930

W B MILLER (QUARRY MASTERS) LTD
LANGSIDE QUARRY   **373**
Kennoway, Fife KY8 5SG
Phone 0333 350306
OS Map 59 **OS Grid Reference** NO345036
Personnel D Philip, Unit Manager

PRODUCT DATA	**Rock type**	Red Rhyolite
	Colour	Red
	Grain	Fine to medium
	Product	Crushed Stone

DRILLING DATA		Hole Diameter 90mm
		Bench Height 20-24m
	Drills	Worthington Rotary Drill

SECONDARY BREAKING		Hydraulic Hammer - Montabert

LOAD AND HAUL	**Face shovel**	Komatsu
	Trucks	Terex
	Loaders	Zettelmeyer

MILLMEAD QUARRY PRODUCTS LTD
Head Office *Scabba Wood Quarry, Cadeby Road, Sprotborough, Doncaster, S Yorks DN5 7SY*
Phone *0302 310018*
Personnel *R Lanni, Director*
Quarries *Scabba Wood*

MILLMEAD QUARRY PRODUCTS LTD
SCABBA WOOD QUARRY   **374**
Cadeby Road, Sprotborough, Doncaster, S Yorks DN5 7SY
Phone 0302 310018
OS Map 111 **OS Grid Reference** SE527015
Personnel J Anderson, Unit Manager

PRODUCT DATA	**Rock type**	Permian Magnesian Limestone
	Colour	Cream to white
	Grain	Coarse
	Product	Crushed Stone

DRILLING DATA		Hole Diameter 102mm
		Bench Height 12m
	Drills	Hausherr HB80RD

SECONDARY BREAKING		Hydraulic Hammer

LOAD AND HAUL	**Trucks**	None
	Loaders	Zettelmeyer

CRUSHING PLANT	**Primary**	Parker 32 × 42 Jaw
	Secondary	(2)Findlay Cone
	Tertiary	(2)Findlay Cone

MINE TRAIN QUARRY
Head Office *39 Upton Gardens, Upton-upon-Severn, Worcester, Hereford & Worcs, WR8 0NU*
Phone *06846 2532*
Personnel *R Tainton, Director*
Quarries *Mine Train*

MINE TRAIN QUARRY

MINE TRAIN QUARRY   375
Parkend, Coleford, Glos GL16 7HX
Phone 06846 2532
OS Map 104   **OS Grid Reference** SE270422
Personnel R Tainton, Director and Unit Manager

PRODUCT DATA	**Rock type**	Carboniferous Pennant Sandstone
	Colour	Greyish blue
	Grain	Medium
	Product	Dressed or Lump Stone

DRILLING DATA		Hand-held equipment

MONE BROS (EXCAVATIONS) LTD
Head Office *Albert Road, Morley, Leeds, W Yorks LS27 8RU*
Phone *0532 523636*
Personnel *J Mone, Director*
          D Mone, Director
Quarries *Blackhill, Rock Cottage*

MONE BROS (EXCAVATIONS) LTD

BLACKHILL QUARRY   376
King Road, Bramhope, Leeds, W Yorks LS16 9JW
Phone 0532 674386   **Telex** 51564
OS Map 104   **OS Grid Reference** SE270422
Personnel M Gowland, Unit Manager

PRODUCT DATA	**Rock type**	Carboniferous Millstone Grit
	Colour	Orangey brown
	Grain	Medium to coarse
	Product	Dressed or Lump Stone

DRILLING DATA		Hand-held equipment

MONE BROS (EXCAVATIONS) LTD

ROCK COTTAGE QUARRY   377
Wormald Green, Harrogate, N Yorks HG3 3NJ
Phone 0765 87271
OS Map 99   **OS Grid Reference** SE300649
Personnel M Gowland, Unit Manager   *entry continues overleaf*

PRODUCT DATA	**Rock type**	Permian Lower Magnesian Limestone
	Colour	Creamy white
	Grain	Medium
	Product	Dressed or Lump Stone

RIPPABLE STRATA

MONTAGUE ESTATE LTD
Head Office *Barn Close House, Itchen Abbas, Winchester, Hants SO21 1BJ*
Phone *096278 464*
Personnel *D E Phelps, Owner*
Quarries *Ham Hill*

MONTAGUE ESTATE LTD

HAM HILL QUARRY   378

Ham Hill, Stoke-sub-Hamdon, Yeovil, Somerset TA14 6RW
Phone 096278 464
OS Map 183   **OS Grid Reference** ST477173
Personnel R Harvey, Unit Manager

PRODUCT DATA	**Rock type**	Jurassic Oolitic Limestone
	Colour	Honey brown
	Grain	Fine to coarse
	Product	Dressed or Lump Stone

| DRILLING DATA | | Hole Diameter 38–40mm |
| | | Hand-held equipment |

| COMPRESSORS | | Ingersoll-Rand |

| LOAD AND HAUL | **Loaders** | JCB 3CX |

MORAY STONE CUTTERS
Head Office *Spynie Quarry, Birnie, Elgin, Grampian IV30 3SW*
Phone *034386 244*
Personnel *A S Baillie, Managing Director*
Quarries *Clashach, Spynie*

MORAY STONE CUTTERS

CLASHACH QUARRY   379

Hopeman, Nr Elgin, Grampian IV30 2NF
Phone 034386 244
OS Map 28   **OS Grid Reference** NJ162701
Personnel A S Baillie, Managing Director and Unit Manager

PRODUCT DATA	**Rock type**	Triassic Sandstone
	Colour	Yellow to pale brown
	Grain	Fine
	Product	Dressed or Lump Stone

| DRILLING DATA | | Hand-held equipment |

MORAY STONE CUTTERS

SPYNIE QUARRY

380

Birnie, Elgin, Grampian IV30 3SW
Phone 034386 244
OS Map 28　**OS Grid Reference** NJ223657
Personnel A S Baillie, Managing Director and Unit Manager

PRODUCT DATA	**Rock type**	Triassic Spynie Sandstone
	Colour	Cream
	Grain	Fine
	Product	Dressed or Lump Stone

DRILLING DATA		Hand-held equipment

D H MORGAN AND SONS
Head Office *Victoria, Roch, Haverfordwest, Dyfed SA62 6JU*
Phone *0437 710259*
Personnel *K Morgan Scourfield, Managing Director*
Quarries *Rhyndaston*

D H MORGAN AND SONS

RHYNDASTON QUARRY

381

Roch, Haverfordwest, Dyfed SA62 5PT
Phone 0437 701361
OS Map 159　**OS Grid Reference** SM890240
Personnel K Morgan Scourfield, Managing Director and Unit Manager

PRODUCT DATA	**Rock type**	Rhyolite
	Colour	Light brown
	Grain	Fine to very fine
	Product	Crushed Stone

DRILLING DATA		Hole Diameter 110mm
		Bench Height 12m
		Contractor A Jones Rock Drillers

SECONDARY BREAKING		Hydraulic Hammer

LOAD AND HAUL	**Trucks**	Foden
		Euclid
	Loaders	Zettelmeyer

MORRIS AND PERRY (GURNEY SLADE QUARRIES)
Head Office *Gurney Slade Quarry, Bath, Avon BA3 4TE*
Phone *0749 840441*
Personnel *G H Perry, Director*
Quarries *Gurney Slade*

MORRIS AND PERRY (GURNEY SLADE QUARRIES)

GURNEY SLADE QUARRY

382

Bath, Avon BA3 4TE
Phone 0749 840441
OS Map 172　**OS Grid Reference** ST626497
Personnel D Roberts, Unit Manager

entry continues overleaf

PRODUCT DATA	**Rock type**	Carboniferous Limestone
	Product	Crushed Stone
DRILLING DATA		Hole Diameter 108mm
		Bench Height 9m
	Drills	Halco SPD43-27
COMPRESSORS		CompAir 37HP
SECONDARY BREAKING		Hydraulic Hammer
LOAD AND HAUL	**Face shovel**	Hitachi
	Trucks	Terex
		Foden
	Loaders	Komatsu
CRUSHING PLANT	**Primary**	Pegson 60 × 48 Jaw
	Secondary	Hazemag Impactor
	Tertiary	Barmac Duopactor

MORRISON CONSTRUCTION LTD
Head Office *Shandwick House, Chapel Street, Tain, Highland IV19 1JF*
Phone *0542 22273*
Personnel *I Mackie, Director*
Quarries *Cairds Hill*

MORRISON CONSTRUCTION LTD

CAIRDS HILL QUARRY 383

Keith, Grampian AY5 3PB
Phone 0542 22273
OS Map 28 **OS Grid Reference** NJ425475
Personnel R Chambers, Unit Manager

PRODUCT DATA	**Rock type**	Basalt
	Colour	Greyish black
	Grain	Medium
	Product	Crushed Stone
DRILLING DATA		Hole Diameter 127mm
		Bench Height 21m
		Contractor Rockblast
LOAD AND HAUL	**Face shovel**	O&K
CRUSHING PLANT	**Primary**	Kue Ken
	Secondary	None
	Tertiary	Allis Chalmers

JOHN MOWLEM AND COMPANY plc
Head Office *Foundation House, Eastern Road, Bracknell, Berks, RG12 2UZ*
Phone *0344 426826* **Fax** *0344 485779* **Telex** *847476*
Personnel *W N Kenrick, Chief Executive*
Companies *Rattee and Kett Ltd* — see separate entry

MULCAIR CONTRACTING COMPANY
Head Office *Cibyn Industrial Estate, Caernarfon, Gwynedd LL55 2BB*
Phone *0286 4928*
Personnel *J Brooks, Director*
        R Jacob, Contracts Engineer
Quarries *Bryn Engan, Hengae*

MULCAIR CONTRACTING COMPANY

BRYN ENGAN QUARRY

384

Llanbedrgoch, Anglesey, Gwynedd LL76 8TZ
Phone 0248 714562
OS Map 115 **OS Grid Reference** SH507815
Personnel R Jacob, Contracts Engineer

PRODUCT DATA	**Rock type**	Carboniferous Limestone
	Colour	Not commercially significant
	Product	Crushed Stone

DRILLING DATA		Hole Diameter 105mm
		Bench Height 12-15m
		Contractor K&H Rockdrillers

| CRUSHING PLANT | **Primary** | Jaw |

MULCAIR CONTRACTING COMPANY

HENGAE QUARRY

385

Llangaffo, Anglesey, Gwynedd LL60 6NE
Phone 0248 714562
OS Map 115 **OS Grid Reference** SH440686
Personnel R Jacob, Contracts Engineer

| PRODUCT DATA | **Rock type** | Granite |
| | **Product** | Crushed Stone |

DRILLING DATA		Hole Diameter 105mm
		Bench Height 12-15m
		Contractor K&H Rockdrillers

| CRUSHING PLANT | **Primary** | Jaw |

MULTI-AGG LTD
Head Office *Shellingford Quarry, Stanford Road, Stanford in the Vale, Faringdon, Oxon SN7 8NE*
Phone *03677 8949/8986*
Personnel *C Puffett, Director*
Quarries *Shellingford*

MULTI-AGG LTD

SHELLINGFORD QUARRY

386

Stanford Road, Stanford in the Vale, Faringdon, Oxon SN7 8NE
Phone 03677 8949/8986
OS Map 151 **OS Grid Reference** SP350434
Personnel C Puffett, Director and Unit Manager

entry continues overleaf

PRODUCT DATA   **Rock type**   Jurassic Limestone
                      Product   Dressed or Lump Stone

RIPPABLE STRATA

NANHORON GRANITE QUARRY LTD

Head Office *Nanhoron Quarry, Pwllheli, Gwynedd LL53 8PR*
Phone *075883 221*
Personnel *R Jones, Managing Director*
           A Jones, Director
Quarries *Nanhoron*

NANHORON GRANITE QUARRY LTD

NANHORON QUARRY

387

Pwllheli, Gwynedd LL53 8PR
Phone 075883 221
OS Map 123   **OS Grid Reference** SH287330
Personnel J Thomas, Unit Manager

PRODUCT DATA	**Rock type**	Microgranite
	Colour	Greyish brown
	Grain	Fine
	Products	Dressed or Lump Stone
		Crushed Stone

DRILLING DATA		Hole Diameter 105-110mm
		Bench Height 24-30m
		Contractor K&H Rockdrillers

SECONDARY BREAKING   Hydraulic Hammer

| LOAD AND HAUL | **Face shovel** | O&K RH90 |
| | **Trucks** | Aveling Barford RD25 |

CRUSHING PLANT	**Primary**	Kue Ken 36X42 Jaw
	Secondary	Kue Ken 42 × 10 Jaw
	Tertiary	Kue Ken Jaw

W NANKIVELL AND SONS LTD

Head Office *Tor Down Quarry, St Breward, Bodmin, Cornwall PL30 4LZ*
Phone *0208 850216*
Personnel *W Nankivell, Managing Director*
Quarries *Tor Down*

W NANKIVELL AND SONS LTD

TOR DOWN QUARRY

388

St Breward, Bodmin, Cornwall PL30 4LZ
Phone 0208 850216
OS Map 200   **OS Grid Reference** SX093767
Personnel B Nottle, Unit Manager

PRODUCT DATA	**Rock type**	Bodmin Moor Granite
	Colour	Silver grey
	Grain	Coarse
	Product	Dressed or Lump Stone

| DRILLING DATA | | Hand-held equipment |

NANT NEWYDD QUARRY

Head Office *Nant Newydd Quarry, Brynteg, Anglesey, Gwynedd LL4 8RE*
Phone *0248 852644*
Personnel *W Owen, Director*
Quarries *Nant Newydd*

NANT NEWYDD QUARRY

NANT NEWYDD QUARRY   389

Brynteg, Anglesey, Gwynedd LL4 8RE
Phone 0248 852644
OS Map 115   **OS Grid Reference** SH481811
Personnel W Owen, Director and Unit Manager

| PRODUCT DATA | **Rock type** | Carboniferous Limestone |
| | **Product** | Dressed or Lump Stone |

RIPPABLE STRATA

| DRILLING DATA | | Hand-held equipment |

| LOAD AND HAUL | **Trucks** | Aveling Barford |
| | **Loaders** | Caterpillar |

NASH ROCKS STONE AND LIME COMPANY

Head Office *P O Box 1, Kington, Hereford & Worcs HR5 3LQ*
Phone *0544 230711*   **Fax** *0544 231406*
Personnel *H Ellam, Managing Director*
     R Lister, Director
Quarries *Dolyhir and Strinds, Tonfanau*

NASH ROCKS STONE AND LIME COMPANY

DOLYHIR AND STRINDS QUARRY   390

Nash, Presteigne, Powys LD8 2RW
Phone 0544 230711
OS Map 148   **OS Grid Reference** SO301623
Personnel J Sinclair, Unit Manager

PRODUCT DATA	**Rock type**	Silurian Limestone and Greywacke
	Colour	Mottled dark and light grey
	Grain	Fine
	Products	Crushed Stone
		Asphalt or Tarmacadam
		Ready-mixed Concrete
		Industrial Limestone

DRILLING DATA		Hole Diameter 105mm
		Bench Height 12.5m
		Contractor W C D Sleeman

entry continues overleaf

SECONDARY BREAKING		Dropball on Ruston-Bucyrus 22-RB
LOAD AND HAUL	**Face shovel**	Ruston-Bucyrus
	Trucks	Volvo
		Aveling Barford
	Loaders	Volvo
CRUSHING PLANT	**Primary**	Baxter RB36 × 30 Jaw
	Secondary	Hazemag Impactor
	Tertiary	Pegson Autocone

NASH ROCKS STONE AND LIME COMPANY
TONFANAU QUARRY
391

Tywyn, Gwynedd LL36 9LP
Phone 0654 710240
OS Map 135 **OS Grid Reference** SH030571
Personnel R Hollyoak, Unit Manager

PRODUCT DATA	**Rock type**	Carboniferous Limestone
	Products	Crushed Stone
		Asphalt or Tarmacadam
		Ready-mixed Concrete
DRILLING DATA		Contractor
		No drilling information provided
CRUSHING PLANT	**Primary**	Baxter RB36 × 30 Jaw
	Secondary	Hazemag Impactor
	Tertiary	Pegson Autocone

NATURAL STONE PRODUCTS LTD (TARMAC BUILDING MATERIALS LTD)
Head Office *Bellmoor, Retford, Notts DN22 8SG*
Phone *0777 708771*
Personnel *M Poland, Director*
Quarries *Blaxter, Darney, Doddington, Hantergantick, High Nick, Merivale, Springwell*

NATURAL STONE PRODUCTS LTD (TARMAC BUILDING MATERIALS LTD)
BLAXTER QUARRY
392

Elsdon, Nr Otterburn, Northumb NE19 1BN
Phone 0777 708771
OS Map 80 **OS Grid Reference** NY932900

QUARRY ON STAND

ROCK DATA	**Rock type**	Lower Carboniferous Sandstone
	Colour	Cream
	Grain	Fine to medium

NATURAL STONE PRODUCTS LTD (TARMAC BUILDING MATERIALS LTD)
DARNEY QUARRY
393

West Woodburn, Hexham, Northumb NE23 4SY
Phone 0830 234
OS Map 80 **OS Grid Reference** NY909871

QUARRY ON STAND

ROCK DATA	**Rock type**	Lower Carboniferous Sandstone
	Colour	White to cream
	Grain	Fine

NATURAL STONE PRODUCTS LTD (TARMAC BUILDING MATERIALS LTD)

DODDINGTON HILL QUARRY

394

Doddington, Wooler, Northumb NE71 6AN
Phone 091 4877842
OS Map 67 **OS Grid Reference** NU008327
Personnel R Forder, Unit Manager

PRODUCT DATA	**Rock type**	Carboniferous Fell Sandstone
	Colour	Pink, cream
	Grain	Fine to medium
	Product	Dressed or Lump Stone

DRILLING DATA		Hand-held equipment

NATURAL STONE PRODUCTS LTD (TARMAC BUILDING MATERIALS LTD)

HANTERGANTICK QUARRY

395

St Breward, Bodmin, Cornwall PL30 4NH
Phone 0208 850483
OS Map 200 **OS Grid Reference** SX103757
Personnel D G Champion, Unit Manager

PRODUCT DATA	**Rock type**	Bodmin Moor Granite
	Colour	Silver grey
	Grain	Fine
	Product	Dressed or Lump Stone

DRILLING DATA		Hole Diameter 45-75mm
		Bench Height 3m
	Drills	Atlas Copco BVB14

COMPRESSORS		Atlas Copco

LOAD AND HAUL	**Trucks**	Aveling Barford RD25
	Loaders	Komatsu

CRUSHING PLANT		None

NATURAL STONE PRODUCTS LTD (TARMAC BUILDING MATERIALS LTD)

HIGH NICK QUARRY

396

Otterburn, Northumb BD23 4SY
Phone 0777 708771
OS Map 80 **OS Grid Reference** NY931873

QUARRY ON STAND

ROCK DATA	**Rock type**	Lower Carboniferous Sandstone
	Colour	Cream
	Grain	Medium

NATURAL STONE PRODUCTS LTD (TARMAC BUILDING MATERIALS LTD)

MERIVALE QUARRY

397

Tavistock, Devon PL20 6NS
Phone 0822 89227
OS Map 191 **OS Grid Reference** SX490756
Personnel R Cooper, Unit Manager

PRODUCT DATA	**Rock type**	Granite
	Colour	Light grey
	Grain	Coarse
	Product	Dressed or Lump Stone
DRILLING DATA		Hole Diameter 75mm
		Bench Height 2–3m
	Drills	Atlas Copco BVB14 Wagon Drill
COMPRESSORS		Atlas Copco
LOAD AND HAUL	**Trucks**	Aveling Barford RD25
CRUSHING PLANT		None

NATURAL STONE PRODUCTS LTD (TARMAC BUILDING MATERIALS LTD)

SPRINGWELL QUARRY

398

Gateshead, Tyne & Wear NE9 7SQ
Phone 0914 877842
OS Map 88 **OS Grid Reference** NZ294586
Personnel P E Longstaff, General Manager

PRODUCT DATA	**Rock type**	Carboniferous Middle Coal Measures Sandstone
	Colour	Fawn
	Grain	Fine to medium
	Product	Dressed or Lump Stone

NORTH WEST AGGREGATES LTD (RMC ROADSTONE PRODUCTS LTD)
Head Office *North West House, Spring Street, Widnes, Cheshire WA8 0NG*
*Phone 051 4236699 **Fax** 051 4951667 **Telex** 628065*
Personnel P J Owen, Managing Director
    G R C Smith, Divisional Director
    J A Beerli, Director and General Manager
    A Armstrong, Production Manager
Quarries Eldon Hill, Halkyn (Pant-y-Pwll Dwr), Raynes

NORTH WEST AGGREGATES LTD (RMC ROADSTONE PRODUCTS LTD)

ELDON HILL QUARRY

399

Sparrowpit, Chapel-en-le-Frith, Derbys SK17 8ER
Phone 0298 812361
OS Map 110 **OS Grid Reference** SK112814
Personnel J Shilcock, Unit Manager

PRODUCT DATA	**Rock type**	Carboniferous Limestone
	Colour	Light grey
	Products	Crushed Stone
		Asphalt or Tarmacadam
		Ready-mixed Concrete
		Industrial Limestone
		Agricultural Lime
DRILLING DATA		Hole Diameter 125mm
		Bench Height 15m
		Contractor A Jones Rock Drillers
SECONDARY BREAKING		Hydraulic Hammer
LOAD AND HAUL	**Trucks**	Aveling Barford RD30
		Aveling Barford 40
		(2)Terex
	Loaders	Michigan
CRUSHING PLANT	**Primary**	Pegson 42 × 48 Jaw
	Secondary	Mansfield No 3 Hammermill
		Mansfield No 4 Hammermill
		Mansfield No 5 Hammermill
	Tertiary	None

NORTH WEST AGGREGATES LTD (RMC ROADSTONE PRODUCTS LTD)

HALKYN QUARRY (PANT-Y-PWLL DWR QUARRY)

400

Pentre Halkyn, Holywell, Clwyd CH8 8HP
Phone 0352 780651
OS Map 116   **OS Grid Reference** SJ190720
Personnel M Wright, Unit Manager

PRODUCT DATA	**Rock type**	Carboniferous Limestone
	Colour	Light creamy grey
	Grain	Fine
	Products	Crushed Stone
		Asphalt or Tarmacadam
		Industrial Limestone
		Agricultural Lime
DRILLING DATA		Hole Diameter 105mm
		Bench Height 15-18m
	Drills	(2)Holman Voltrak
COMPRESSORS		CompAir 650HE
SECONDARY BREAKING		Hydraulic Hammer - Furukawa
LOAD AND HAUL	**Trucks**	Terex
	Loaders	Komatsu WA600
CRUSHING PLANT	**Primary**	Goodwin Barsby RB36 × 30 Jaw
	Secondary	Pegson 36 × 30 Jaw
		Mansfield No 5
		Hammermill
	Tertiary	Mansfield No 3
		Hammermill

NORTH WEST AGGREGATES LTD (RMC ROADSTONE PRODUCTS LTD)

RAYNES QUARRY

401

Abergele Road, Colwyn Bay, Clwyd LL29 9YN
Phone 0492 517564
OS Map 116   **OS Grid Reference** SH890780
Personnel K Andrews, Unit Manager

PRODUCT DATA	**Rock type**	Carboniferous Limestone
	Colour	Pink
	Grain	Fine
	Product	Crushed Stone

DRILLING DATA		Hole Diameter 105mm
		Bench Height 20m
	Drills	(2)Holman Voltrak

| COMPRESSORS | CompAir 650HE |

| SECONDARY BREAKING | Hydraulic Hammer - Montabert on JCB |

LOAD AND HAUL	**Trucks**	(3)Heathfield
	Loaders	Caterpillar
		Komatsu WA600

| CRUSHING PLANT | No information provided |

ARCHIBALD NOTT AND SONS LTD

Head Office *Tinto House, Station Road, South Molton, Devon EX36 3LL*
Phone *07695 2763*
Personnel *C Nott, Managing Director*
S Nott, Director
K J Nott, Director
Quarries *Bray Valley, Plaistow*

ARCHIBALD NOTT AND SONS LTD

BRAY VALLEY QUARRY

402

Brayford, Barnstable, Devon EX32 7QB
Phone 0598 710421
OS Map 180 **OS Grid Reference** SS687338
Personnel S Nott, Director and Unit Manager

PRODUCT DATA	**Rock type**	Devonian Sandstone
	Colour	Not Commercially Significant
	Grain	Medium
	Products	Crushed Stone
		Asphalt or Tarmacadam

DRILLING DATA		Hole Diameter 105mm
		Bench Height 12m
		Contractor Celtic Rock Services

| SECONDARY BREAKING | Dropball on Ruston-Bucyrus 22-RB |

| LOAD AND HAUL | **Trucks** | (2)Deutz 15t |
| | **Loaders** | Terex 7271B |

| CRUSHING PLANT | Babbitless |

ARCHIBALD NOTT AND SONS LTD

PLAISTOW QUARRY

403

Shirewell, Barnstable, Devon EX3l 4EX
Phone 027 182382
OS Map 180   **OS Grid Reference** SS568373
Personnel T Barrowcliffe, Unit Manager

PRODUCT DATA	**Rock type**	Devonian Sandstone
	Colour	Not Commercially Significant
	Grain	Medium
	Product	Crushed Stone

DRILLING DATA		Hole Diameter 105mm
		Bench Height 12m
	Drills	Halco 400C

| COMPRESSORS | | CompAir 650HE |

| SECONDARY BREAKING | | Dropball on Ruston-Bucyrus 22-RB |

| LOAD AND HAUL | **Trucks** | Deutz |
| | **Loaders** | Terex |

CRUSHING PLANT	**Primary**	Goodwin Barsby 36 × 24 Jaw (Mobile)
	Secondary	Goodwin Barsby 20 × 10 Jaw (Single Toggle)
	Tertiary	None

L OATES
Head Office *Squire Hill Quarry, Southowram, Halifax, W Yorks HX3 9SX*
Phone *0484 715407*
Personnel *L Oates, Owner*
Quarries *Squire Hill*

L OATES

SQUIRE HILL QUARRY

404

Southowram, Halifax, W Yorks HX3 9SX
Phone 0484 715407
OS Map 104   **OS Grid Reference** SE133233
Personnel L Oates, Owner and Unit Manager

PRODUCT DATA	**Rock type**	Carboniferous Sandstone
	Colour	Buff, cream
	Grain	Medium
	Product	Dressed or Lump Stone

| DRILLING DATA | | Hand-held equipment |

OGDEN ROADSTONE LTD (EVERED HOLDINGS plc)
Head Office *St Clair Road, Otley, W Yorks LS21 1HX*
Phone *0943 464531* **Fax** *0943 466044* **Telex** *51189*
Personnel *J Ogden, Managing Director*
    J Cresswell, General Manager
Quarries *Ivonbrook, Peckfield*

entry continues overleaf

OGDEN ROADSTONE LTD (EVERED HOLDINGS plc)

IVONBROOK QUARRY

405

Grange Mill, Wirksworth, Derbys DE4 4HY
Phone 062988 275
OS Map 119   **OS Grid Reference** SK241573
Personnel F Fanton, Unit Manager

PRODUCT DATA	**Rock type**	Carboniferous Limestone
	Products	Crushed Stone
		Asphalt or Tarmacadam
DRILLING DATA		Hole Diameter 110mm
		Bench Height 18m
	Drills	Halco 410C
COMPRESSORS		CompAir 650HE
LOAD AND HAUL	**Trucks**	(3)Aveling Barford RD40
	Loaders	Caterpillar 966C
		Caterpillar 968
CRUSHING PLANT	**Primary**	Parker 50 × 36 Jaw (Single Toggle)
	Secondary	Parker 105 Impactor
	Tertiary	Kue Ken PA32 Impactor

OGDEN ROADSTONE LTD (EVERED HOLDINGS plc)

PECKFIELD QUARRY

406

Micklefield, Nr Leeds, W Yorks LS25 4AE
Phone 0532 861087
OS Map 109   **OS Grid Reference** SE460325
Personnel D Holmes, Unit Manager

PRODUCT DATA	**Rock type**	Permian Magnesian Limestone
	Products	Crushed Stone
		Agricultural Lime
DRILLING DATA		Hole Diameter 125mm
		Bench Height 17m
	Drills	Halco 410C
COMPRESSORS		CompAir 650HE
LOAD AND HAUL	**Trucks**	DJB 25
	Loaders	Caterpillar 966C
		Caterpillar 980C
CRUSHING PLANT	**Primary**	Parker 42 × 20 Jaw
	Secondary	Parker 24 × 10 Jaw
	Tertiary	None

J OLDHAM AND COMPANY (STONEMASONS) LTD

Head Office Tearne House, Hollington, Uttoxeter, Staffs ST10 4HR
Phone 088926 353/354/355   **Fax** 088926 212
Personnel J Oldham, Managing Director
Quarries Tearne, Wooton

J OLDHAM AND COMPANY (STONEMASONS) LTD

TEARNE QUARRY

407

Hollington, Uttoxeter, Staffs ST14 5JF
Phone 088926 353
OS Map 128   **OS Grid Reference** SK055390
Personnel M J Donnelly, Works Manager

PRODUCT DATA	**Rock type**	Triassic Sandstone
	Colour	Light reddish pink, mottled red
	Grain	Fine to medium
	Product	Dressed or Lump Stone

| DRILLING DATA | | Hand-held equipment |

J OLDHAM AND COMPANY (STONEMASONS) LTD

WOOTON QUARRY

408

Wooton, Rocester, Derbys DE6 2GW
Phone 088926 353
OS Map 119   **OS Grid Reference** SK114445
Personnel M J Donnelly, Works Manager

| PRODUCT DATA | **Rock type** | Triassic Sandstone |
| | **Product** | Dressed or Lump Stone |

| DRILLING DATA | | Hand-held equipment |

ORKNEY BUILDERS

Head Office *Heddle Hill Quarry, Finstown, Orkney KW17 2JW*
Phone *085676 511*
Personnel *L Firth, Managing Director*
Quarries *Heddle Hill*

ORKNEY BUILDERS

HEDDLE HILL QUARRY

409

Finstown, Orkney KW17 2JW
Phone 085676 511
OS Map 6   **OS Grid Reference** HY358132
Personnel J Breck, Unit Manager

PRODUCT DATA	**Rock type**	Devonian Sandstone
	Colour	Not commercially significant
	Grain	Fine to medium
	Product	Crushed Stone

DRILLING DATA		Hole Diameter 105mm
		Bench Height 17m
	Drills	Halco 400C

| COMPRESSORS | | Ingersoll-Rand |

| SECONDARY BREAKING | | Dropball |

| LOAD AND HAUL | **Trucks** | Caterpillar |
| | **Loaders** | Caterpillar |

entry continues overleaf

CRUSHING PLANT	Primary	Parker 36 × 24 Jaw
	Secondary	Parker Kubitizer
	Tertiary	None

ORKNEY ISLAND COUNCIL
Head Office Council Offices, Kirkwall, Orkney KW15 1NY
Phone 0856 3535
Personnel H L Cross, Director of Engineering
Quarries Curister

ORKNEY ISLAND COUNCIL
CURISTER QUARRY
Finstown, Orkney KW15 1TT
Phone 085676 295
OS Map 6 **OS Grid Reference** HY377126
Personnel W Brown, Unit Manager

410

PRODUCT DATA	Rock type	Devonian Sandstone
	Colour	Not commercially significant
	Grain	Medium
	Product	Crushed Stone

DRILLING DATA		Hole Diameter 105mm
		Bench Height 18-20m
	Drills	Holman Voltrak
		Halco 400C

| SECONDARY BREAKING | | Dropball |

| LOAD AND HAUL | Trucks | Caterpillar |
| | Loaders | Caterpillar |

CRUSHING PLANT	Primary	Kue Ken 36 × 42 Jaw
		Parker 18 × 32 Jaw (Mobile)
	Secondary	Hazemag Impactor

OTS HOLDINGS
Head Office Happylands Quarry, Broadway, Hereford & Worcs WR12 7WD
Phone 0386 853582
Personnel E Woodall, Managing Director
Quarries Happylands

OTS HOLDINGS
HAPPYLANDS QUARRY
Broadway, Hereford & Worcs WR12 7WD
Phone 0386 853582
OS Map 150 **OS Grid Reference** SP128359
Personnel J Lee, Unit Manager

411

PRODUCT DATA	Rock type	Jurassic Oolitic Limestone
	Colour	Yellow to buff
	Grain	Fine
	Products	Dressed or Lump Stone
		Crushed Stone

DRILLING DATA		Contractor occasionally
LOAD AND HAUL	**Trucks**	None
	Loaders	Caterpillar
		Volvo
		JCB
CRUSHING PLANT		None

HARRY PARKER AND SON
Head Office *Chellow Grange Quarry, Haworth Road, Bradford, W Yorks BD9 6NL*
Phone *0274 42439*
Personnel *H Parker, Owner*
Quarries *Chellow Grange*

HARRY PARKER AND SON

CHELLOW GRANGE QUARRY

412

Haworth Road, Bradford, W Yorks BD9 6NL
Phone 0274 42439
OS Map 104 **OS Grid Reference** SE115353
Personnel H Parker, Owner and Unit Manager

PRODUCT DATA	**Rock type**	Carboniferous Sandstone
	Colour	Buff to grey
	Grain	Fine
	Product	Dressed or Lump Stone
DRILLING DATA		Hand-held equipment

PATERSON OF GREENOAK HILL LTD
Head Office *Gartsherrie, Coatbridge, Strathclyde ML5 2EU*
Phone *0236 33351/33356*
Personnel *W Paterson, Director*
       J Sawyers, Office Manager
Quarries *Dunduff*

PATERSON OF GREENOAK HILL LTD

DUNDUFF QUARRY

413

Blackwood, Kirkmuirhill, Lanark, Strathclyde ML11 0JQ
Phone 0555 894134
OS Map 71 **OS Grid Reference** NS779410
Personnel J Thompson, Unit Manager

PRODUCT DATA	**Rock type**	Devonian Greywacke
	Colour	Not commercially significant
	Grain	Fine
	Product	Crushed Stone
DRILLING DATA		Hole Diameter 108mm
		Bench Height 18m
		Contractor Ritchies Equipment
SECONDARY BREAKING		Dropball on Ruston-Bucyrus 22-RB

entry continues overleaf

| LOAD AND HAUL | **Trucks** | Terex |
| | **Loaders** | Caterpillar |

CRUSHING PLANT — No information provided

E J PATERSON
Head Office *Woodside Quarry, Woodside, Auchlee, Longside, Grampian AB4 7UA*
Phone *077982 314*
Personnel *E J Paterson, Owner*
Quarries *Woodside*

E J PATERSON
WOODSIDE QUARRY   414
Woodside, Auchlee, Longside, Grampian AB4 7UA
Phone 077982 314
OS Map 30 **OS Grid Reference** NK050460
Personnel E J Paterson, Owner and Unit Manager

| PRODUCT DATA | **Rock type** | Pre-Cambrian Dalradian Limestone |
| | **Product** | Dressed or Lump Stone |

DRILLING DATA — Hand-held equipment

PAWSON BROS LTD
Head Office *Brittania Quarry, Howley Park, Morley, W Yorks LS27 0SW*
Phone *0532 530464*
Personnel *A Gascoine, Managing Director*
Quarries *Brittania*

PAWSON BROS LTD
BRITTANIA QUARRY   415
Howley Park, Morley, W Yorks LS27 0SW
Phone 0532 530464
OS Map 104 **OS Grid Reference** SE267263
Personnel O Ingham, Unit Manager

PRODUCT DATA	**Rock type**	Carboniferous Lower Coal Measures Sandstone
	Colour	Buff to light brown
	Grain	Fine
	Product	Dressed or Lump Stone

DRILLING DATA — Hand-held equipment

J D PEARCE QUARRY COMPANY
Head Office *Johnstone House, Church Street, Keinton Mandeville, Somerton, Somerset TA11 6ER*
Phone *045822 3326*
Personnel *J D Pearce, Owner*
Quarries *Peddles Lane*

J D PEARCE QUARRY COMPANY

PEDDLES LANE QUARRY

416

Charlton Mackrell, Somerton, Somerset TA11 7LE
Phone 045822 3326
OS Map 183   **OS Grid Reference** ST520280
Personnel J D Pearce, Owner and Unit Manager

PRODUCT DATA   **Rock type**   Jurassic Oolitic Limestone
                        Grain   Fine
                      Product   Dressed or Lump Stone

RIPPABLE STRATA

PENHILL QUARRY AND HAULAGE COMPANY LTD
Head Office *Jubilee House, Kilhampton, Bude, Cornwall EX23 9RS*
Phone *028882 326*
Personnel *R Pengilley, Director*
Quarries *Hertbury, Pigsden*

PENHILL QUARRY AND HAULAGE COMPANY LTD

HERTBURY QUARRY

417

Ivy Leaf Bush, Stratton, Bude, Cornwall EX23 9LQ
Phone 028882 326
OS Map 190   **OS Grid Reference** SS206066

QUARRY ON STAND

ROCK DATA   **Rock type**   Carboniferous Sandstone (Gritstone)

PENHILL QUARRY AND HAULAGE COMPANY LTD

PIGSDEN QUARRY

418

Launcells, Kilkhampton, Bude, Cornwall EX23 9RF
Phone 028882 489
OS Map 190   **OS Grid Reference** SS248115
Personnel R Pengilley, Director and Unit Manager

PRODUCT DATA   **Rock type**   Carboniferous Sandstone (Gritstone)
                      Product   Crushed Stone

DRILLING DATA   Contractor Saxton Co (Deep Drillers) occasionally

NO FURTHER INFORMATION PROVIDED

PENRYN GRANITE LTD (CHARTER CONSOLIDATED plc)
Head Office *15–17 Fairmantle St, Truro, Cornwall TR1 2EH*
Phone *0872 75626*
Personnel *J Haskins, Production Director*
          P Gerrish, General Manager
Quarries *Black Hill, Carnsew (Penryn), Castle-an-Dinas, Dairy, Kessel Downs Penlee*

entry continues overleaf

PENRYN GRANITE LTD (CHARTER CONSOLIDATED plc)

BLACK HILL QUARRY

419

Lewannick, Launceston, Cornwall PL15 7PZ
Phone 0566 86309
OS Map 201   **OS Grid Reference** SX267818
Personnel D Trice, Unit Manager

PRODUCT DATA   **Rock type**   Dolerite

QUARRY UNDER DEVELOPMENT

PENRYN GRANITE LTD (CHARTER CONSOLIDATED plc)

CARNSEW QUARRY (PENRYN QUARRY)

420

Mabe Burnthouse, Penryn, Cornwall TR10 9DH
Phone 0326 73851
OS Map 204   **OS Grid Reference** SW764355
Personnel D Toy, Unit Manager

PRODUCT DATA	**Rock type**	Carnmenellis Granite
	Colour	Silver grey
	Grain	Coarse
	Products	Dressed or Lump Stone
		Crushed Stone
		Asphalt or Tarmacadam

DRILLING DATA		Hole Diameter 89mm
		Bench Height 10–12m
	Drills	Ingersoll-Rand CM351

SECONDARY BREAKING		(2)Dropball on Ruston-Bucyrus 22-RB

LOAD AND HAUL	**Trucks**	Caterpillar
	Loaders	Caterpillar 966C
		Caterpillar 968

CRUSHING PLANT	**Primary**	Pegson 36 × 42 Jaw
	Secondary	Pegson 4′ Gyratory
	Tertiary	(2)Pegson 3′ Hydrocone

PENRYN GRANITE LTD (CHARTER CONSOLIDATED plc)

CASTLE-AN-DINAS QUARRY

421

Nancledra, Penzance, Cornwall TR20 8NA
Phone 0736 65155
OS Map 203   **OS Grid Reference** SW489341
Personnel W Gilbard, Unit Manager

PRODUCT DATA	**Rock type**	Lands End Granite
	Colour	Buff, light grey
	Grain	Coarse
	Product	Crushed Stone

DRILLING DATA		Hole Diameter 105mm
		Bench Height 20m
	Drills	Ingersoll-Rand CM351

SECONDARY BREAKING		Dropball on Ruston-Bucyrus 22-RB
LOAD AND HAUL	**Trucks**	Caterpillar
	Loaders	Caterpillar 966C
		Caterpillar 968
CRUSHING PLANT	**Primary**	Goodwin Barsby Goliath 42 × 24 Jaw (Mobile)
	Secondary	Symons Jaw
	Tertiary	Symons Cone

PENRYN GRANITE LTD (CHARTER CONSOLIDATED plc)

DAIRY QUARRY

422

Pentewan, St Austell, Cornwall PL26 6DQ
Phone 0872 75626
OS Map 204 **OS Grid Reference** SX017468

QUARRY ON STAND

ROCK DATA	**Rock type**	Carboniferous Sandstone (Gritstone)

PENRYN GRANITE LTD (CHARTER CONSOLIDATED plc)

KESSEL DOWNS QUARRY

423

Kessel Downs, Longdowns, Penryn, Cornwall TR10 9DH
Phone 0326 73851
OS Map 204 **OS Grid Reference** SW749327
Personnel D Toy, Unit Manager

PRODUCT DATA	**Rock type**	Carnmenellis Granite
	Grain	Coarse
	Products	Crushed Stone
		Ready-mixed Concrete
DRILLING DATA		Hole Diameter 105mm
		Bench Height 20m
	Drills	Ingersoll-Rand CM351
SECONDARY BREAKING		Dropball on Ruston-Bucyrus 38-RB
		Dropball on Smith 21
LOAD AND HAUL	**Trucks**	Caterpillar
	Loaders	Caterpillar 966C
		Caterpillar 968
CRUSHING PLANT	**Primary**	Kue Ken 36 × 42 Jaw (Double Toggle)
	Secondary	Symons 3' Cone
	Tertiary	Symons Cone

PENRYN GRANITE LTD (CHARTER CONSOLIDATED plc)

PENLEE QUARRY

424

Newlyn, Penzance, Cornwall TR18 5AA
Phone 0736 62378
OS Map 203 **OS Grid Reference** SW469279
Personnel D Cook, Unit Manager

entry continues overleaf

PRODUCT DATA	**Rock type**	Hornfels Diabase
	Colour	Light speckled grey
	Grain	Medium to fine
	Products	Crushed Stone
		Asphalt or Tarmacadam
		Ready-mixed Concrete
DRILLING DATA		Hole Diameter 110mm
		Bench Height 20–30m
	Drills	Holman Voltrak
SECONDARY BREAKING		Dropball on Ruston-Bucyrus 22-RB
LOAD AND HAUL	**Trucks**	(3)Caterpillar
	Loaders	Caterpillar 966C
		Caterpillar 968
CRUSHING PLANT	**Primary**	Kue Ken 36 × 24 Jaw (Double Toggle)
	Secondary	Allis Chalmers Cone
	Tertiary	(2)Allis Chalmers 3′ Cone

PETERBOROUGH QUARRIES LTD (EVERED HOLDINGS plc)
Head Office *West End, Maxey, Peterborough, Cambs PE6 9HA*
Phone *0778 342355*
Personnel *A Crowson, Managing Director and Production Director*
Quarries *Crossleys, Greetham, Thornaugh*

PETERBOROUGH QUARRIES LTD (EVERED HOLDINGS plc)
CROSSLEYS QUARRY 425
Stamford, Lincs PE9 9ZZ
Phone 0780 783170
OS Map 141 **OS Grid Reference** TF032048

QUARRY ON STAND

| ROCK DATA | **Rock type** | Jurassic Oolitic Limestone |

PETERBOROUGH QUARRIES LTD (EVERED HOLDINGS plc)
GREETHAM QUARRY 426
Ketton, Stamford, Lincs PE9 9ZZ
Phone 0778 342355
OS Map 141 **OS Grid Reference** SK933147
Personnel L Townsend, Unit Manager

PRODUCT DATA	**Rock type**	Jurassic Oolitic Limestone
	Grain	Fine
	Product	Crushed Stone
DRILLING DATA		Hole Diameter 105mm
		Bench Height 12m
	Drills	Ingersoll-Rand CM351
COMPRESSORS		Ingersoll-Rand

LOAD AND HAUL	**Trucks**	(2)Aveling Barford RD40
	Loaders	Aveling Barford WXL113
		Volvo 860

| CRUSHING PLANT | **Primary** | (2)Goodwin Barsby 24 × 10 Jaw (Mobile) |
| | **Secondary** | (2)Lightning Hammermill |

PETERBOROUGH QUARRIES LTD (EVERED HOLDINGS plc)

THORNAUGH QUARRY 427

Stamford, Lincs PE8 6NJ
Phone 0780 782210
OS Map 141   **OS Grid Reference** TL110035
Personnel L Townsend, Unit Manager

PRODUCT DATA	**Rock type**	Jurassic Oolitic Limestone
	Colour	Cream
	Grain	Fine
	Product	Crushed Stone

NO FURTHER INFORMATION PROVIDED

PHILPOTS QUARRIES LTD
Head Office *West Hoathly,East Grinstead, W Sussex RH19 4PS*
Phone *0342 810428*
Personnel *L A Hannah, Managing Director*
Quarries *Philpots*

PHILPOTS QUARRIES LTD

PHILPOTS QUARRY 428

West Hoathly, East Grinstead, W Sussex RH19 4PS
Phone 0342 810428
OS Map 187   **OS Grid Reference** TQ353323
Personnel L A Hannah, Managing Director and Unit Manager

PRODUCT DATA	**Rock type**	Cretaceous Wealden Sandstone
	Colour	Honey brown
	Grain	Medium
	Product	Dressed or Lump Stone

| DRILLING DATA | | Hand-held equipment |

PERCY PICKARD (SALES) LTD
Head Office *Fagley Quarries, Fagley Lane, Eccleshill, Bradford, W Yorks BD2 3NT*
Phone *0274 637307* **Fax** *0274 626146*
Personnel *L Bodill, Director*
        P Gilmore, Sales Manager
Quarries *Bolton Woods, Deep Lane, Fagley*

PERCY PICKARD (SALES) LTD

BOLTON WOODS QUARRY 429

Bolton Hall Road, Bradford, W Yorks BD2 1BQ
Phone 0274 595534
OS Map 104   **OS Grid Reference** SE163363
Personnel Names not provided

entry continues overleaf

PRODUCT DATA	**Rock type**	Carboniferous Sandstone (Gritstone)
	Colour	Dark buff
	Grain	Fine
	Product	Dressed or Lump Stone

| DRILLING DATA | | Hand-held equipment |

PERCY PICKARD (SALES) LTD

DEEP LANE QUARRY

430

Deep Lane, Bradford, W Yorks BD2 1BQ
Phone 0274 637307
OS Map 104 **OS Grid Reference** SE120329

QUARRY ON STAND

ROCK DATA	**Rock type**	Carboniferous Sandstone (Gritstone)
	Colour	Dark buff
	Grain	Fine

PERCY PICKARD (SALES) LTD

FAGLEY QUARRY

431

Fagley Lane, Eccleshill, Bradford, W Yorks BD2 3NT
Phone 0274 637307
OS Map 104 **OS Grid Reference** SE179360
Personnel Names not provided

PRODUCT DATA	**Rock type**	Carboniferous Sandstone (Hard York Gritstone)
	Colour	Dark buff
	Grain	Fine
	Product	Dressed or Lump Stone

| DRILLING DATA | | Hand-held equipment |

PILKINGTON QUARRIES LTD
Head Office *The Dolomite Quarry, Warmsworth, Doncaster, S Yorks DN14 9RG*
Phone *0302 853092/853354* **Fax** *0302 310259*
Personnel *P A Littlewood, Director*
Quarries *The Dolomite*

PILKINGTON QUARRIES LTD

THE DOLOMITE QUARRY

432

Warmsworth, Doncaster, S Yorks DN14 9RG
Phone 0302 853092/853354
OS Map 111 **OS Grid Reference** SE537006
Personnel G L Blount, Unit Manager

PRODUCT DATA	**Rock type**	Permian Magnesian Limestone
	Colour	Cream with dark spots of iron oxide
	Grain	Coarse
	Products	Crushed Stone
		Asphalt or Tarmacadam
		Industrial Limestone
		Agricultural Lime

DRILLING DATA		Hole Diameter 105–115mm Bench Height 13,21,34m Contractor R&B Rock Drillers
SECONDARY BREAKING		Hydraulic Hammer - Gullick Dropball on Ruston-Bucyrus 22-RB
LOAD AND HAUL	**Face shovel**	Ruston-Bucyrus
	Trucks	Terex R25
CRUSHING PLANT	**Primary**	Pegson 36 × 24 Jaw (Single Toggle)
	Secondary	BJD Pickbreaker
	Tertiary	BJD Swinghammer
		BJD Hammermill

PIONEER AGGREGATES (UK) LTD
Corporate Head Office 59-60 Northolt Road, South Harrow, Middx HA2 0EY
Phone 081 423 3066 **Fax** 081 423 3845 **Telex** 928153
Personnel F S C Manson, Managing Director
M Ogden, Regional Director - North
P Quinn, Regional Director - South
T Pittams, Production Manager - North
D Wanklyn, Production Manager - South
Quarries Aberduna, Coldstones, Forest Wood, Gelligaer (Pencaemawr), Griff, Grove, Longwood, Syke, Tam's Loup, Westbury

PIONEER AGGREGATES (UK) LTD
ABERDUNA QUARRY   **433**
Maeshafn, Mold, Clwyd CH7 5LE
Phone 0352 85591
OS Map 117 **OS Grid Reference** SJ204617
Personnel D Mills, Unit Manager

PRODUCT DATA	**Rock type**	Carboniferous Limestone
	Colour	Light creamy grey
	Products	Crushed Stone
		Asphalt or Tarmacadam
DRILLING DATA		Hole Diameter 105mm Bench Height 18m Contractor A Jones Rock Drillers
SECONDARY BREAKING		Dropball on Ruston-Bucyrus 22-RB
LOAD AND HAUL	**Trucks**	Aveling Barford RD30
		Terex 30t
	Loaders	Liebherr
CRUSHING PLANT		No information provided

PIONEER AGGREGATES (UK) LTD
COLDSTONES QUARRY   **434**
Greenhow Hill, Pateley Bridge, Harrogate, N Yorks HG3 5JQ
Phone 0423 711356
OS Map 99 **OS Grid Reference** SE126642
Personnel R Orange, Unit Manager

entry continues overleaf

PRODUCT DATA	**Rock type**	Carboniferous Limestone
	Colour	Light to medium grey
	Products	Crushed Stone
		Asphalt or Tarmacadam
DRILLING DATA		Hole Diameter 108mm
		Bench Height 12–25m
	Drills	Drillmark Rig
SECONDARY BREAKING		Dropball on Ruston-Bucyrus 30-RB
LOAD AND HAUL	**Trucks**	(2)Pellini 40t
	Loaders	Poclain 350
CRUSHING PLANT		No information provided

PIONEER AGGREGATES (UK) LTD

FOREST WOOD QUARRY 435

Pontyclun, Mid Glam CF7 9XD
Phone 0443 225000
OS Map 170   **OS Grid Reference** ST016797
Personnel A Talling, Unit Manager

PRODUCT DATA	**Rock type**	Carboniferous Limestone
	Colour	Brownish grey
	Grain	Medium
	Products	Dressed or Lump Stone
		Crushed Stone
		Asphalt or Tarmacadam
DRILLING DATA		Hole Diameter 110mm
		Bench Height 18m
		Contractor Drill Quip UK
LOAD AND HAUL	**Trucks**	Aveling Barford RD30
		(2)Aveling Barford RD40
		Terex
	Loaders	Caterpillar 950
		Caterpillar 966C
		Caterpillar 980C
		Terex
CRUSHING PLANT	**Primary**	BJD 55 Rotary
	Secondary	Hazemag APK50 Impactor

PIONEER AGGREGATES (UK) LTD

GELLIGAER QUARRY (PENCAEMAWR QUARRY) 436

Gelligaer Common, Treharris, Trelewis, Mid Glam CS46 6TA
Phone 0443 410886
OS Map 171   **OS Grid Reference** ST117995
Personnel P Boniface, Unit Manager

PRODUCT DATA	**Rock type**	Carboniferous Pennant Sandstone
	Colour	Dark grey, blue, rustic brown
	Grain	Medium
	Products	Dressed or Lump Stone
		Crushed Stone

DRILLING DATA		Hole Diameter 110mm Bench Height 15m Contractor Drill Quip UK
LOAD AND HAUL	**Trucks** **Loaders**	(2)Aveling Barford (2)Caterpillar 966C
CRUSHING PLANT	**Primary** **Secondary**	Kue Ken (Mobile) Kue Ken (Mobile)

PIONEER AGGREGATES (UK) LTD

GRIFF QUARRY   **437**

Gypsy Lane, Nuneaton, Warwicks CV10 7PJ
Phone 0203 345777
OS Map 140   **OS Grid Reference** SP362889
Personnel R Bush, Unit Manager

PRODUCT DATA	**Rock type** **Colour** **Grain** **Products**	Diorite Light to medium grey Coarse Crushed Stone Asphalt or Tarmacadam Ready-mixed Concrete
DRILLING DATA		Hole Diameter 105mm Bench Height 15m Contractor S & R
SECONDARY BREAKING		Dropball on Ruston-Bucyrus 22-RB
LOAD AND HAUL	**Face shovel** **Trucks** **Loaders**	Poclain 350CK (3)Aveling Barford RD40 Caterpillar 950 Caterpillar 966C Caterpillar 980C
CRUSHING PLANT	**Primary** **Secondary** **Tertiary**	Pegson Jaw Pegson Cone Pegson Gyratory

---

PIONEER AGGREGATES (UK) LTD

GROVE QUARRY   **438**

South Cornelly, Nr Pyle, Mid Glam CF33 4RP
Phone 0656 740522
OS Map 170   **OS Grid Reference** SS822799
Personnel C Wood, Unit Manager

PRODUCT DATA	**Rock type** **Colour** **Grain** **Product**	Carboniferous Limestone Medium brownish grey Medium Crushed Stone
DRILLING DATA		Hole Diameter 105mm Bench Height 15m Contractor Drill Quip UK

entry continues overleaf

SECONDARY BREAKING		Dropball on Ruston-Bucyrus 22-RB
LOAD AND HAUL	**Trucks**	Terex
	Loaders	Terex
CRUSHING PLANT	**Primary**	Baxter Jaw
	Secondary	BJD Hammermill

PIONEER AGGREGATES (UK) LTD

LONGWOOD QUARRY

439

Long Ashton, Bristol, Avon BS18 9DW
Phone 0272 392471
OS Map 172 **OS Grid Reference** ST537714
Personnel D Howarth, Unit Manager

PRODUCT DATA	**Rock type**	Carboniferous Limestone
	Colour	Greyish brown
	Products	Crushed Stone
		Ready-mixed Concrete
DRILLING DATA		Hole Diameter 105–125mm
		Bench Height 12–18m
	Drills	Holman Voltrak
COMPRESSORS		CompAir 650HE
SECONDARY BREAKING		Hydraulic Hammer
LOAD AND HAUL	**Trucks**	(2)Terex 3305B
	Loaders	Caterpillar 980C
CRUSHING PLANT	**Primary**	Baioni Impactor
	Secondary	Baioni Impactor
	Tertiary	Baioni Impactor

PIONEER AGGREGATES (UK) LTD

SYKE QUARRY

440

Walwyns Castle, Haverfordwest, Dyfed SA62 3DZ
Phone 0437 781261
OS Map 158 **OS Grid Reference** SM872106
Personnel G Banner, Unit Manager

PRODUCT DATA	**Rock type**	Silurian Sandstone
	Colour	Dark blue, grey, brown
	Grain	Medium
	Product	Crushed Stone
DRILLING DATA		Hole Diameter 105–110mm
		Bench Height 15–20m
		Contractor A Jones Rock Drillers
LOAD AND HAUL	**Trucks**	Aveling Barford RD30
		Aveling Barford RD35
	Loaders	Caterpillar 950
		Terex 7261

CRUSHING PLANT	**Primary**	Pegson 48×27 Jaw (Single Toggle)
	Secondary	Pegson 900 Autocone
	Tertiary	Pegson 2′ Gyrasphere

PIONEER AGGREGATES (UK) LTD

TAM'S LOUP QUARRY

441

Harthill, Strathclyde ML7 5TN
Phone 0501 51325
OS Map 65   **OS Grid Reference** NS885640
Personnel K Walton, Unit Manager

PRODUCT DATA	**Rock type**	Quartzitic Dolerite
	Colour	Dark grey
	Grain	Medium
	Products	Dressed or Lump Stone
		Crushed Stone
		Asphalt or Tarmacadam
DRILLING DATA		Hole Diameter 105mm
		Bench Height 17m
		Contractor
LOAD AND HAUL	**Loaders**	Caterpillar
CRUSHING PLANT	**Primary**	Boliden Allis 120×100 Jaw
	Secondary	Boliden Allis 45″ Gyratory
	Tertiary	(3)Boliden Allis 36″ Hydrocone

PIONEER AGGREGATES (UK) LTD

WESTBURY QUARRY

442

Westbury-sub-Mendip, Wells, Somerset BA5 3BZ
Phone 0749 870472
OS Map 183   **OS Grid Reference** ST507505
Personnel R Mawdley, Unit Manager

PRODUCT DATA	**Rock type**	Carboniferous Limestone
	Colour	Medium grey
	Products	Crushed Stone
		Asphalt or Tarmacadam
DRILLING DATA		Hole Diameter 105mm
		Bench Height 12-18m
		Contractor Celtic Rock Services
LOAD AND HAUL	**Trucks**	Terex
		Foden
	Loaders	Caterpillar
CRUSHING PLANT	**Primary**	Baxter 45×32 Jaw
	Secondary	BJD 33 Impactor
	Tertiary	Hazemag APK40 Impactor

I C PIPER
Head Office *4 Caradon View, Minions, Liskeard, Cornwall PL14 5LL*
Phone *0579 62016*
Personnel *I C Piper, Owner*
Quarries *Bearrah Tor*

I C PIPER

BEARRAH TOR QUARRY   443

Panorama, Minions, Liskeard, Cornwall PL14 5LM
Phone 0579 62452
OS Map 201 **OS Grid Reference** SX261746
Personnel I C Piper, Owner and Unit Manager

PRODUCT DATA		
	Rock type	Granite
	Colour	Silver grey
	Grain	Coarse
	Product	Dressed or Lump Stone

DRILLING DATA		
		Hand-held equipment

PREMIER LIME AND STONE COMPANY
Head Office *Creeton Quarry, Little Bytham, Grantham, Lincs NG33 4QC*
Phone *078081 202*
Personnel *J W Moseley, Director*
Quarries *Creeton*

PREMIER LIME AND STONE COMPANY

CREETON QUARRY   444

Little Bytham, Grantham, Lincs NG33 4QC
Phone 078081 202
OS Map 130 **OS Grid Reference** SK002203
Personnel J W Moseley, Director and Unit Manager

PRODUCT DATA		
	Rock type	Jurassic Limestone
	Colour	Cream, bluish grey
	Grain	Fine
	Product	Crushed Stone

DRILLING DATA		
		Hole Diameter 105mm
		Bench Height 16-21m
	Drills	Holman Voltrak

CRUSHING PLANT	**Primary**	Goodwin Barsby Goliath 42 × 24 Jaw

NO FURTHER INFORMATION PROVIDED

P & H RADLEY LTD
Head Office *Stone Quarry, Windy Ridge, Cartworth Moor, Holmfirth, W Yorks HD7 1QS*
Phone *0484 683536*
Personnel *P Radley, Owner*
Quarries *Stone*

P & H RADLEY LTD

STONE QUARRY

445

Windy Ridge, Cartworth Moor, Holmfirth, W Yorks HD7 1QS
Phone 0484 683536
OS Map 110 **OS Grid Reference** SE072146
Personnel P Radley, Owner and Unit Manager

PRODUCT DATA	**Rock type**	Carboniferous Millstone Grit
	Colour	Buff with brown graining
	Grain	Medium
	Product	Dressed or Lump Stone

RIPPABLE STRATA

SECONDARY BREAKING   Dropball on Smith

RATTEE AND KETT LTD (JOHN MOWLEM AND COMPANY plc)
Head Office Purbeck Road, Cambridge, Cambs CB2 2PF
Phone 0223 248061 **Fax** 0223 410284
Personnel B W May, Managing Director
        M Hill, Marketing Manager
Quarries Stamford

RATTEE AND KETT LTD (JOHN MOWLEM AND COMPANY plc)

STAMFORD QUARRY

446

Little Casterton Road, Stamford, Lincs PE9 4DA
Phone 0223 248061
OS Map 144 **OS Grid Reference** TF501308
Personnel Not appointed, Unit Manager

PRODUCT DATA	**Rock type**	Jurassic Inferior Oolitic Limestone (Stamford Freestone)
	Colour	Light buff to very light brown
	Grain	Fine to medium
	Product	Dressed or Lump Stone
DRILLING DATA		Hand-held equipment
	Drills	Holman drills

COMPRESSORS   CompAir

RBS BROOKLYNS LTD (TARMAC BUILDING MATERIALS LTD)
Head Office Shaft Road, Combe Down, Bath, Avon BA2 7HP
Phone 0225 835650 **Fax** 0225 835950
Personnel R Purshouse, Managing Director
        M J King, Director
Quarries Mount Pleasant

RBS BROOKLYNS LTD (TARMAC BUILDING MATERIALS LTD)

MOUNT PLEASANT QUARRY

447

Shaft Road, Combe Down, Bath, Avon BA2 7HP
Phone 0225 835650
OS Map 183 **OS Grid Reference** ST772624
Personnel A Deacon, Unit Manager

entry continues overleaf

PRODUCT DATA	**Rock type**	Jurassic Great Oolitic Limestone
	Colour	Cream
	Grain	Medium
	Product	Dressed or Lump Stone
DRILLING DATA		No drilling

GEORGE M READ
Head Office *Wilderness Quarry, Mitcheldean, Glos GL17 0DS*
Phone *0452 830395*
Personnel *G M Read, Director*
Quarries *Wilderness*

GEORGE M READ
WILDERNESS QUARRY   448
Mitcheldean, Glos GL17 0DS
Phone 0452 830395
OS Map 162 **OS Grid Reference** SO672185

QUARRY ON STAND

ROCK DATA	**Rock type**	Devonian Sandstone
	Colour	Dull red
	Grain	Fine

REALSTONE LTD
Head Office *Wingerworth, Chesterfield, Derbys S42 6RG*
Phone *0246 270244* **Fax** *0246 201060* **Telex** *547779*
Personnel *D Gregory, Managing Director*
Quarries *Bolehill, Dukes, Prince of Wales, Stancliffe, Stanton Moor Stoneraise, Watts Cliff*

REALSTONE LTD
BOLEHILL QUARRY   449
Wingerworth, Chesterfield, Derbys S42 6RG
Phone 0246 270244
OS Map 119 **OS Grid Reference** SK367662

QUARRY ON STAND

ROCK DATA	**Rock type**	Carboniferous Sandstone
	Colour	Dark buff, dark green to grey
	Grain	Fine

REALSTONE LTD
DUKES QUARRY   450
Whatstandwell, Matlock, Derbys DE4 5LG
Phone 0246 270244
OS Map 119 **OS Grid Reference** SK335545
Personnel B Bailey, Unit Manager

PRODUCT DATA	**Rock type**	Carboniferous Sandstone
	Colour	Lilac, buff, yellow
	Grain	Medium
	Product	Dressed or Lump Stone
DRILLING DATA		Hand-held equipment

REALSTONE LTD

PRINCE OF WALES QUARRY **451**

Lower Penpethy, Tintagel, Cornwall PL34 0HH
Phone 0840 213225
OS Map 200 **OS Grid Reference** SX075862
Personnel G Pearce, Unit Manager

PRODUCT DATA	**Rock type**	Devonian Slate
	Colour	Rustic brown
	Grain	Fine
	Product	Dressed or Lump Stone
DRILLING DATA		Hand-held equipment
		Contractor Saxton Co (Deep Drillers)

REALSTONE LTD

STANCLIFFE QUARRY **452**

Darley Dale, Matlock, Derbys DE4 2FZ
Phone 0246 270244
OS Map 119 **OS Grid Reference** SK268639

QUARRY ON STAND

ROCK DATA	**Rock type**	Carboniferous Sandstone
	Colour	Light buff
	Grain	Fine

REALSTONE LTD

STANTON MOOR QUARRY **453**

Stanton in the Peak, Matlock, Derbys DE4 2LX
Phone 0246 270244
OS Map 119 **OS Grid Reference** SK251645

QUARRY ON STAND

ROCK DATA	**Rock type**	Carboniferous Sandstone
	Colour	Light brown to pink
	Grain	Fine to medium

REALSTONE LTD

STONERAISE QUARRY **454**

Great Salkeld, Penrith, Cumbria CA11 9LN
Phone 0768 65811
OS Map 90 **OS Grid Reference** NY523352
Personnel M Dixon, Unit Manager

entry continues overleaf

PRODUCT DATA	**Rock type**	Permian Sandstone
	Colour	Red, dark pink
	Grain	Fine to medium
	Product	Dressed or Lump Stone

DRILLING DATA		Hand-held equipment

REALSTONE LTD

WATTS CLIFF QUARRY

455

Elton, Matlock, Derbys CE4 2LZ
Phone 0246 270244
OS Map 119 **OS Grid Reference** SU222610
Personnel P Bailey, Unit Manager

PRODUCT DATA	**Rock type**	Carboniferous Sandstone
	Colour	Lilac
	Grain	Medium
	Product	Dressed or Lump Stone

RIPPABLE STRATA

LOAD AND HAUL	**Face shovel**	Akermans 16C

REDLAND AGGREGATES LTD
Head Office Bradgate House, Groby, Nr Leicester, Leics LE6 0FA
Phone 0530 242151 *Fax 0530 245396* *Telex 34481*
Personnel G Phillipson, Chairman
I Reid, Managing Director
S Savery, Deputy Managing Director
J Close, Director
R Flack, Production Manager
Quarries Buddon Wood (Mountsorrel), Dean, Dry Rigg, Graig, Hafod, Southorpe, Wardlow

---

REDLAND AGGREGATES LTD

BUDDON WOOD QUARRY (MOUNTSORREL QUARRY)

456

Mountsorrel, Loughborough, Leics LE12 7AT
Phone 0537 243881
OS Map 129 **OS Grid Reference** SK562150
Personnel J Close, Director and General Manager
G Rudin, Unit Manager

PRODUCT DATA	**Rock type**	Granite
	Colour	Speckled medium pink
	Grain	Coarse
	Products	Crushed Stone
		Asphalt or Tarmacadam
		Ready-mixed Concrete
		Pre-cast Concrete Products

DRILLING DATA		Hole Diameter 127-132mm
		Bench Height 18-30m
	Drills	Halco 450H
		(2)Halco TS75-35
		Bohler DCT122

LOAD AND HAUL	**Face shovel**	O&K RH120C
		Caterpillar
	Trucks	(10)Caterpillar 777
		(4)Volvo BM A25
	Loaders	Caterpillar 960D
		(2)Caterpillar 980C
		Caterpillar 992C

CRUSHING PLANT	**Primary**	Nordberg 60-180 Gyratory
	Secondary	(5)Nordberg 7' Cone (Standard)
	Tertiary	(5)Nordberg 4.25' Cone (Shorthead)
		Nordberg 560 Omnicone

REDLAND AGGREGATES LTD

DEAN QUARRY   457

Dean Point, St Keverne, Helston, Cornwall TR12 6MY
Phone 0326 280472
OS Map 204   **OS Grid Reference** SW202800
Personnel B Retallack, Unit Manager

PRODUCT DATA	**Rock type**	Lizard Gabbro
	Colour	Bluish grey
	Grain	Coarse
	Products	Dressed or Lump Stone
		Crushed Stone
		Asphalt or Tarmacadam

DRILLING DATA		Hole Diameter 110mm
		Bench Height 12.5m
	Drills	Holman Voltrak

| COMPRESSORS | | CompAir 700HE |

| SECONDARY BREAKING | | Dropball on Ruston-Bucyrus 22-RB |

LOAD AND HAUL	**Face shovel**	Caterpillar 245
	Trucks	Aveling Barford
	Loaders	Aveling Barford WXL118

CRUSHING PLANT	**Primary**	Pegson 44 × 48 Jaw (Single Toggle)
	Secondary	Nordberg 4' Cone
	Tertiary	Nordberg 4.25' Cone (Shorthead)
		(5)Nordberg 3' Cone

REDLAND AGGREGATES LTD

DRY RIGG QUARRY   458

Horton in Ribblesdale, Settle, N Yorks BD23 0EL
Phone 07296 411
OS Map 90   **OS Grid Reference** SD803695
Personnel E Ralph, Unit Manager

PRODUCT DATA	**Rock type**	Silurian Sandstone (Horton Formation)
	Colour	Dark grey
	Grain	Medium
	Products	Crushed Stone
		Asphalt or Tarmacadam

entry continues overleaf

DRILLING DATA		Hole Diameter 108mm
		Bench Height 14m
	Drills	Halco SPD43-22
		Halco 400C

SECONDARY BREAKING		Dropball on Ruston-Bucyrus 30-RB

LOAD AND HAUL	**Trucks**	(2)Aveling Barford RD40
	Loaders	Caterpillar 988B

CRUSHING PLANT	**Primary**	Newel Dumford Jaw
	Secondary	Nordberg 5.5' Cone
	Tertiary	(2)Nordberg 4'Cone
		(3)Nordberg 3'Cone

REDLAND AGGREGATES LTD

GRAIG QUARRY

459

Llanarmon-yn-Ial, Mold, Clwyd CH7 4QA
Phone 082 43643
OS Map 117   **OS Grid Reference** SJ192555
Personnel B Pugh, Unit Manager

PRODUCT DATA	**Rock type**	Carboniferous Limestone
	Colour	Medium grey
	Products	Crushed Stone
		Asphalt or Tarmacadam

DRILLING DATA		Hole Diameter 105mm
		Bench Height 15m
	Drills	Atlas Copco ROC404-04

COMPRESSORS		Atlas Copco XA350

SECONDARY BREAKING		Hydraulic Hammer - Furukawa

LOAD AND HAUL	**Trucks**	Aveling Barford RD50
	Loaders	Komatsu WA500

CRUSHING PLANT		No information provided

REDLAND AGGREGATES LTD

HAFOD QUARRY

460

Abercarn, Newbridge, Newport, Gwent NP1 5LH
Phone 0495 247459
OS Map 171   **OS Grid Reference** ST225965
Personnel A Wells, Unit Manager

PRODUCT DATA	**Rock type**	Carboniferous Pennant Sandstone
	Colour	Dark greyish blue
	Grain	Fine
	Product	Crushed Stone

DRILLING DATA		Hole Diameter 105mm
		Bench Height 12-15m
		Contractor EMK Drilling & Blasting

SECONDARY BREAKING		Dropball
LOAD AND HAUL	**Trucks**	Caterpillar
	Loaders	Volvo BM 4600
CRUSHING PLANT	**Primary**	Kue Ken 48 × 42 Jaw (Double Toggle)
	Secondary	Nordberg 4.25' Cone (Standard)
	Tertiary	Nordberg 3' Cone (Shorthead)

REDLAND AGGREGATES LTD

SOUTHORPE QUARRY
461

High Farm, Southorpe, Stamford, Lincs PE9 9BZ
Phone 0780 782750
OS Map 142   **OS Grid Reference** TF095026
Personnel A Pack, Unit Manager

PRODUCT DATA	**Rock type**	Jurassic Limestone
	Colour	Light brown
	Grain	Fine
	Product	Crushed Stone
DRILLING DATA		Hole Diameter 105mm
		Bench Height 15m
		Contractor S.E.S.
LOAD AND HAUL	**Loaders**	Caterpillar
CRUSHING PLANT	**Primary**	Kue Ken 48 × 36 Jaw
	Secondary	Kue Ken 36 × 24 Jaw

REDLAND AGGREGATES LTD

WARDLOW QUARRY
462

Cauldon Low, Waterhouse, Stoke-on-Trent, Staffs ST10 3HA
Phone 0538 702366
OS Map 119   **OS Grid Reference** SK083475
Personnel F Thompson, Unit Manager

PRODUCT DATA	**Rock type**	Carboniferous Milldale Limestone
	Colour	Light creamy grey
	Products	Crushed Stone
		Asphalt or Tarmacadam
		Industrial Limestone
		Agricultural Lime
DRILLING DATA		Hole Diameter 110mm
		Bench Height 18m
	Drills	Halco SPD43-22
SECONDARY BREAKING		Dropball on Ruston-Bucyrus 30-RB
LOAD AND HAUL	**Trucks**	(3)Aveling Barford RD40
	Loaders	Komatsu
		Caterpillar 950
		Caterpillar 988B

entry continues overleaf

CRUSHING PLANT	**Primary**	Sheepbridge Twin Rotor
	Secondary	BJD Slugger
	Tertiary	BJD Swinghammer

JAMES REID AND COMPANY (1947)
Head Office Lugton Lime Works, Lugton, Kilmarnock, Strathclyde KA3 4EB
Phone 050585 435
Personnel W Iley, Managing Director
Quarries Hessilhead, Middleton Mine, Trearne

JAMES REID AND COMPANY (1947)

HESSILHEAD QUARRY

463

Beith, Strathclyde KA15 1HT
Phone 05055 2536
OS Map 63 **OS Grid Reference** NS374535
Personnel W Edmunds, Unit Manager

| PRODUCT DATA | **Rock type** | Carboniferous Limestone |
| | **Product** | Dressed or Lump Stone |

RIPPABLE STRATA

JAMES REID AND COMPANY (1947)

MIDDLETON MINE

464

16 Borthwick Castle Rd, Middleton, Gorebridge, Lothian EH23 4QP
Phone 0875 20339
OS Map 66 **OS Grid Reference** NT355580
Personnel W Redpath, Unit Manager

PRODUCT DATA	**Rock type**	Carboniferous Lower Limestone
	Colour	Greyish blue
	Products	Crushed Stone
		Industrial Limestone
		Agricultural Lime
DRILLING DATA		Hole Diameter 105mm
		Bench Height 10-30m
	Drills	Halco 410C
COMPRESSORS		CompAir 650HE
SECONDARY BREAKING		Dropball
LOAD AND HAUL	**Trucks**	Aveling Barford RD30
	Loaders	Caterpillar
CRUSHING PLANT		No information provided

JAMES REID AND COMPANY (1947)

TREARNE QUARRY

465

Beith, Strathclyde KA15 1HT
Phone 05055 2536
OS Map 63 **OS Grid Reference** NS376533
Personnel W Edmunds, Unit Manager

PRODUCT DATA	**Rock type**	Carboniferous Dockra Limestone
	Products	Crushed Stone
		Agricultural Lime

DRILLING DATA		Hole Diameter 95mm
		Bench Height 12m
	Drills	Wagon Drill

| SECONDARY BREAKING | | Dropball |

| LOAD AND HAUL | **Trucks** | Aveling Barford |
| | **Loaders** | JCB |

CRUSHING PLANT	**Primary**	Kue Ken 24 × 36 Jaw
		Pegson 25 × 40 Jaw
	Secondary	Kue Ken Jaw
		(2)Pegson Gyratory

RHUDDLAN STONE LTD (C & M PARRY PLANT HIRE)
Head Office Rhuddlan Bach Quarry, Brynteg, Anglesey, Gwynedd LL78 8QA
Phone 0248 852282
Personnel C Parry, Director
Quarries Rhuddlan Bach

RHUDDLAN STONE LTD (C & M PARRY PLANT HIRE)
RHUDDLAN BACH QUARRY   **466**
Brynteg, Anglesey, Gwynedd LL78 8QA
Phone 0248 852282
OS Map 115   **OS Grid Reference** SH486807
Personnel C Parry, Director and Unit Manager

PRODUCT DATA	**Rock type**	Carboniferous Limestone
	Colour	Light grey
	Grain	Fine
	Product	Dressed or Lump Stone

| DRILLING DATA | | Hand-held equipment |

RISKEND QUARRYING COMPANY
Head Office Riskend Quarry, Tak Ma Doon Road, Kilsyth, Glasgow, Strathclyde G65 9JY
Phone 0236 821486
Personnel W Clelland, Director
Quarries Riskend

RISKEND QUARRYING COMPANY
RISKEND QUARRY   **467**
Tak Ma Doon Road, Kilsyth, Glasgow, Strathclyde G65 9JY
Phone 0236 821486
OS Map 64   **OS Grid Reference** NS797792
Personnel R Moodie, Unit Manager

entry continues overleaf

PRODUCT DATA	**Rock type**	Quartzitic Dolerite Sill
	Colour	Dark grey
	Grain	Medium
	Products	Crushed Stone
		Ready-mixed Concrete

DRILLING DATA		Hole Diameter 110mm
		Bench Height 25m
	Drills	Atlas Copco ROC304

| COMPRESSORS | | Ingersoll-Rand |

| SECONDARY BREAKING | | Dropball on Ruston-Bucyrus 22-RB |

LOAD AND HAUL	**Trucks**	Aveling Barford RD40
	Loaders	Caterpillar
		Broyt

CRUSHING PLANT	**Primary**	Kue Ken 160
	Secondary	Kue Ken 108
	Tertiary	(2)Allis Chalmers 36″ Cone

RMC GROUP plc
Corporate Head Office *RMC House, Coldharbour Lane, Thorpe, Egham, Surrey TW20 8TD*
Phone *0932 568833* **Fax** *0932 568933*
Personnel *J Camden, Chairman*
J W Gauntlett, Deputy Chairman
P J Owen, Managing Director
Companies *Butterley Aggregates Ltd, RMC Dimensional Stone Ltd, RMC Industrial Minerals Ltd, RMC Roadstone Products Ltd,* Scottish Aggregates Ltd   ✱ — see separate entry

RMC DIMENSIONAL STONE LTD (RMC GROUP plc)
Head Office *Ann Twyford Quarry, Birchover, Matlock, Derbys DE4 2BN*
Phone *0629 88287*
Personnel *J F Brighton, General Manager*
Quarries *Ann Twyford, De Lank*

RMC DIMENSIONAL STONE LTD (RMC GROUP plc)

ANN TWYFORD QUARRY   **468**
Birchover, Matlock, Derbys DE4 2BN
Phone 0629 88287
OS Map 119   **OS Grid Reference** SK242624
Personnel J F Brighton, General Manager

PRODUCT DATA	**Rock type**	Carboniferous Millstone Grit
	Colour	Pink to buff
	Grain	Medium to coarse
	Product	Dressed or Lump Stone

| DRILLING DATA | | Hole Diameter 38mm |
| | | Hand-held equipment |

| COMPRESSORS | | Ingersoll-Rand |

| LOAD AND HAUL | **Trucks** | Terex |
| | **Loaders** | Terex |

CRUSHING PLANT　　　　　　　None

RMC DIMENSIONAL STONE LTD (RMC GROUP plc)

DE LANK QUARRY　　　　　　　　　　469

St Breward, Bodmin, Cornwall PL30 4NQ
Phone 0208 850217
OS Map 200　**OS Grid Reference** SX102754
Personnel J F Brighton, General Manager

PRODUCT DATA	**Rock type**	Granite
	Colour	Mottled silver grey
	Grain	Fine to medium
	Product	Dressed or Lump Stone

DRILLING DATA		Hole Diameter 80mm
		Bench Height 10m
		Hand-held equipment

RMC INDUSTRIAL MINERALS LTD (RMC GROUP plc)
Head Office *Hindlow, Buxton, Derbys SK17 0EL*
Phone *0298 71155*
Personnel *P J Hutchins, General Manager*
Quarries *Brierlow*

RMC INDUSTRIAL MINERALS LTD (RMC GROUP plc)

BRIERLOW QUARRY　　　　　　　　470

Hindlow, Buxton, Derbys SK17 9PY
Phone 0298 71155
OS Map 119　**OS Grid Reference** SK086691
Personnel C Weston, Unit Manager

PRODUCT DATA	**Rock type**	Carboniferous Limestone
	Colour	Not commercially significant
	Products	Industrial Limestone
		Agricultural Lime

DRILLING DATA		Hole Diameter 127mm
		Bench Height 17m
	Drills	Halco 400C

| COMPRESSORS | | CompAir RO160-70 |

| SECONDARY BREAKING | | Hydraulic Hammer |

LOAD AND HAUL	**Trucks**	Aveling Barford
	Loaders	Caterpillar 966C
		Caterpillar 968

CRUSHING PLANT	**Primary**	Goodwin Barsby 54 × 43 Jaw
	Secondary	BJD Flextooth
	Tertiary	None

RMC ROADSTONE PRODUCTS LTD (RMC GROUP plc)
Head Office *RMC House, Church Lane, Bromsgrove, Hereford & Worcs B61 8RA*
Phone *0527 575777* **Fax** *0527 575263*
Personnel *G R C Smith, Managing Director*
G G Cooper, Marketing Director
Companies *McLaren Roadstone Ltd, North West Aggregates Ltd, Shap Granite Company Ltd, Thomas W Ward Roadstone Ltd, Western Roadstone Ltd, Wotton Roadstone Ltd*
Quarries *Dove Holes*

RMC ROADSTONE PRODUCTS LTD (RMC GROUP plc)

DOVE HOLES QUARRY

471

Dale Road, Dove Holes, Buxton, Derbys SK17 8BH
Phone 0298 77531
OS Map 119 **OS Grid Reference** SK090775
Personnel D Wall, Unit Manager

PRODUCT DATA	**Rock type**	Carboniferous Bee Low Limestone
	Products	Crushed Stone
		Asphalt or Tarmacadam
		Pre-cast Concrete Products
		Industrial Limestone
		Burnt Limestone
		Agricultural Lime
DRILLING DATA		Hole Diameter 125-200mm
		Bench Height 10-20m
	Drills	Drillmark DM25
SECONDARY BREAKING		Dropball on Ruston-Bucyrus 22-RB
LOAD AND HAUL	**Trucks**	Aveling Barford RD40
		Caterpillar 773B
	Loaders	(3)Caterpillar 988B
		(2)Komatsu WA450
		Komatsu WA500
CRUSHING PLANT	**Primary**	Sheepbridge 72 × 27 Double Impactor
	Secondary	FBM Wagenader
		Sheepbridge Swing Impactor
	Tertiary	Sheepbridge 3826

ROBINSON ROCK
Head Office *Holmescales Quarry, Old Hutton, Kendal, Cumbria LA8 0NB*
Phone *0539 32255/33018*
Personnel *W L Robinson, Director*
Quarries *Holmescales*

ROBINSON ROCK

HOLMESCALES QUARRY

472

Old Hutton, Kendal, Cumbria LA8 0NB
Phone 0539 32255
OS Map 97 **OS Grid Reference** SD540782
Personnel A Woodson, Unit Manager

PRODUCT DATA	**Rock type**	Permian Sandstone
	Colour	Red
	Grain	Coarse
	Product	Dressed or Lump Stone

| DRILLING DATA | Contractor |

NO FURTHER INFORMATION PROVIDED

ROSELAND AGGREGATES LTD
Head Office *Lean Quarry, Horningtops, Liskeard, Cornwall PL14 3QD*
Phone *0579 42342*
Personnel *W D Crocker, Director*
Quarries *Lean*

ROSELAND AGGREGATES LTD

LEAN QUARRY   **473**

Horningtops, Liskeard, Cornwall PL14 3QD
Phone 0579 42342
OS Map 201   **OS Grid Reference** SX263614
Personnel K Williams, Unit Manager

PRODUCT DATA	**Rock type**	Basalt
	Colour	Reddish brown, brown, grey
	Grain	Fine
	Products	Crushed Stone
		Asphalt or Tarmacadam
		Ready-mixed Concrete

DRILLING DATA	Hole Diameter 105mm
	Bench Height 17m
	Contractor Saxton Co (Deep Drillers)

| LOAD AND HAUL | **Loaders** | Zettelmeyer |
| | | Hannomag |

NO FURTHER INFORMATION

RTZ CORPORATION plc
Corporate Head Office *6 St James's Square, London SW17 4LD*
Phone *071 930 2399* **Fax** *071 930 3249* **Telex** *24639*
Personnel *A E Buxton, Director of Mining*
Companies *RTZ Mining and Exploration Ltd Delabole Slate Division*

RTZ MINING AND EXPLORATION LTD DELABOLE SLATE DIVISION (RTZ CORPORATION plc)
Head Office *Pengelly House, Delabole, Cornwall PL33 9AZ*
Phone *0840 212242* **Fax** *0840 212948* **Telex** *45157*
Personnel *J Dowling, General Manager*
Quarries *Delabole, Laneast*

---

entry continues overleaf

RTZ MINING AND EXPLORATION LTD, DELABOLE SLATE DIVISION
(RTZ CORPORATION plc)

474

DELABOLE QUARRY

Pengelly House, Delabole, Cornwall PL33 9AZ
Phone 0840 212242
OS Map 200 **OS Grid Reference** SX073840
Personnel J Dowling, General Manager and Unit Manager

PRODUCT DATA	**Rock type**	Devonian Slate
	Colour	Bluish grey
	Grain	Fine
	Product	Dressed or Lump Stone

QUARRY UNDER DEVELOPMENT

RTZ MINING AND EXPLORATION LTD, DELABOLE SLATE DIVISION
(RTZ CORPORATION plc)

475

LANEAST QUARRY

Laneast, Launceston, Cornwall PL15 7TS
Phone 0840 212242
OS Map 201 **OS Grid Reference** SX223833

QUARRY ON STAND

ROCK DATA	**Rock type**	Basalt

RUGBY PORTLAND CEMENT plc

Corporate Head Office *Crown House, Rugby, Warwicks CV21 2DT*
Phone *0788 542111* **Fax** *0788 540256* **Telex** *31523*
Personnel *A H Teare, Managing Director*
Quarries *Barrington, Chinnor, Greys, Kensworth, Middlegate*

RUGBY PORTLAND CEMENT plc

476

BARRINGTON QUARRY

Barrington Works, 41 Slingfield Road, Barrington, Cambs CB2 5RG
Phone 0223 870781
OS Map 154 **OS Grid Reference** TL390510
Personnel J Drayton, Unit Manager

PRODUCT DATA	**Rock type**	Cretaceous Gault and Lower Chalk
	Colour	White
	Grain	Fine
	Product	Industrial

RIPPABLE STRATA

NO FURTHER INFORMATION

RUGBY PORTLAND CEMENT plc

CHINNOR QUARRY

477

Chinnor Works, Chinnor, Oxford, Oxon OX9 8DP
Phone 0844 51477 **Fax** 0788 540256 **Telex** 31523
OS Map 164 **OS Grid Reference** SP756002
Personnel D Howlett, Unit Manager

PRODUCT DATA	**Rock type**	Cretaceous Lower Chalk
	Colour	White
	Grain	Fine
	Product	Dressed or Lump Stone

RIPPABLE STRATA

LOAD AND HAUL	**Face shovel**	Poclain
		Case 360B
	Trucks	(3)Caterpillar D30D
	Loaders	None

CRUSHING PLANT

NO FURTHER INFORMATION PROVIDED

RUGBY PORTLAND CEMENT plc

GREYS PIT

478

Rochester Works, Halling-on-the-Medway, Rochester, Kent, ME2 1AW
Phone 0634 240261
OS Map 178 **OS Grid Reference** TQ698652
Personnel G Fuller, General Works Manager
　　　　　Vacant, Unit Manager

PRODUCT DATA	**Rock type**	Cretaceous Lower-Middle Chalk
	Colour	White
	Grain	Fine
	Product	Industrial

RIPPABLE STRATA

NO FURTHER INFORMATION

RUGBY PORTLAND CEMENT plc

KENSWORTH QUARRY

479

PO Box 6, Kensworth, Dunstable, Bedfords LU6 2PR
Phone 0582 872156
OS Map 166 **OS Grid Reference** TL017198
Personnel J Barton, Unit Manager

PRODUCT DATA	**Rock type**	Cretaceous Middle-Upper Chalk
	Colour	White
	Grain	Fine
	Product	Industrial

RIPPABLE STRATA

NO FURTHER INFORMATION

RUGBY PORTLAND CEMENT plc

MIDDLEGATE QUARRIES

480

South Ferriby Works, Barton-on-Humber, Humbers DN18 6JL
Phone 0724 732434
OS Map 1112   **OS Grid Reference** SE990203
Personnel A Kitchen, Unit Manager

PRODUCT DATA | **Rock type** Cretaceous Ferriby Welton Chalk
Colour White
Grain Fine
Product Industrial

RIPPABLE STRATA

NO FURTHER INFORMATION

A SANDERS LTD
Head Office *Tuckingmill Quarry, Bow, Crediton, Devon EX17 6HN*
Phone *03633 315*
Personnel *A Sanders, Owner*
Quarries *Tuckingmill*

A SANDERS LTD

TUCKINGMILL QUARRY

481

Bow, Crediton, Devon EX17 6HN
Phone 03633 315
OS Map 191   **OS Grid Reference** SS724034
Personnel A Sanders, Owner and Unit Manager

PRODUCT DATA | **Rock type** Carboniferous Sandstone
Products Dressed or Lump Stone
Crushed Stone

RIPPABLE STRATA

ALEXANDER SANDISON AND SONS LTD
Head Office *Northside, Baltasound, Unst, Shetland ZE2 9DS*
Phone *095781 444*
Personnel *A Sandison, Owner*
Quarries *Setters*

ALEXANDER SANDISON AND SONS LTD

SETTERS QUARRY

482

Northside, Baltasound, Unst, Shetland ZE2 9DS
Phone 095781 444
OS Map 1   **OS Grid Reference** HP623104
Personnel A Sandison, Owner and Unit Manager

PRODUCT DATA | **Rock type** Palaeozoic Schist
Colour Grey
Grain Fine to medium
Product Crushed Stone

DRILLING DATA   Hole Diameter 110mm
Bench Height 17–20m
Contractor Lewis Land Services

SANTIME LTD
Head Office *Pilkington Quarry, Makinson Lane, Horwich, Bolton, Greater Manchester BL6 6NA*
Phone *0204 693679*
Personnel *D Pickavan, Director*
Quarries *Pilkington*

SANTIME LTD
PILKINGTON QUARRY   **483**
Makinson Lane, Horwich, Bolton, Greater Manchester BL6 6NA
Phone 0204 693679
OS Map 109   **OS Grid Reference** SD655124
Personnel W Harrison, Unit Manager

PRODUCT DATA	**Rock type**	Carboniferous Lower Coal Measure Sandstone
	Colour	Not commercially significant
	Product	Crushed Stone
DRILLING DATA		Hole Diameter 105mm
		Bench Height 12–20m
	Drills	Halco 410C
COMPRESSORS		Consolidated Pneumatic
SECONDARY BREAKING		Dropball on Ruston-Bucyrus 22-RB
		Dropball on Ransomes & Rapier NCK-Rapier 605
LOAD AND HAUL	**Face shovel**	Ruston-Bucyrus
	Trucks	Foden
		Caterpillar 769
	Loaders	Caterpillar 966C
		Caterpillar 980
CRUSHING PLANT	**Primary**	Parker Jaw
	Secondary	(2)Parker Kubitizer

SAXONMOOR LTD
Head Office *Hillhouse Edge Quarry, Cartworth Moor, Holmfirth, Huddersfield, W Yorks HD7 1RL*
Phone *0484 683239*
Personnel *B Graham, Director*
Quarries *Hillhouse Edge*

SAXONMOOR LTD
HILLHOUSE EDGE QUARRY   **484**
Cartworth Moor, Holmfirth, Huddersfield, W Yorks HD7 1RL
Phone 0484 683239
OS Map 110   **OS Grid Reference** SE130060
Personnel S Graham, Unit Manager

entry continues overleaf

PRODUCT DATA	**Rock type**	Carboniferous Sandstone
	Colour	Fawn with brown speckles
	Grain	Medium
	Product	Dressed or Lump Stone

| DRILLING DATA | Hand-held equipment |

SCOTTISH AGGREGATES LTD (RMC GROUP plc)

Head Office 6 Union Street, Bridge of Allan, Stirling, Central FK9 4NT
Phone 0786 834055
Personnel R Davies, General Manager
Quarries Hillend, Kilbarchan

SCOTTISH AGGREGATES LTD (RMC GROUP plc)

HILLEND QUARRY   485

Caldercruix, Strathclyde ML6 6NB
Phone 07715 441
OS Map 65   **OS Grid Reference** NS823673
Personnel

PRODUCT DATA	**Rock type**	Basalt
	Colour	Dark grey
	Grain	Fine
	Product	Crushed Stone

| DRILLING DATA | Contractor Rocklift |

NO FURTHER INFORMATION PROVIDED

SCOTTISH AGGREGATES LTD (RMC GROUP plc)

KILBARCHAN QUARRY   486

Kilbarchan, Strathclyde PA10 2AD
Phone 0505 75422
OS Map 64   **OS Grid Reference** NS407636
Personnel J Hamilton, Unit Manager

PRODUCT DATA	**Rock type**	Basalt
	Colour	Dark grey
	Grain	Fine
	Products	Crushed Stone
		Asphalt or Tarmacadam

DRILLING DATA	Hole Diameter 105mm
	Bench Height 12m
	Contractor Rocklift

| SECONDARY BREAKING | Dropball on Ruston-Bucyrus 22-RB |

| LOAD AND HAUL | **Trucks** | (2)Terex |
| | **Loaders** | Komatsu WA470 |

| CRUSHING PLANT | No information provided |

SCOTTISH NATURAL STONES LTD (BALFOUR BEATTY LTD)
Head Office *Edinburgh Road, Springhill, Shotts, Strathclyde ML7 5DT*
Phone *0501 23248*
Personnel *I McDonald, Director and General Manager*
Quarries *Black Pasture, Dunmore, Gatelawbridge, Kingoodie, Newbiggin Ross of Mull*

SCOTTISH NATURAL STONES LTD (BALFOUR BEATTY LTD)

BLACK PASTURE QUARRY 487

Hexham, Northumb NE46 4EQ
Phone 0501 23248
OS Map 87   **OS Grid Reference** NY930700
Personnel Names not provided

PRODUCT DATA	**Rock type**	Carboniferous Sandstone
	Colour	Light brown
	Grain	Medium
	Product	Dressed or Lump Stone

DRILLING DATA		Hand-held equipment

SCOTTISH NATURAL STONES LTD (BALFOUR BEATTY LTD)

DUNMORE QUARRY 488

Stirling, Central FK2 0SX
Phone 0501 23248
OS Map 58   **OS Grid Reference** NS894896
Personnel Names not provided

PRODUCT DATA	**Rock type**	Carboniferous Sandstone
	Colour	Cream
	Grain	Fine
	Product	Dressed or Lump Stone

DRILLING DATA		Hand-held equipment

SCOTTISH NATURAL STONES LTD (BALFOUR BEATTY LTD)

GATELAWBRIDGE QUARRY 489

Gatelawbridge, Thornhill, Dumfries, Dumfries & Galloway, DG1 1QQ
Phone 0501 23248
OS Map 78   **OS Grid Reference** NX902965
Personnel Names not provided

PRODUCT DATA	**Rock type**	Silurian Sandstone
	Grain	Medium
	Product	Dressed or Lump Stone

DRILLING DATA		Hand-held equipment

SCOTTISH NATURAL STONES LTD (BALFOUR BEATTY LTD)

KINGOODIE QUARRY 490

Kingoodie, Dundee, Fife DD2 5DE
Phone 0501 23248
OS Map 53   **OS Grid Reference** NO335294

entry continues overleaf

QUARRY ON STAND

ROCK DATA   **Rock type**   Devonian Sandstone

SCOTTISH NATURAL STONES LTD (BALFOUR BEATTY LTD)

NEWBIGGIN QUARRY

491

Burntisland, Fife KY3 0AQ
Phone 0501 23248
OS Map 66 **OS Grid Reference** NT211864

QUARRY ON STAND

ROCK DATA   **Rock type**   Carboniferous Sandstone
                Colour   White, cream, buff
                Grain   Medium

SCOTTISH NATURAL STONES LTD (BALFOUR BEATTY LTD)

ROSS OF MULL QUARRY

492

Isle of Mull, Strathclyde PA65 6BD
Phone 0501 23248
OS Map 48 **OS Grid Reference** NM305235
Personnel Names not provided

PRODUCT DATA   **Rock type**   Muscovite Granite
                Colour   Pink, red with pale grey felspar
                Grain   Coarse
                Product   Dressed or Lump Stone

DRILLING DATA   Hand-held equipment

SCOTTS (BAMPTON)LTD

Head Office *Kersdown Quarry, Tiverton Road Works, Bampton, Tiverton, Devon EX16 9DX*
Phone *0398 31466*
Personnel *M Hellier, Managing Director*
Quarries *Kersdown*

SCOTTS (BAMPTON)LTD

KERSDOWN QUARRY

493

Tiverton Road Works, Bampton, Tiverton, Devon EX16 9DX
Phone 0398 31466
OS Map 181 **OS Grid Reference** SS963222
Personnel D Denscome, Unit Manager

PRODUCT DATA   **Rock type**   Carboniferous Limestone
                Product   Crushed Stone

RIPPABLE STRATA

SECONDARY BREAKING   Dropball on Smith 21

NO FURTHER INFORMATION PROVIDED

THOMAS SCOURFIELD AND SONS
Head Office *Kiln Quarry, Carew Cheriton, Tenby, Dyfed SA70 8SR*
Phone *0646 651326*
Personnel *K Morgan Scourfield, Director*
Quarries *Carew, Kiln*

THOMAS SCOURFIELD AND SONS

CAREW QUARRIES 494

Carew Cheriton, Tenby, Dyfed SA70 8SR
Phone 0646 651326
OS Map 158 **OS Grid Reference** SN049045
Personnel K Morgan Scourfield, Director and Unit Manager

PRODUCT DATA	**Rock type**	Carboniferous Limestone
	Colour	Not commercially significant
	Product	Crushed Stone

DRILLING DATA		Hole Diameter 110mm
		Bench Height 12–15m
		Contractor A Jones Rock Drillers

| SECONDARY BREAKING | | Dropball on Ruston-Bucyrus 22-RB |

LOAD AND HAUL	**Face shovel**	Poclain 90
	Trucks	Foden
		Euclid
	Loaders	Bray

| CRUSHING PLANT | **Primary** | Pegson 25 × 40 Jaw |
| | **Secondary** | Parker 103 Kubitizer |

THOMAS SCOURFIELD AND SONS

KILN QUARRY 495

Carew Cheriton, Tenby, Dyfed SA70 8SL
Phone 0646 651326
OS Map 158 **OS Grid Reference** SN039043
Personnel K Morgan Scourfield, Director and Unit Manager

PRODUCT DATA	**Rock type**	Carboniferous Limestone
	Colour	Not commercially significant
	Product	Crushed Stone

DRILLING DATA		Hole Diameter 108mm
		Bench Height 15m
		Contractor A Jones Rock Drillers

| SECONDARY BREAKING | | Hydraulic Hammer |

| LOAD AND HAUL | **Trucks** | Foden |
| | **Loaders** | Zettelmeyer |

| CRUSHING PLANT | | No information provided |

SHAP GRANITE COMPANY LTD (RMC ROADSTONE PRODUCTS LTD)
Head Office Shap, Penrith, Cumbria CA10 3QQ
Phone 09316 787 **Fax** *09316 775*
Personnel J P Daglish, General Manager
Quarries Shap Blue, Shap Pink

SHAP GRANITE COMPANY LTD (RMC ROADSTONE PRODUCTS LTD)

SHAP BLUE QUARRY

496

Shap, Penrith, Cumbria, CA10 3QQ
Phone 093 16787
OS Map 90   **OS Grid Reference** NY558084
Personnel D Nethven, Unit Manager

PRODUCT DATA	**Rock type**	Hornfels
	Colour	Dark pink to brown
	Product	Dressed or Lump Stone
		Crushed Stone

NO FURTHER INFORMATION PROVIDED

SHAP GRANITE COMPANY LTD (RMC ROADSTONE PRODUCTS LTD)

SHAP PINK QUARRY

497

Shap, Penrith, Cumbria, CA10 3QQ
Phone 093 16787
OS Map 90   **OS Grid Reference** NY564106
Personnel D Nethven, Unit Manager

PRODUCT DATA	**Rock type**	Biotite Granite
	Colour	Light grey, pink
	Product	Dressed or Lump Stone
		Crushed Stone

NO FURTHER INFORMATION PROVIDED

R SHEPHERD
Head Office 3 Raise Hamlet, Ward Way, Alston, Cumbria CA9 3AS
Phone 0498 81288
Personnel R Shepherd, Owner
Quarries Leipsic

R SHEPHERD

LEIPSIC QUARRY

498

Alston, Cumbria CA9 3NQ
Phone 0498 81288
OS Map 86   **OS Grid Reference** NY735504
Personnel R Shepherd, Owner and Unit Manager

PRODUCT DATA	**Rock type**	Carboniferous Sandstone
	Product	Dressed or Lump Stone

DRILLING DATA		Hand-held equipment

SHERBURN STONE COMPANY LTD
Head Office *15 Front Street, Sherburn Hill, Durham DH6 1PA*
Phone *0913 720636*
Personnel *J Allison, Managing Director*
G Ascough, Production Manager
Quarries *Barton, Crime Rig, Hart, Shadforth, Witch Hill*

SHERBURN STONE COMPANY LTD

BARTON QUARRY
499

Barton, Richmond, N Yorks DL10 6NF
Phone 0325 77864
OS Map 93   **OS Grid Reference** NY220078
Personnel J Allison, Managing Director and Unit Manager

PRODUCT DATA	**Rock type**	Permian Limestone
	Colour	Buffish grey
	Products	Dressed or Lump Stone
		Crushed Stone
DRILLING DATA		Hole Diameter 105mm
		Bench Height 10m
	Drills	Hausherr HBM60
SECONDARY BREAKING		Hydraulic Hammer - Furukawa on Komatsuu PC200
LOAD AND HAUL	**Trucks**	Aveling Barford RD30
	Loaders	Komatsu WA200
		Komatsu WA300
CRUSHING PLANT	**Primary**	Parker Impactor (Mobile)

SHERBURN STONE COMPANY LTD

CRIME RIG QUARRY
500

Shadforth, Durham DH6 1LA
Phone 0913 720232
OS Map 88   **OS Grid Reference** NZ344418
Personnel J Allison, Managing Director and Unit Manager

PRODUCT DATA	**Rock type**	Carboniferous Limestone
	Colour	Not commercially significant
	Product	Crushed Stone
DRILLING DATA		Hole Diameter 105mm
		Bench Height 12m
	Drills	Atlas Copco ROC404
COMPRESSORS		Atlas Copco XA350DD
SECONDARY BREAKING		Hydraulic Hammer - Furukawa
LOAD AND HAUL	**Trucks**	Komatsu
	Loaders	Komatsu
CRUSHING PLANT	**Primary**	Parker Impactor

SHERBURN STONE COMPANY LTD

HART QUARRY

501

Hart Village, Hartlepool, Cleveland TS27 3BP
Phone 0429 274177
OS Map 93 **OS Grid Reference** NZ475345
Personnel G Ascough, Production Manager and Unit Manager

PRODUCT DATA	**Rock type**	Permian Magnesian Limestone
	Colour	Not commercially significant
	Products	Crushed Stone
		Agricultural Lime
DRILLING DATA		Hole Diameter 105mm
		Bench Height 10m
	Drills	Hausherr HNB60
SECONDARY BREAKING		Hydraulic Hammer - Furukawa
LOAD AND HAUL	**Trucks**	Aveling Barford RD30
	Loaders	Komatsu
CRUSHING PLANT	**Primary**	(2)Parker Impactor (Mobile)

SHERBURN STONE COMPANY LTD

SHADFORTH QUARRY

502

Shadforth, Durham DH1 4JA
Phone 0385 720232
OS Map 88 **OS Grid Reference** NZ340422
Personnel G Ascough, Production Manager and Unit Manager

PRODUCT DATA	**Rock type**	Permian Upper Magnesian Limestone
	Colour	Not commercially significant
	Products	Crushed Stone
		Agricultural Lime
DRILLING DATA		Hole Diameter 105mm
		Bench Height 8m
	Drills	Atlas Copco ROC404
COMPRESSORS		Atlas Copco XA350DD
SECONDARY BREAKING		Dropball on Ruston-Bucyrus 22-RB
LOAD AND HAUL	**Trucks**	Aveling Barford RD40
	Loaders	Komatsu
CRUSHING PLANT	**Primary**	(2)Parker (Mobile)

SHERBURN STONE COMPANY LTD

WITCH HILL QUARRY

503

Silent Bank, Thornley, Durham DH6 2QA
Phone 0429 280956
OS Map 93 **OS Grid Reference** NZ365395
Personnel G Ascough, Production Manager and Unit Manager

PRODUCT DATA	**Rock type**	Permian Magnesian Limestone
	Colour	Not commercially significant
	Products	Crushed Stone
		Agricultural Lime

DRILLING DATA		Hole Diameter 105mm
		Bench Height 15m
	Drills	Atlas Copco ROC404

| COMPRESSORS | | Atlas Copco XA350DD |

| SECONDARY BREAKING | | Dropball on Ruston-Bucyrus 22-RB |

LOAD AND HAUL	**Trucks**	Aveling Barford RD40
	Loaders	Komatsu WA200
		Komatsu WA300

| CRUSHING PLANT | **Primary** | (2)Parker Impactor (Mobile) |
| | **Secondary** | Cones |

NO FURTHER INFORMATION PROVIDED

SHETLAND ISLES COUNCIL

Head Office *Council Offices, Grantfield, Lerwick, Shetland ZE1 0NT*
Phone *0595 3535* **Fax** *0595 3535 Ext 223*
Personnel *B Davidson, Technical Director*
Quarries *Scord*

SHETLAND ISLES COUNCIL

SCORD QUARRY

504

Scalloway, Shetland ZE2 9PD
Phone 0595 882279
OS Map 4 **OS Grid Reference** HU400390
Personnel D Adamson, Unit Manager

PRODUCT DATA	**Rock type**	Palaeozoic Schist
	Colour	Grey
	Grain	Fine to medium
	Product	Crushed Stone

DRILLING DATA		Hole Diameter 105mm
		Bench Height 13m
		Contractor Lewis

| SECONDARY BREAKING | | Dropball |

| LOAD AND HAUL | **Trucks** | (2)Heathfield |
| | **Loaders** | Caterpillar |

CRUSHING PLANT	**Primary**	Pegson 42 × 30 Jaw
		Parker 36 × 24 Jaw
	Secondary	Nordberg 3' Cone (Standard)
		Barmac Duopactor

SHIPLEY BANK QUARRIES

Head Office *Rose Cottage, Lartington, Barnard Castle, Durham DL12 8BP*
Phone *0833 50529*
Personnel *T H Cross, Owner*
Quarries *Shipley Bank*

SHIPLEY BANK QUARRIES

SHIPLEY BANK QUARRY

505

Marwood, Barnard Castle, Durham DL12 9BP
Phone 0833 50529
OS Map 92   **OS Grid Reference** NY015209
Personnel N Cross, Unit Manager

PRODUCT DATA	**Rock type**	Carboniferous Sandstone (Gritstone)
	Colour	Buff with faint grey marking
	Grain	Fine
	Product	Dressed or Lump Stone
DRILLING DATA		Hand-held equipment

SINGLETON BIRCH LTD

Head Office *Melton Ross Quarry, Melton Ross, Brigg, Lincs DN38 6AE*
Phone *0652 688386*
Personnel *K O Aldrich, Production Manager*
Quarries *Melton Ross*

SINGLETON BIRCH LTD

MELTON ROSS QUARRY

506

Melton Ross, Brigg, Lincs DN38 6AE
Phone 0652 688386
OS Map 112   **OS Grid Reference** TA085113
Personnel R R Stansfield, Unit Manager

PRODUCT DATA	**Rock type**	Cretaceous Melton Chalk
	Colour	White
	Grain	Fine
	Products	Dressed or Lump Stone
		Crushed Stone
		Industrial Limestone
		Burnt Limestone
		Agricultural Lime
DRILLING DATA		Hole Diameter 108mm
		Bench Height 14m
	Drills	(2)Holman Voltrak
COMPRESSORS		(2)CompAir 700HE
LOAD AND HAUL	**Trucks**	Terex
	Loaders	Komatsu
CRUSHING PLANT		No information

SMITH AND SONS (BLETCHINGTON) LTD
Head Office *Enslow, Bletchington, Oxon OX5 3AY*
Phone *086983 281*
Personnel *J Smith, Managing Director*
                N Bailey, Operations Director
Quarries *Ardley, Fish Hill*

SMITH AND SONS (BLETCHINGTON) LTD

ARDLEY QUARRY   507

Ardley, Bicester, Oxon OX6 9PH
Phone 0869 89221
OS Map 164   **OS Grid Reference** SP532271
Personnel J P Walker, Unit Manager

PRODUCT DATA	**Rock type**	Jurassic Great Oolitic Limestone
	Colour	Not commercially significant
	Grain	Fine
	Products	Crushed Stone
		Pre-cast Concrete Products
DRILLING DATA		Hole Diameter 110mm
		Bench Height 18m
	Drills	Halco mast on Cat
COMPRESSORS		CompAir 650HE
LOAD AND HAUL	**Face shovel**	Akermans H16
	Trucks	None
	Loaders	(2)Caterpillar 966D
		Caterpillar 980C
		Caterpillar 988B
		Ferguson 560
CRUSHING PLANT	**Primary**	Parker 32 × 42 Jaw (Mobile)
	Secondary	Parker (Mobile)

SMITH AND SONS (BLETCHINGTON) LTD

FISH HILL QUARRY   508

Fish Hill, Broadway, Hereford & Worcs WR12 7LL
Phone 086983 281
OS Map 150   **OS Grid Reference** SP118370
Personnel K Dowding, Unit Manager

PRODUCT DATA	**Rock type**	Jurassic Limestone
	Colour	Cream to yellow
	Grain	Fine
	Product	Dressed or Lump Stone
RIPPABLE STRATA		
LOAD AND HAUL	**Loaders**	Caterpillar
		Volvo
		JCB
CRUSHING PLANT		None

SPAR SUPPLY COMPANY
Head Office *Slieau Chiarn House, Braaid, Marown, Isle of Man*
Phone *0624 851756/851904*
Personnel *A T Jones, Director*
        G B Corfield, Director
Quarries *Granite Mountain*

SPAR SUPPLY COMPANY

GRANITE MOUNTAIN QUARRY   509

Granite Mountain, Foxdale, Douglas, Isle of Man
Phone 0624 851756
OS Map 95   **OS Grid Reference** SC291760
Personnel G B Corfield, Director and Unit Manager

PRODUCT DATA	**Rock type**	Granite
	Colour	Not commercially significant
	Grain	Coarse
	Product	Crushed Stone

DRILLING DATA		Hole Diameter 105mm
		Bench Height 20m
		Contractor A Jones Rock Drillers

SECONDARY BREAKING   Dropball

NO FURTHER INFORMATION PROVIDED

SPEYSIDE SAND AND GRAVEL
Head Office *Rothes Glen Quarry, Rothes, Grampian AB3 9CB*
Phone *03403 700*
Personnel *D Carmichael, Director*
Quarries *Rothes Glen*

SPEYSIDE SAND AND GRAVEL

ROTHES GLEN QUARRY   510

Rothes, Grampian AB3 9CB
Phone 03403 700
OS Map 28   **OS Grid Reference** NJ279495
Personnel D Ross, Unit Manager

PRODUCT DATA	**Rock type**	Pre-Cambrian Moine Schist
	Products	Dressed or Lump Stone
		Crushed Stone

NO FURTHER INFORMATION

C W SPRONSTON (LIME) LTD
Head Office *Afonwen Quarry, Riverview, Townfield, Frodsham, Cheshire WA6 7RL*
Phone *0928 32250*
Personnel *C W Spronston, Managing Director*
Quarries *Afonwen*

C W SPRONSTON (LIME) LTD

AFONWEN QUARRY

511

Riverview, Townfield, Frodsham, Cheshire WA6 7RL
Phone 0928 32250
OS Map 117 **OS Grid Reference** SJ520725
Personnel C W Spronston, Managing Director and Unit Manager

PRODUCT DATA	**Rock type**	Carboniferous Limestone
	Colour	Whitish brown
	Grain	Open
	Products	Dressed or Lump Stone
		Agricultural Lime

RIPPABLE STRATA

LOAD AND HAUL   **Face shovel**   Hymac

NO FURTHER INFORMATION

STAFFORDSHIRE STONE (HOLLINGTON) LTD
Head Office Institute Road, Kings Heath, Birmingham B14 7EG
Phone 021 444 1087
Personnel C Shaw, Director
Quarries Red Hole (Great Gate)

STAFFORDSHIRE STONE (HOLLINGTON) LTD

RED HOLE QUARRY (GREAT GATE QUARRY)

512

Quarry Bank Road, Hollington, Stoke-on-Trent, Staffs ST10 4HR
Phone 088926 435
OS Map 128 **OS Grid Reference** SK054398
Personnel C Shaw, Director and Unit Manager

PRODUCT DATA	**Rock type**	Triassic Sandstone (Gritstone)
	Colour	Dull red
	Grain	Fine to medium
	Product	Dressed or Lump Stone

DRILLING DATA   Hand-held equipment

COMPRESSORS   CompAir

S D & M STAINSBY
Head Office Parkside, The Avenue, High Shincliffe, Durham DH1 2PT
Phone 0913 867098
Personnel S D Stainsby, Owner
Quarries Dead Friars

S D & M STAINSBY

DEAD FRIARS QUARRY

513

Blanchland, Consett, Durham DH8 9XD
Phone 0913 867098
OS Map 87 **OS Grid Reference** NY969453
Personnel S D Stainsby, Owner and Unit Manager

entry continues overleaf

PRODUCT DATA	**Rock type**	Carboniferous Millstone Grit
	Product	Dressed or Lump Stone

DRILLING DATA		Hand-held equipment

STANCLIFFE STONE COMPANY
Head Office *Cressbrook Mill, Cressbrook, Buxton, Derbys SK17 8SY*
Phone *0298 871228*
Personnel *C D Holmes, Director*
Quarries *Palmers*

STANCLIFFE STONE COMPANY

PALMERS QUARRY   514

Stanton in Peak, Matlock, Derbys DE4 2NG
Phone 0298 871228
OS Map 119   **OS Grid Reference** SK251643
Personnel C D Holmes, Director and Unit Manager

PRODUCT DATA	**Rock type**	Carboniferous Sandstone (Gritstone)
	Colour	Buff, light pink
	Grain	Fine to medium
	Product	Dressed or Lump Stone

DRILLING DATA		Hole Diameter 32mm
		Hand-held equipment

STEETLEY plc
Corporate Head Office *PO Box 53, Brownsover Road, Rugby, Warwicks CV21 2UT*
Phone *0788 535621* **Fax** *0788 535361* **Telex** *311677*
Personnel *D L Donne, Chairman*
     R M Miles, Group Managing Director
Companies *Steetley Quarry Products Ltd*

STEETLEY QUARRY PRODUCTS LTD
Head Office *Coventry Point, Market Way, Coventry, Warwicks CY1 1EA*
Phone *0203 633888* **Fax** *0203 26788*
Personnel *R M Miles, Chairman*
     M J Lodge, Managing Director

STEETLEY QUARRIES DIVISION
Head Office *Kiveton Lane, Kiveton Park, Sheffield, S Yorks S31 8NN*
Phone *0909 770581* **Fax** *0909 773628* **Telex** *54276*
Personnel *Vacant, Managing Director*
     B Brown, Divisional Director

STEETLEY QUARRIES DIVISION *Southern Region*
Regional Head Office *Gibbet Lane, Shawell, Lutterworth, Leics LE17 6AA*
Phone *0788 74841* **Fax** *0788 536226*
Personnel *P Bowden, Regional Manager*

STEETLEY QUARRIES DIVISION Southern Region (Eastern Area)
Area Office *Gibbet Lane, Shawell, Lutterworth, Leics LE17 6AA*
Phone *0788 74841* **Fax** *0788 536226*
Personnel *J Gordon, Production Manager*
         G W Parkin, Commercial Manager
Quarries *Corby (Cowltick), Great Ponton, South Witham, Worsham*

STEETLEY QUARRIES DIVISION Southern Region (Eastern Area)

CORBY QUARRY (COWLTICK QUARRY)   515

121 Kettering Road, Corby, Northants NN17 3JG
Phone 0536 66660
OS Map 141   **OS Grid Reference** SP916885
Personnel A Simpson, Unit Manager

PRODUCT DATA	**Rock type**	Jurassic Oolitic Limestone
	Colour	Not commercially significant
	Product	Crushed Stone
DRILLING DATA		Hole Diameter 105mm
		Bench Height 10m
		Contractor S.E.S
SECONDARY BREAKING		Hydraulic Hammer - Krupp
LOAD AND HAUL	**Trucks**	None
	Loaders	Caterpillar 360
		Caterpillar 966C
		Zettelmeyer ZL1801
CRUSHING PLANT	**Primary**	Parker 32 × 42 Jaw

NO FURTHER INFORMATION

STEETLEY QUARRIES Southern Region (Eastern Area)

GREAT PONTON QUARRY   516

Station Yard, Dallygate, Grantham, Lincs NG33 5DZ
Phone 0476 83415
OS Map 130   **OS Grid Reference** SK935303
Personnel No information

PRODUCT DATA	**Rock type**	Jurassic Inferior Oolitic Limestone
	Colour	Not commercially significant
	Products	Crushed Stone
		Agricultural Lime

NO FURTHER INFORMATION

STEETLEY QUARRIES DIVISION Southern Region (Eastern Area)

SOUTH WITHAM QUARRY   517

Mill Lane, South Witham, Grantham, Lincs NG33 5OL
Phone 0572 83272   **Fax** 0788 536226
OS Map 130   **OS Grid Reference** SK920191
Personnel R Fletcher, Unit Manager

entry continues overleaf

PRODUCT DATA	**Rock type**	Jurassic Limestone
	Colour	Not commercially significant
	Product	Crushed Stone

DRILLING DATA		Hole Diameter 105mm
		Bench Height 15m
		Contractor S E S

LOAD AND HAUL	**Trucks**	None
	Loaders	Caterpillar 950E
		(2)Zettelmeyer ZL5000

| CRUSHING PLANT | **Primary** | Goodwin Barsby 36 × 24 Jaw |
| | | Pegson Jaw |

STEETLEY QUARRIES DIVISION Southern Region (Eastern Area)

WORSHAM QUARRY

518

Burford Road, Asthall Leigh, Witney, Oxon OX8 5SS
Phone 0604 491138
OS Map 164   **OS Grid Reference** SP295103

QUARRY ON STAND

| ROCK DATA | **Rock type** | Jurassic Great Oolitic Limestone |

STEETLEY QUARRIES DIVISION Southern Region (Western Area)
Area Office *Heol Goch, Pentyrch, Taffs Well, Cardiff, S Glam CF4 8XB*
Phone *0222 810549*   **Fax** *0222 810747*
Personnel *D R Lewis, Production Manager*
　　　　　　C J Law, Commercial Manager
Quarries *Ammanford, Llynclys, Shadwell, Taffs Well, Woodbury*

STEETLEY QUARRIES DIVISION Southern Region (Western Area)

AMMANFORD QUARRY

519

Craig-y-Odyn, Trap, Llandeilo, Dyfed SA19 6TT
Phone 0269 850225   **Fax** 0269 851204
OS Map 159   **OS Grid Reference** SN634178
Personnel S Milne, Unit Manager

PRODUCT DATA	**Rock type**	Carboniferous Limestone
	Colour	Not commercially significant
	Product	Crushed Stone

DRILLING DATA		Hole Diameter 105mm
		Bench Height 15m
	Drills	Halco 400C

| COMPRESSORS | | CompAir 650HE |

| LOAD AND HAUL | **Trucks** | (2)Aveling Barford RD30 |
| | **Loaders** | Caterpillar 988B |

| CRUSHING PLANT | **Primary** | Goodwin Barsby 40 × 32 Jaw (Double Toggle) |
| | **Secondary** | Goodwin Impactor (Reversible) |

STEETLEY QUARRIES DIVISION Southern Region (Western Area)

LLYNCLYS QUARRY

520

Oswestry, Shrops SY10 8LW
Phone 0691 830642
OS Map 126 **OS Grid Reference** SJ265245
Personnel P Breeze, Unit Manager

PRODUCT DATA	**Rock type**	Carboniferous Dolomitic Limestone
	Colour	Not commercially significant
	Grain	Coarse
	Products	Crushed Stone
		Industrial Limestone
DRILLING DATA		Hole Diameter 105-125mm
		Bench Height 20m
	Drills	Halco 410C
COMPRESSORS		CompAir 700HE
SECONDARY BREAKING		Dropball on Ruston-Bucyrus 22-RB
LOAD AND HAUL	**Face shovel**	Caterpillar 245
	Trucks	Aveling Barford RD25
		Aveling Barford RD30
	Loaders	Caterpillar 966D
CRUSHING PLANT	**Primary**	Pegson 1100 × 800 Jaw (Mobile)
		Sheepbridge 38 × 36 Jaw
	Secondary	Hazemag SAP5

STEETLEY QUARRIES DIVISION Southern Region (Western Area)

SHADWELL QUARRY

521

Farley Road, Much Wenlock, Shrops TF13 6PF
Phone 0952 727539
OS Map 127 **OS Grid Reference** SO625008
Personnel D McGill, Unit Manager

PRODUCT DATA	**Rock type**	Silurian Wenlock Limestone
	Colour	Not commercially significant
	Product	Crushed Stone
DRILLING DATA		Hole Diameter 110mm
		Bench Height 15m
		Contractor Ritchies Equipment
	Drills	Halco 410C
COMPRESSORS		Atlas Copco XRH350DD
SECONDARY BREAKING		Dropball on Ruston-Bucyrus 22-RB
LOAD AND HAUL	**Trucks**	(2)Terex 33-05B
	Loaders	Caterpillar 980
CRUSHING PLANT	**Primary**	Pegson 30 × 42 Jaw
		Pegson 25 × 40 Jaw (Mobile)
	Secondary	Pegson 4′ Autocone

TAFFS WELL QUARRY

522

Heol Goch, Pentyrch, Taffs Well, Cardiff, S Glam CF4 8XB
Phone 0222 810525
OS Map 171 **OS Grid Reference** ST122822
Personnel R Hines, Unit Manager

PRODUCT DATA	**Rock type**	Carboniferous Dolomitic Limestone
	Colour	Pinkish grey
	Products	Crushed Stone
		Ready-mixed Concrete
		Agricultural Lime
DRILLING DATA		Hole Diameter 110-125mm
		Bench Height 17.5m
	Drills	Halco 425X
		Halco 410C
		Halco 400C
COMPRESSORS		(2)CompAir 650HE
SECONDARY BREAKING		Hydraulic Hammer - Furukawa
LOAD AND HAUL	**Trucks**	Terex
		(3)Aveling Barford RD30
	Loaders	Caterpillar 966C
		Caterpillar 966D
		Caterpillar 988B
CRUSHING PLANT	**Primary**	Kennedy P36 Cone
	Secondary	Hazemag Impactor
	Tertiary	(2)Nordberg 48″ Gyradisc
		Nordberg 36″ Cone

WOODBURY QUARRY

523

Shelsley Beauchamp, Worcester, Hereford & Worcs WR6 6RE
Phone 08865 648
OS Map 150 **OS Grid Reference** SO743637
Personnel S Badelek, Unit Manager

PRODUCT DATA	**Rock type**	Silurian Limestone
	Colour	Not commercially significant
	Product	Crushed Stone
DRILLING DATA		Hole Diameter 105mm
		Bench Height 15m
		Contractor K&H Rockdrillers
SECONDARY BREAKING		Dropball
LOAD AND HAUL	**Trucks**	None
	Loaders	Caterpillar

NO FURTHER INFORMATION

STEETLEY QUARRIES DIVISION Northern Region
Regional Head Office *Northern Region House, Copgrove, Harrogate, N Yorks HG3 3TB*
Phone *0423 340781* **Fax** *0423 340750*
Personnel *N G Bentley, Regional Manager*
M Turner, Regional Production Manager
G Bell, Production Manager - North
Vacant, Commercial Manager - North
Vacant, Production Manager - South
M Hazlehurst, Commercial Manager - South
Quarries *Cadeby, Dowlow, Gebdykes, Harrycroft, Potgate Ruston Parva (Lowthorpe), Smaws, Thrislington, Wath, Whitwell Wormersley*

STEETLEY QUARRIES DIVISION Northern Region

CADEBY QUARRY
524

PO Box I, Cadeby Works, Conisbrough, Doncaster, S Yorks DN12 3AP
Phone 0709 867474
OS Map 111 **OS Grid Reference** SK521999
Personnel D Gavin, Unit Manager

PRODUCT DATA	**Rock type**	Permian Magnesian Limestone
	Colour	Light cream
	Products	Dressed or Lump Stone
		Crushed Stone
		Industrial Limestone
		Agricultural Lime
DRILLING DATA		Hole Diameter 120mm
		Bench Height 40m
	Drills	Hausherr HBM120 RB
SECONDARY BREAKING		Dropball on Ruston-Bucyrus 22-RB
LOAD AND HAUL	**Trucks**	(2)Terex R25
	Loaders	Caterpillar 950
		Caterpillar 966C
		Caterpillar 988B
CRUSHING PLANT	**Primary**	Pegson 48 × 60 Jaw
	Secondary	Sheepbridge Cone

STEETLEY QUARRIES DIVISION Northern Region

DOWLOW QUARRY
525

Sterndale Moor, Buxton, Derbys SK17 9QP
Phone 0298 77611
OS Map 119 **OS Grid Reference** SK097677
Personnel P Spivey, Unit Manager

PRODUCT DATA	**Rock type**	Carboniferous Bee Low Limestone
	Colour	Not commercially significant
	Products	Crushed Stone
		Agricultural Lime
DRILLING DATA		Hole Diameter 150–165mm
		Bench Height 20m
	Drills	Halco MPD75-30T
		Halco 650H

entry continues overleaf

COMPRESSORS		CompAir 75–170
SECONDARY BREAKING		Dropball on Ruston-Bucyrus 22-RB
LOAD AND HAUL	**Trucks**	(3)Terex
		Volvo
	Loaders	Caterpillar 960
		Caterpillar 966C
		Caterpillar 988B
CRUSHING PLANT	**Primary**	Sheepbridge 60 × 50 Impactor
	Secondary	Sheepbridge 36 × 30 Impactor
		Parker Impactor (Reversible)

STEETLEY QUARRIES DIVISION Northern Region

GEBDYKES QUARRY 526

Masham, Ripon, N Yorks HG4 4BT
Phone 0765 85434
OS Map 99 **OS Grid Reference** SE237823
Personnel D Dunn, Unit Manager

PRODUCT DATA	**Rock type**	Permian Lower Magnesian Limestone
	Colour	Grey to buff
	Products	Dressed or Lump Stone
		Crushed Stone
		Industrial Limestone
		Agricultural Lime
DRILLING DATA		Hole Diameter 108mm
		Bench Height 12m
		Contractor R&B Rock Drillers
SECONDARY BREAKING		Dropball on Ruston-Bucyrus 22-RB
LOAD AND HAUL	**Trucks**	Aveling Barford RD40
	Loaders	Caterpillar
CRUSHING PLANT	**Primary**	Parker Jaw
	Secondary	Parker No 4

STEETLEY QUARRIES DIVISION Northern Region

HARRYCROFT QUARRY 527

Kiveton Park Station, Sheffield, S Yorks S31 8NP
Phone 0909 472623
OS Map 120 **OS Grid Reference** SK508827
Personnel D Fox, Unit Manager

PRODUCT DATA	**Rock type**	Permian Magnesian Limestone
	Colour	Not commercially significant
	Product	Crushed Stone
DRILLING DATA		Hole Diameter 110mm
		Bench Height 8m
		Contractor R&B Rock Drillers

| LOAD AND HAUL | **Trucks** | None |
| | **Loaders** | Caterpillar 966C |

| CRUSHING PLANT | | No information |

STEETLEY QUARRIES DIVISION Northern Region

POTGATE QUARRY

528

North Stainley, Ripon, N Yorks HG4 3JN
Phone 0765 85341
OS Map 99 **OS Grid Reference** SE285744
Personnel D Dunn, Unit Manager

PRODUCT DATA	**Rock type**	Permian Lower Magnesian Limestone
	Colour	Not commercially significant
	Product	Crushed Stone

DRILLING DATA		Hole Diameter 108mm
		Bench Height 20m
		Contractor R&B Rock Drillers

| SECONDARY BREAKING | | Dropball on Ruston-Bucyrus 22-RB |

| LOAD AND HAUL | **Trucks** | Terex |
| | **Loaders** | Caterpillar |

| CRUSHING PLANT | **Primary** | Baxter Jaw |
| | **Secondary** | Parker 105 |

STEETLEY QUARRIES DIVISION Northern Region

RUSTON PARVA QUARRY (LOWTHORPE QUARRY)

529

Ruston Parva, Driffield, Humbers YO25 0DF
Phone 0377 44215
OS Map 101 **OS Grid Reference** TA071615

PRODUCT DATA	**Rock type**	Cretaceous Flamborough Chalk
	Colour	Not Commercially Significant
	Grain	Fine
	Product	Crushed Stone

NO FURTHER INFORMATION

STEETLEY QUARRIES DIVISION Northern Region

SMAWS QUARRY

530

Rudgate, Tadcaster, N Yorks LS24 9LY
Phone 0937 832993
OS Map 105 **OS Grid Reference** SE462430
Personnel M Kendel, Unit Manager

PRODUCT DATA	**Rock type**	Permian Limestone
	Colour	Not commercially significant
	Product	Crushed Stone

entry continues overleaf

DRILLING DATA		Hole Diameter 108mm Bench Height 17m Contractor R&B Rock Drillers
SECONDARY BREAKING		Hydraulic Hammer
LOAD AND HAUL	**Trucks** **Loaders**	Foden Volvo
CRUSHING PLANT	**Primary** **Secondary** **Tertiary**	Pegson Jaw None None

STEETLEY QUARRIES DIVISION Northern Region

THRISLINGTON QUARRY   531

West Cornforth, Durham DL17 9EU
Phone 0740 654461
OS Map 93   **OS Grid Reference** NZ311328
Personnel M Candlish, Unit Manager

PRODUCT DATA	**Rock type** **Colour** **Products**	Permian Magnesian Limestone Not commercially significant Crushed Stone Ready-mixed Concrete Industrial Limestone Agricultural Lime
DRILLING DATA	 **Drills**	Hole Diameter 125mm Bench Height 12–25m (2)Hausherr HBM80
LOAD AND HAUL	**Trucks** **Loaders**	Aveling Barford RD40 Komatsu WA600
CRUSHING PLANT	**Primary** **Secondary**	Pegson 60 × 48 Jaw (2)Pegson 4' Gyrasphere

STEETLEY QUARRIES DIVISION Northern Region

WATH QUARRY   532

Malton Road, Hovingham, Malton, N Yorks YO6 7LT
Phone 065382 204
OS Map 100   **OS Grid Reference** SE679747
Personnel B Downs, Unit Manager

PRODUCT DATA	**Rock type** **Colour** **Products**	Jurassic Limestone White Dressed or Lump Stone Crushed Stone Agricultural Lime
DRILLING DATA		Hole Diameter 105mm Bench Height 8m Contractor S.E.S.

LOAD AND HAUL	**Trucks**	Volvo
	Loaders	Volvo

CRUSHING PLANT	**Primary**	Sheepbridge 42 × 30 Jaw
	Secondary	Findlay Impactor

STEETLEY QUARRIES DIVISION Northern Region

WHITWELL QUARRY

533

South Field Lane, Whitwell, Worksop, Notts S80 3LJ
Phone 0909 720751
OS Map 120   **OS Grid Reference** SK535759
Personnel J E Ward, General Manager
           M Beckett, Unit Manager

PRODUCT DATA	**Rock type**	Permian Magnesian Limestone
	Colour	Not commercially significant
	Products	Crushed Stone
		Ready-mixed Concrete
		Industrial Limestone
		Agricultural Lime

DRILLING DATA		Hole Diameter 115–145mm
		Bench Height 20–27m
	Drills	Hausherr HBM120 RD

LOAD AND HAUL	**Trucks**	Aveling Barford RD50
	Loaders	Caterpillar

CRUSHING PLANT	**Primary**	Pegson Jaw
		(2)Pegson Gyratory
	Secondary	(2)Gyratory
	Tertiary	Gyratory

NO FURTHER INFORMATION

STEETLEY QUARRIES DIVISION Northern Region

WORMERSLEY QUARRY

534

Wormersley, Doncaster, S Yorks DN6 9AX
Phone 0977 620593
OS Map 111   **OS Grid Reference** SE520200
Personnel G Pollard, Unit Manager

PRODUCT DATA	**Rock type**	Permian Upper Magnesian Limestone
	Colour	Not commercially significant
	Product	Crushed Stone

RIPPABLE STRATA

CRUSHING PLANT	**Primary**	Kue Ken 110 Jaw
	Secondary	Mansfield No 1 Hammermill
	Tertiary	Mansfield No 3 Hammermill

GR STEIN REFRACTORIES LTD
Head Office *Manuel Works, Linlithgow, Lothian EH49 6LH*
Phone *0506 843333*
Personnel *A Clarke, Director*
Quarries *Harthorpe East, Tippetcraig*

GR STEIN REFRACTORIES LTD

HARTHORPE EAST QUARRY

535

St John's Chapel, Bishop Auckland, Durham DL13 3QJ
Phone 0506 843333
OS Map 92   **OS Grid Reference** NY869350
Personnel J Duncan, Unit Manager

| PRODUCT DATA | **Rock type** | Carboniferous Sandstone (Gritstone) |
| | **Product** | Dressed or Lump Stone |

| DRILLING DATA | | Hand-held equipment |

GR STEIN REFRACTORIES LTD

TIPPETCRAIG QUARRY

536

High Bonnybridge, Central FK4 2EU
Phone 0506 843333   **Fax** 0423 340750
OS Map 65   **OS Grid Reference** NS825791

QUARRY ON STAND

| ROCK DATA | **Rock type** | Carboniferous Shale |
| | **Grain** | Fine |

J H L STEPHENS
Head Office *Callywith Quarry, Penquite Lane, Bodmin, Cornwall PL31 2AZ*
Phone *0208 72029*
Personnel *J H L Stephens, Owner*

J H L STEPHENS

CALLYWITH QUARRY

537

Penquite Lane, Bodmin, Cornwall PL31 2AZ
Phone 0208 72029
OS Map 200   **OS Grid Reference** SX085680
Personnel J H L Stephens, Owner and Unit Manager

PRODUCT DATA	**Rock type**	Middle Devonian Slate
	Colour	Buff to grey
	Grain	Fine
	Product	Dressed or Lump Stone

| DRILLING DATA | | Hand-held equipment |
| | | Contractor occasionally |

STOKE HALL QUARRY (STONE SALES) LTD
Head Office *Grindleford, Sheffield, S Yorks S30 1HW*
Phone *0433 30313*
Personnel *D Haywood, Director*
         P C Haywood, Director
Quarries *Stoke Hall*

STOKE HALL QUARRY (STONE SALES) LTD

STOKE HALL QUARRY

538

Grindleford, Sheffield, S Yorks S30 1HW
Phone 0433 30313
OS Map 119   **OS Grid Reference** SD236770
Personnel N Bater, Unit Manager

PRODUCT DATA	**Rock type**	Carboniferous Sandstone
	Colour	Buff
	Grain	Fine
	Product	Dressed or Lump Stone

DRILLING DATA		Hole Diameter 32mm
		Hand-held equipment

STONEGRAVE AGGREGATES LTD
Head Office *Aycliffe Quarry, Aycliffe, Durham DL5 6NB*
Phone *0325 313129*
Personnel *B Wade, Director*
Quarries *Aycliffe*

STONEGRAVE AGGREGATES LTD

AYCLIFFE QUARRY

539

Aycliffe, Durham DL5 6NB
Phone 0325 313129
OS Map 93   **OS Grid Reference** NZ225286
Personnel B Wade, Director and General Manager

PRODUCT DATA	**Rock type**	Carboniferous Dolomitic Limestone
	Colour	Grey
	Grain	Coarse
	Products	Crushed Stone
		Industrial Limestone

DRILLING DATA		Hole Diameter 105mm
		Bench Height 25m
	Drills	Halco 400C

COMPRESSORS	CompAir 650HE

SECONDARY BREAKING	Hydraulic Hammer - Furukawa

LOAD AND HAUL	**Face shovel**	Komatsu
	Trucks	Aveling Barford RD35
	Loaders	Komatsu WA120
		Komatsu WA400
		Komatsu WA450

CRUSHING PLANT	No information provided

STOWE HILL QUARRIES
Head Office *Stowe Hill Quarry, Stowe, St Briavels, Lydney, Glos GL15 6QH*
Phone *0594 530100*
Personnel *G Simms, Owner*
Quarries *Stowe Hill*

STOWE HILL QUARRIES

STOWE HILL QUARRY 540
Stowe, St Briavels, Lydney, Glos GL15 6QH
Phone 0594 530100
OS Map 162 **OS Grid Reference** SO564065
Personnel G Simms, Owner and Unit Manager

PRODUCT DATA	**Rock type**	Carboniferous Limestone
	Products	Dressed or Lump Stone
		Crushed Stone
DRILLING DATA		Hole Diameter 80mm
		Bench Height 10m
	Drills	Holman Universal Wagon Drill
COMPRESSORS		CompAir RO37HP
LOAD AND HAUL	**Loaders**	Fiatallis
CRUSHING PLANT	**Primary**	Goodwin Barsby 42 × 30 Jaw
	Secondary	None
	Tertiary	None

STOWEY STONE LTD
Head Office *2 Market Place, Radstock, Bath, Avon BA3 2AW*
Phone *0761 33167*
Personnel *G J Bissex, Managing Director*
Quarries *Stowey*

STOWEY STONE LTD

STOWEY QUARRY 541
Bishop Sutton, Bristol, Avon BS18 4UJ
Phone 0761 52356
OS Map 172 **OS Grid Reference** ST598587
Personnel G J Bissex, Managing Director and Unit Manager

PRODUCT DATA	**Rock type**	Jurassic Lias Limestone
	Colour	White, blue
	Grain	Fine
	Product	Dressed or Lump Stone
DRILLING DATA		Hand-held equipment

STRATHCLYDE REGION COUNCIL
Head Office *Department of Roads, Hamilton, Strathclyde ML3 0AL*
Phone *0698 454444* **Fax** *0698 454294*
Personnel *D Carruthers, Technical Director*
Quarries *Cairngryffe, Duntilland, Tincornhill*

STRATHCLYDE REGION COUNCIL

CAIRNGRYFFE QUARRY

542

Lanark, Strathclyde ML11 8SW
Phone 0698 454444
OS Map 65   **OS Grid Reference** NS943411

QUARRY ON STAND

ROCK DATA	**Rock type**	Microgranite
	Colour	Red
	Grain	Fine to medium

STRATHCLYDE REGION COUNCIL

DUNTILLAND QUARRY

543

Salsburgh, Shotts, Strathclyde ML7 4NZ
Phone 0698 87211
OS Map 65   **OS Grid Reference** NS843633
Personnel J McWell, Unit Manager

PRODUCT DATA	**Rock type**	Basalt
	Colour	Black
	Grain	Fine
	Products	Crushed Stone
		Asphalt or Tarmacadam
DRILLING DATA		Hole Diameter 105mm
		Bench Height 19m
		Contractor Rocklift
SECONDARY BREAKING		Dropball
LOAD AND HAUL	**Trucks**	Aveling Barford
	Loaders	(2)Caterpillar
CRUSHING PLANT	**Primary**	Robey-Farrel 42 × 40 Jaw
	Secondary	Kennedy 38″ Cone
	Tertiary	Kennedy 36″ Cone

STRATHCLYDE REGION COUNCIL

TINCORNHILL QUARRY

544

Sorn, Mauchline, Strathclyde KA5 5JE
Phone 0698 282828
OS Map 70   **OS Grid Reference** TS576275

QUARRY ON STAND

ROCK DATA	**Rock type**	Granodiorite

SULLOM QUARRIES
Head Office Gulberwick, Lerwick, Shetland ZE2 9EX
Phone 0595 2401
Personnel J Scoals, Director
Quarries Brindister

entry continues overleaf

BRINDISTER QUARRY

545

Gulberwick, Lerwick, Shetland ZE2 9EX
Phone 0595 2401
OS Map 4 **OS Grid Reference** HU479410
Personnel J Scoals, Unit Manager

PRODUCT DATA	**Rock type**	Devonian Sandstone
	Colour	Not commercially significant
	Grain	Fine
	Products	Crushed Stone
		Ready-mixed Concrete
		Pre-cast Concrete Products

DRILLING DATA		Hole Diameter 108mm
		Contractor McKay

CRUSHING PLANT	**Primary**	Parker Rocksizer
	Secondary	Granulator
	Tertiary	(2)Lokomo G1810 Cone

A & D SUTHERLAND

Head Office 69 Princes Street, Thurso, Highland KW14 7DJ
Phone 0847 63224
Personnel A Sutherland, Owner
Quarries Spittal

A & D SUTHERLAND

SPITTAL QUARRY

546

Westerdale, Halkirk, Thurso, Highland KW12 6UP
Phone 0847 84239
OS Map 12 **OS Grid Reference** ND172542
Personnel A Sutherland, Owner and Unit Manager

PRODUCT DATA	**Rock type**	Devonian Sandstone
	Colour	Dark grey
	Grain	Fine
	Product	Dressed or Lump Stone

NO INFORMATION PROVIDED

J SUTTLE

Head Office Swanage Quarry, Off Panorama Road, Swanage, Dorset BH19 2QS
Phone 0929 423576
Personnel C J Suttle, Director
Quarries Swanage

J SUTTLE

SWANAGE QUARRY

547

Off Panorama Road, Swanage, Dorset BH19 2QS
Phone 0929 423576
OS Map 195 **OS Grid Reference** SZ015785
Personnel C J Suttle, Director and Unit Manager

PRODUCT DATA | **Rock type** | Jurassic Limestone
Colour | Dark blue, bluish grey
Grain | Shelly
Products | Dressed or Lump Stone
Crushed Stone

RIPPABLE STRATA

NO FURTHER INFORMATION

SUTTON LIMES LTD
Head Office *Sutton Quarry, Ripon, N Yorks HG4 3JZ*
Phone *0765 3841*
Personnel *B Hunter, Director*
R Hunter, Director
Quarries *Sutton*

SUTTON LIMES LTD

SUTTON QUARRY   548
Ripon, N Yorks HG4 3JZ
Phone 0765 3841
OS Map 99   **OS Grid Reference** SE285744
Personnel R Hulliah, Unit Manager

PRODUCT DATA | **Rock type** | Permian Lower Magnesian Limestone
Colour | Light grey
Product | Crushed Stone

DRILLING DATA | **Drills** | Holman Voltrak

LOAD AND HAUL | | No information

CRUSHING PLANT | **Primary** | Goodwin Barsby 36 × 24 Jaw
Secondary | Lanway Hammermill
Tertiary | None

M & L SYMONDS
Head Office *Pilsamoor Quarry, Egloskerry, Launceston, Cornwall PL15 8SH*
Phone *0566 86362*
Personnel *L Symonds, Owner*
Quarries *Pilsamoor*

M & L SYMONDS

PILSAMOOR QUARRY   549
Egloskerry, Launceston, Cornwall PL15 8SH
Phone 0566 86362
OS Map 201   **OS Grid Reference** SX316858
Personnel L Symonds, Owner and Unit Manager

PRODUCT DATA | **Rock type** | Devonian Sandstone (Gritstone)
Colour | Light buff to dark brown
Grain | Medium
Product | Dressed or Lump Stone

DRILLING DATA | | Hand-held equipment

TADCASTER BUILDING LIMESTONE LTD
Head Office *Highmoor Quarry, Warren Lane, Toulson, Tadcaster, N Yorks LS24 9NU*
Phone *0937 833956*
Personnel *S Smith, Director*
Quarries *Highmoor*

TADCASTER BUILDING LIMESTONE LTD
HIGHMOOR QUARRY
Warren Lane, Toulson, Tadcaster, N Yorks LS24 9NU
Phone 0937 833956
OS Map 105   **OS Grid Reference** SE449425
Personnel A Parnaby, Unit Manager

550

PRODUCT DATA		
	Rock type	Permian Magnesian Limestone
	Colour	Creamy white
	Grain	Coarse
	Products	Crushed Stone
		Pre-cast Concrete Products

RIPPABLE STRATA

NO INFORMATION PROVIDED

TARMAC plc
Corporate Head Office *Hilton Hall, Essington, Wolverhampton, W Midlands WV11 2BQ*
Phone *0902 307407*   **Fax** *0902 307408*   **Telex** *338544*
Personnel *Sir E Pountain, Chairman*
B W Baker, Group Managing Director
J Mawdsley, Director
J D Sims, Director
Companies *Tarmac Building Materials Ltd, Tarmac Quarry Products Ltd*

TARMAC BUILDING MATERIALS LTD (TARMAC plc)
Head Office *Millfield Road, Ettingshall, Wolverhampton, W Midlands WV4 6JP*
Phone *0902 353522* **Fax** *0902 353920* **Telex** *339825*
Personnel *J D Sims, Chief Executive*
Companies *Natural Stone Products Ltd, RBS Brooklyns Ltd*

TARMAC QUARRY PRODUCTS LTD (TARMAC plc)
Head Office *Millfield Road, Ettingshall, Wolverhampton, W Midlands WV4 6JP*
Phone *0902 353522* **Fax** *0902 353920* **Telex** *339825*
Personnel *J Mawdsley, Chief Executive*
D Andrews, Assistant Managing Director
C Nototny, Assistant Managing Director
L Falkner, Director
Companies *Cumbria Stone Quarries Ltd, Tarmac Roadstone (Eastern) Ltd,*
Tarmac Roadstone (East Midlands) Ltd, Tarmac Roadstone (North West) Ltd,
Tarmac Roadstone (Scotland) Ltd, Tarmac Roadstone (Southern) Ltd,
Tarmac Roadstone (Western) Ltd

TARMAC ROADSTONE (EASTERN) LTD (TARMAC QUARRY PRODUCTS LTD)
Head Office John Hadfield House, Dale Road, Matlock, Derbys DE4 3PL
Phone 0629 580300 *Fax* 0629 580204
Personnel G Cliffe, Managing Director
J Williams, Production Manager
A B Wiper, Production Manager
Vacant, Production Manager
Quarries Barrasford, Bolsover Moor, Cowdor and Hall Dale, Dene, Holme Hall Howick, Middle Peak, Mootlaw, Old Quarrington, Raisby, Redmire, Threshfield

---

TARMAC ROADSTONE (EASTERN) LTD (TARMAC QUARRY PRODUCTS LTD)

BARRASFORD QUARRY | 551

Nr Hexham, Northumb NE48 4DJ
Phone 0434 681443
OS Map 87 **OS Grid Reference** NY914745
Personnel R Henderson, Unit Manager

PRODUCT DATA	**Rock type**	Dolerite Whin Sill
	Colour	Dark grey
	Grain	Medium
	Products	Crushed Stone
		Asphalt or Tarmacadam
		Pre-cast Concrete Products
DRILLING DATA		Hole Diameter 125mm
		Bench Height 14m
	Drills	Bohler DCT122
COMPRESSORS		Ingersoll-Rand
SECONDARY BREAKING		Dropball
LOAD AND HAUL	**Trucks**	Aveling Barford
	Loaders	Caterpillar
CRUSHING PLANT		No information given

TARMAC ROADSTONE (EASTERN) LTD (TARMAC QUARRY PRODUCTS LTD)

BOLSOVER MOOR QUARRY | 552

Whatley Road, Bolsover, Chesterfield, Derbys S44 6XE
Phone 0246 823141/822123
OS Map 120 **OS Grid Reference** SK503714
Personnel G Knapper, Unit Manager

PRODUCT DATA	**Rock type**	Permian Magnesian Limestone
	Colour	Cream
	Grain	Coarse
	Product	Crushed Stone
DRILLING DATA		Hole Diameter 105mm
		Bench Height 15m
	Drills	Halco 410C
COMPRESSORS		CompAir 650HE

entry continues overleaf

SECONDARY BREAKING		Dropball on Ruston-Bucyrus 38-RB
LOAD AND HAUL	**Trucks**	(3)Aveling Barford RD30
	Loaders	Caterpillar 980C
		Caterpillar 986A
		Komatsu WA500
CRUSHING PLANT	**Primary**	Sheepbridge 4850
		Sheepbridge 3024

TARMAC ROADSTONE (EASTERN) LTD (TARMAC QUARRY PRODUCTS LTD)

COWDOR AND HALL DALE QUARRY 553

Matlock, Derbys DE4 2HJ
Phone 0629 822104
OS Map 119 **OS Grid Reference** SK288604

QUARRY ON STAND

ROCK DATA	**Rock type**	Carboniferous Monsal Dale Limestone
	Colour	Light grey

TARMAC ROADSTONE (EASTERN) LTD (TARMAC QUARRY PRODUCTS LTD)

DENE QUARRY 554

Cromford, Nr Matlock, Derbys DE4 3QS
Phone 0629 822104
OS Map 119 **OS Grid Reference** SK291563
Personnel D Pargeter, Unit Manager

PRODUCT DATA	**Rock type**	Carboniferous Monsal Dale and Eyam Limestone
	Colour	Grey
	Products	Crushed Stone
		Asphalt or Tarmacadam
		Industrial Limestone
		Agricultural Lime
DRILLING DATA		Hole Diameter 110mm
		Bench Height 15m
	Drills	Bohler DTC122
		Halco 410C
COMPRESSORS		Atlas Copco XRH350S
SECONDARY BREAKING		Dropball on Ruston-Bucyrus 22-RB
LOAD AND HAUL	**Trucks**	(2)Aveling Barford RD028
		(4)Aveling Barford RD30
		(4)Aveling Barford RD40
		Volvo BM 4400
	Loaders	Caterpillar 950E
		Caterpillar 966C
		Caterpillar 986C
		(2)Caterpillar 988B
CRUSHING PLANT		No information given

TARMAC ROADSTONE (EASTERN) LTD (TARMAC QUARRY PRODUCTS LTD)

HOLME HALL QUARRY
555

Stainton, Maltby, Rotherham, S Yorks S66 7RD
Phone 0709 814491
OS Map 111   **OS Grid Reference** SK545940
Personnel J Lewis, Unit Manager

PRODUCT DATA	**Rock type**	Permian Magnesian Limestone
	Colour	Cream, white
	Products	Crushed Stone
		Asphalt or Tarmacadam
		Ready-mixed Concrete
		Industrial Limestone
		Agricultural Lime
DRILLING DATA		Hole Diameter 105mm
		Bench Height 10–20m
	Drills	(2)Hausherr HBM 80RD
LOAD AND HAUL	**Trucks**	Aveling Barford RD44
		Aveling Barford RD55
	Loaders	Komatsu WA450
		(2)Komatsu WA500
		(2)Komatsu WA600
CRUSHING PLANT	**Primary**	(2)Pegson 30×42 Jaw
		Goodwin 30×24 Jaw
		BJD 55
		Wageneder Impactor
	Secondary	(3)Mansfield No 4 Hammermill
		(2)Mansfield No 5 Hammermill
	Tertiary	Mansfield No 4 Hammermill

TARMAC ROADSTONE (EASTERN) LTD (TARMAC QUARRY PRODUCTS LTD)

HOWICK QUARRY
556

Little Houghton, Alnwick, Northumb NE66 3JY
Phone 0665 77641
OS Map 81   **OS Grid Reference** NU238169
Personnel T Thompson, Unit Manager

PRODUCT DATA	**Rock type**	Basalt
	Colour	Greyish black
	Grain	Fine
	Products	Crushed Stone
		Asphalt or Tarmacadam
DRILLING DATA		Hole Diameter 125mm
		Bench Height 15m
	Drills	Bohler DTC122 DHD
COMPRESSORS		CompAir 650HE
SECONDARY BREAKING		Hydraulic Hammer
		Dropball on Ruston-Bucyrus 22-RB

entry continues overleaf

LOAD AND HAUL	**Trucks**	Terex
		Aveling Barford RD35
	Loaders	Komatsu WA500
		Caterpillar 980C

| CRUSHING PLANT | | No information given |

TARMAC ROADSTONE (EASTERN) LTD (TARMAC QUARRY PRODUCTS LTD)

MIDDLE PEAK QUARRY

557

Wirksworth, Derbys DE4 4BU
Phone 0629 822207
OS Map 119   **OS Grid Reference** SK285543
Personnel K Marshall, Unit Manager

PRODUCT DATA	**Rock type**	Carboniferous Monsal Dale and Hopton Limestone
	Colour	Cream, grey
	Products	Crushed Stone
		Asphalt or Tarmacadam
		Industrial Limestone
		Agricultural Lime

DRILLING DATA		Hole Diameter 110–125mm
		Bench Height 15m
	Drills	Ingersoll-Rand CM350
		Atlas Copco ROC404

| COMPRESSORS | | Atlas Copco XRH350S |

| SECONDARY BREAKING | | Dropball on Ruston-Bucyrus 38-RB |

LOAD AND HAUL	**Trucks**	(4)Aveling Barford RD30
		(4)Aveling Barford RD40
		(2)Aveling Barford RD50
	Loaders	Caterpillar 950
		(2)Caterpillar 966D
		Caterpillar 980C
		(2)Caterpillar 988B
		Volvo BM 4400

CRUSHING PLANT	**Primary**	Pegson Jaw
		Hadfield Jaw
	Secondary	(2)Hazemag APK60
	Tertiary	Mansfield No 4 Hammermill
		Hazemag APK30
		Parker
		Barmac Duopactor

TARMAC ROADSTONE (EASTERN) LTD (TARMAC QUARRY PRODUCTS LTD)

MOOTLAW QUARRY

558

Mootlaw, Matfen, Ryall, Newcastle upon Tyne, Northumb NE20 0TB
Phone 0661 6222
OS Map 88   **OS Grid Reference** NY024746
Personnel I Cambell, Unit Manager

PRODUCT DATA	**Rock type**	Carboniferous Limestone
	Colour	Medium grey
	Products	Crushed Stone
		Industrial Limestone
		Agricultural Lime
DRILLING DATA		Hole Diameter 90mm
		Bench Height 15m
	Drills	Ingersoll-Rand CM340
SECONDARY BREAKING		Hydraulic Hammer
LOAD AND HAUL	**Trucks**	Terex
	Loaders	Komatsu
CRUSHING PLANT		No information given

TARMAC ROADSTONE (EASTERN) LTD (TARMAC QUARRY PRODUCTS LTD)

OLD QUARRINGTON QUARRY 559

Old Quarrington, Bowburn, Durham DH6 5NN
Phone 0913 770852
OS Map 93 **OS Grid Reference** NZ330378
Personnel A Barrett, Unit Manager

PRODUCT DATA	**Rock type**	Permian Magnesian Limestone
	Colour	Not commercially significant
	Grain	Coarse
	Products	Crushed Stone
		Pre-cast Concrete Products
		Agricultural Lime
DRILLING DATA		Hole Diameter 125mm
		Bench Height 30m
	Drills	Hausherr HB80 RD
SECONDARY BREAKING		Dropball
LOAD AND HAUL	**Trucks**	Aveling Barford
	Loaders	Komatsu WA450
CRUSHING PLANT	**Primary**	Sheepbridge Jaw
	Secondary	Cat Hammermill
	Tertiary	None

TARMAC ROADSTONE (EASTERN) LTD (TARMAC QUARRY PRODUCTS LTD)

RAISBY QUARRY 560

Coxhoe, Durham DH6 4BB
Phone 0913 770611
OS Map 93 **OS Grid Reference** NZ340353
Personnel A Barrett, Unit Manager

entry continues overleaf

PRODUCT DATA	**Rock type**	Permian Magnesian Limestone
	Colour	Not commercially significant
	Grain	Coarse
	Products	Crushed Stone
		Asphalt or Tarmacadam
		Industrial Limestone
		Agricultural Lime
DRILLING DATA		Hole Diameter 125mm
		Bench Height 30m
	Drills	Hausherr HB80 RD
SECONDARY BREAKING		Dropball on Ruston-Bucyrus 22-RB
LOAD AND HAUL	**Trucks**	Aveling Barford
		Terex
	Loaders	Komatsu WA400
		Komatsu WA450
		Komatsu WA500
CRUSHING PLANT	**Primary**	Baxter ST42×32 Jaw (Single Toggle)
		Sheepbridge Impactor
	Secondary	(2)Mansfield No 2 Hammermill
		(2)Mansfield No 5 Hammermill
	Tertiary	Nordberg 4.25′ Cone

TARMAC ROADSTONE (EASTERN) LTD (TARMAC QUARRY PRODUCTS LTD)

REDMIRE QUARRY

561

Leyburn, N Yorks DL8 4HD
Phone 0969 22342
OS Map 98 **OS Grid Reference** SE050928
Personnel T Robson, Unit Manager

PRODUCT DATA	**Rock type**	Carboniferous Great Scar Limestone
	Colour	Not commercially significant
	Product	Crushed Stone
DRILLING DATA		Hole Diameter 110mm
		Bench Height 9–18m
	Drills	Ingersoll-Rand CM300
COMPRESSORS		Atlas Copco XR350H
SECONDARY BREAKING		Dropball on Ruston-Bucyrus 30-RB
LOAD AND HAUL	**Trucks**	Aveling Barford RD30
		Euclid
	Loaders	Caterpillar 920
		Michigan 125
CRUSHING PLANT	**Primary**	Sheepbridge Impactor
	Secondary	Mansfield No 4 Hammermill
	Tertiary	None

TARMAC ROADSTONE (EASTERN) LTD (TARMAC QUARRY PRODUCTS LTD)

THRESHFIELD QUARRY

562

Skyrethornes Lane, Threshfield, Grassington, Skipton, N Yorks BD23 5PA
Phone 0756 752318
OS Map 98   **OS Grid Reference** SD982638
Personnel B Gill, Unit Manager

PRODUCT DATA	**Rock type**	Carboniferous Great Scar Limestone
	Colour	Medium grey
	Product	Crushed Stone
DRILLING DATA		Hole Diameter 125mm
		Bench Height 20–25m
	Drills	Halco 410C
SECONDARY BREAKING		Dropball on Ruston-Bucyrus 30-RB
LOAD AND HAUL	**Trucks**	Aveling Barford RD25
	Loaders	(2)Komatsu WA450
		Komatsu WA500
CRUSHING PLANT	**Primary**	Hadfield 55 × 36 Jaw
		Sheepbridge 48 × 42 Jaw
	Secondary	(2)Sheepbridge 36 × 30 Jaw
	Tertiary	Parker Kubitizer

TARMAC ROADSTONE (EAST MIDLANDS) LTD (TARMAC QUARRY PRODUCTS LTD)
Head Office *John Hadfield House, Dale Road, Matlock, Derbys DE4 3PL*
Phone 0629 580300
Personnel C Shaw, Managing Director
A Jones, Production Manager
Quarries Ashwood Dale, Cliffe Hill, Hopton Wood, Old Cliffe Hill

---

TARMAC ROADSTONE (EAST MIDLANDS) LTD (TARMAC QUARRY PRODUCTS LTD)

ASHWOOD DALE QUARRY

563

Buxton, Derbys SK17 9TB
Phone 0536 67311
OS Map 119   **OS Grid Reference** SK080726

QUARRY ON STAND

| ROCK DATA | **Rock type** | Carboniferous Limestone |

TARMAC ROADSTONE (EAST MIDLANDS) LTD (TARMAC QUARRY PRODUCTS LTD)

CLIFFE HILL QUARRY

564

Battleflat Lane, Ellistown, Leics LE6 1FA
Phone 0530 230530
OS Map 129   **OS Grid Reference** SK459113
Personnel M Bills, Unit Manager

PRODUCT DATA	**Rock type**	Diorite
	Colour	Mottled grey
	Grain	Coarse
	Products	Crushed Stone
		Asphalt or Tarmacadam

entry continues overleaf

DRILLING DATA		Contractor A Jones Rock Drillers
		No further information

CRUSHING PLANT	**Primary**	(2)Pegson 30 × 42 Jaw (Mobile)
	Secondary	Gyrasphere (Mobile)
	Tertiary	Pegson Impactor (Mobile)

TARMAC ROADSTONE (EAST MIDLANDS) LTD (TARMAC QUARRY PRODUCTS LTD)

HOPTON WOOD QUARRY

565

Middleton, Derbys DE4 4BU
Phone 0536 67311
OS Map 119 **OS Grid Reference** SK261555

QUARRY ON STAND

ROCK DATA	**Rock type**	Carboniferous Limestone
	Colour	Grey

TARMAC ROADSTONE (EAST MIDLANDS) LTD (TARMAC QUARRY PRODUCTS LTD)

OLD CLIFFE HILL QUARRY

566

Markfield, Leics LE6 0TW
Phone 0530 242561
OS Map 129 **OS Grid Reference** SK473106
Personnel K Toon, Unit Manager

PRODUCT DATA	**Rock type**	Diorite
	Colour	Mottled grey, pink
	Grain	Coarse
	Products	Crushed Stone
		Asphalt or Tarmacadam
		Ready-mixed Concrete

ALL WORK SUB-CONTRACTED

TARMAC ROADSTONE (NORTH WEST) LTD (TARMAC QUARRY PRODUCTS LTD)
Head Office Moorcroft, Lismore Road, Buxton, Derbys SK17 9AP
Phone 0298 25441 Fax 0298 72413
Personnel J H Glaves, Managing Director
P Markham, Production Manager
G Clarke, Production Manager
Quarries Arcow, Bankfield, Bold Venture, Denbigh, Dunald Mill, Hillhead, Horton, Kendal Fell, Minera,
Sandside, Tendley, Topley Pike, Water Swallows

TARMAC ROADSTONE (NORTH WEST) LTD (TARMAC QUARRY PRODUCTS LTD)

ARCOW QUARRY

567

Helwith Bridge, Settle, N Yorks BD24 0EW
Phone 072 96310
OS Map 98 **OS Grid Reference** SD804704
Personnel R Mason, Unit Manager

PRODUCT DATA	**Rock type**	Silurian Sandstone (Austwick Formation)
	Colour	Not commercially significant
	Products	Crushed Stone
		Asphalt or Tarmacadam

DRILLING DATA		Hole Diameter 105mm
		Bench Height 14m
	Drills	(2)Halco 410C

COMPRESSORS		CompAir RO60–250
		Ingersoll-Rand DXL700H

SECONDARY BREAKING		Dropball on Ruston-Bucyrus 22-RB

LOAD AND HAUL	**Trucks**	Volvo
	Loaders	Caterpillar

CRUSHING PLANT	**Primary**	Allis Chalmers Jaw
	Secondary	Kue Ken 108
		Nordberg 3' Cone (Shorthead)
		Nordberg 4' Cone (Shorthead)
	Tertiary	(2)Barmac Duopactor

TARMAC ROADSTONE (NORTH WEST) LTD (TARMAC QUARRY PRODUCTS LTD)

BANKFIELD QUARRY   568

Clitheroe, Lancs BB7 4QA
Phone 0200 22371
OS Map 103   **OS Grid Reference** SD755426
Personnel G Longden, Unit Manager

PRODUCT DATA	**Rock type**	Carboniferous Chatburn Limestone
	Colour	Light brown
	Products	Crushed Stone
		Asphalt or Tarmacadam
		Industrial Limestone
		Agricultural Lime

DRILLING DATA		Hole Diameter 105mm
		Bench Height 6–15m
	Drills	Bohler DTC122

SECONDARY BREAKING		Hydraulic Hammer - Montabert

LOAD AND HAUL	**Trucks**	Aveling Barford RD40
	Loaders	Caterpillar 966C
		Caterpillar 980C
		Volvo BM 4500

CRUSHING PLANT	**Primary**	Sheepbridge Jaw
		Sheepbridge 650 Double Impactor
	Secondary	Hazemag APK60
	Tertiary	Mansfield No 4 Hammermill

TARMAC ROADSTONE (NORTH WEST) LTD (TARMAC QUARRY PRODUCTS LTD)

BOLD VENTURE QUARRY   569

Chatburn, Clitheroe, Lancs BB7 4LA
Phone 0200 41355
OS Map 103   **OS Grid Reference** SD761437
Personnel C Widdas, Works Manager

entry continues overleaf

PRODUCT DATA   **Rock type**   Carboniferous Chatburn Limestone
                   Colour   Medium grey
                Products   Crushed Stone
                                    Asphalt or Tarmacadam
                                    Industrial Limestone

CRUSHING AND COATING PLANT ONLY

NO FURTHER INFORMATION

TARMAC ROADSTONE (NORTH WEST) LTD (TARMAC QUARRY PRODUCTS LTD)

DENBIGH QUARRY   570

Craig Road, Ruthin, Clwyd LL16 3YE
Phone 0745 714551
OS Map 116   **OS Grid Reference** SJ053668
Personnel J I Jones, Unit Manager

PRODUCT DATA   **Rock type**   Carboniferous Limestone
                   Colour   Brownish grey
                Products   Crushed Stone
                                    Asphalt or Tarmacadam
                                    Ready-mixed Concrete

DRILLING DATA   Hole Diameter 105mm
                                    Bench Height 20m
                                    Contractor D & H Rockdriller

SECONDARY BREAKING   Dropball

LOAD AND HAUL   **Trucks**   Aveling Barford
                                      Volvo
                  Loaders   Komatsu WA500

CRUSHING PLANT   No information given

TARMAC ROADSTONE (NORTH WEST) LTD (TARMAC QUARRY PRODUCTS LTD)

DUNALD MILL QUARRY   571

Nether Kellet, Carnforth, Lancs LA6 1HE
Phone 0524 734911
OS Map 97   **OS Grid Reference** SD515679
Personnel E Hodgson, Unit Manager

PRODUCT DATA   **Rock type**   Carboniferous Limestone
                   Colour   Medium grey
                Products   Crushed Stone
                                    Asphalt or Tarmacadam
                                    Industrial Limestone

DRILLING DATA   Hole Diameter 105mm
                                    Bench Height 18m
                    Drills   Halco 410C

SECONDARY BREAKING   Hydraulic Hammer - Montabert

LOAD AND HAUL	**Trucks**	Aveling Barford RD40
	Loaders	Caterpillar 966C
		Caterpillar 988B
CRUSHING PLANT	**Primary**	Sheepbridge Jaw
	Secondary	(2)Mansfield No 3 Hammermill
	Tertiary	None

TARMAC ROADSTONE (NORTH WEST) LTD (TARMAC QUARRY PRODUCTS LTD)

HILLHEAD QUARRY   572

Wormhill, Buxton, Derbys SK17 8EJ
Phone 0298 23638
OS Map 119   **OS Grid Reference** SK129680
Personnel W Renshaw, Unit Manager

PRODUCT DATA	**Rock type**	Carboniferous Bee Low Limestone
	Colour	Not commercially significant
	Products	Crushed Stone
		Agricultural Lime
DRILLING DATA		Hole Diameter 110mm
		Bench Height 22m
	Drills	Halco 625H
SECONDARY BREAKING		Hydraulic Hammer
		Dropball on Ruston-Bucyrus 22-RB
LOAD AND HAUL	**Trucks**	Aveling Barford
		Volvo
	Loaders	Komatsu WA500
CRUSHING PLANT	**Primary**	Pegson Jaw
	Secondary	Hazemag APK60
	Tertiary	(2)Mansfield Hammermill

TARMAC ROADSTONE (NORTH WEST) LTD (TARMAC QUARRY PRODUCTS LTD)

HORTON QUARRY   573

Horton in Ribblesdale, Settle, N Yorks BD24 0HR
Phone 072 96301
OS Map 98   **OS Grid Reference** SD796772
Personnel N Bond, Unit Manager

PRODUCT DATA	**Rock type**	Carboniferous Great Scar Limestone
	Colour	Bluish grey
	Products	Crushed Stone
		Industrial Limestone
		Burnt Limestone
		Agricultural Lime
DRILLING DATA		Hole Diameter 125mm
		Bench Height 30m
	Drills	Halco SDP43-22
SECONDARY BREAKING		Dropball on Ruston-Bucyrus 22-RB

entry continues overleaf

LOAD AND HAUL	**Trucks**	Aveling Barford
		Terex
	Loaders	Caterpillar

| CRUSHING PLANT | | No information provided |

TARMAC ROADSTONE (NORTH WEST) LTD (TARMAC QUARRY PRODUCTS LTD)

KENDAL FELL QUARRY

574

Underbarrow Road, Kendal, Cumbria LA9 5RT
Phone 0539 721280
OS Map 97   **OS Grid Reference** SD505225
Personnel C Craystone, Unit Manager

PRODUCT DATA	**Rock type**	Carboniferous Great Scar Limestone
	Colour	Light grey
	Products	Crushed Stone
		Ready-mixed Concrete
		Pre-cast Concrete Products

DRILLING DATA		Hole Diameter 105mm
		Bench Height 9–10m
	Drills	Holman Voltrak
		Halco 410C

| SECONDARY BREAKING | | Hydraulic Hammer |

LOAD AND HAUL	**Trucks**	Volvo
	Loaders	Caterpillar 966C
		Caterpillar 966D

CRUSHING PLANT	**Primary**	Parker 32 × 42 Jaw
	Secondary	Mansfield No 5 Hammermill
	Tertiary	None

TARMAC ROADSTONE (NORTH WEST) LTD (TARMAC QUARRY PRODUCTS LTD)

MINERA QUARRY

575

Nr Wrexham, Clwyd LL11 3DS
Phone 0978 720123
OS Map 117   **OS Grid Reference** SJ253521
Personnel E Jones, Unit Manager

PRODUCT DATA	**Rock type**	Carboniferous Limestone
	Colour	Brownish grey
	Products	Crushed Stone
		Asphalt or Tarmacadam
		Industrial Limestone
		Agricultural Lime

DRILLING DATA		Hole Diameter 105mm
		Bench Height 18m
	Drills	Holman Voltrak

| SECONDARY BREAKING | | Dropball on Ruston-Bucyrus 22-RB |

LOAD AND HAUL	**Trucks**	Aveling Barford
		Heathfield
		(2)Volvo
	Loaders	Komatsu WA300

| CRUSHING PLANT | | No information provided |

TARMAC ROADSTONE (NORTH WEST) LTD (TARMAC QUARRY PRODUCTS LTD)

SANDSIDE QUARRY **576**

Milnthorpe, Cumbria LA7 7HW
Phone 0448 23587
OS Map 96 **OS Grid Reference** SD482808
Personnel M Ellison, Unit Manager

PRODUCT DATA	**Rock type**	Carboniferous Limestone
	Colour	Grey
	Products	Crushed Stone
		Asphalt or Tarmacadam
		Industrial Limestone
		Agricultural Lime

DRILLING DATA		Hole Diameter 108mm
		Bench Height 13-21m
	Drills	Halco 410C

| SECONDARY BREAKING | | Hydraulic Hammer - Krupp |

LOAD AND HAUL	**Trucks**	Caterpillar
		Volvo
	Loaders	Caterpillar
		Komatsu WA250

CRUSHING PLANT	**Primary**	Sheepbridge 48 × 50 Jaw
	Secondary	Mansfield No 3 Hammermill
		Mansfield No 4 Hammermill

TARMAC ROADSTONE (NORTH WEST) LTD (TARMAC QUARRY PRODUCTS LTD)

TENDLEY QUARRY **577**

Brigham, Cockermouth, Cumbria CA13 0FE
Phone 0900 823962
OS Map 89 **OS Grid Reference** NY088287
Personnel J Collister, Unit Manager

PRODUCT DATA	**Rock type**	Carboniferous Limestone
	Colour	Light grey
	Products	Crushed Stone
		Asphalt or Tarmacadam
		Agricultural Lime

DRILLING DATA		Hole Diameter 105mm
		Bench Height 11-15m
	Drills	Halco 410C

| LOAD AND HAUL | **Trucks** | (2)Heathfield |
| | **Loaders** | (2)Caterpillar 966D |

entry continues overleaf

CRUSHING PLANT	**Primary**	Blake 38 × 48 Jaw
	Secondary	Mansfield No 4 Hammermill
	Tertiary	Mansfield No 2 Hammermill

TARMAC ROADSTONE (NORTH WEST) LTD (TARMAC QUARRY PRODUCTS LTD)

TOPLEY PIKE QUARRY

578

Nr Buxton, Derbys SK17 9TG
Phone 0298 212351
OS Map 119 **OS Grid Reference** SK104722
Personnel P Mallet, Unit Manager

PRODUCT DATA	**Rock type**	Carboniferous Woo Dale Limestone
	Colour	Not commercially significant
	Products	Crushed Stone
		Asphalt or Tarmacadam
		Industrial Limestone
		Agricultural Lime

DRILLING DATA		Hole Diameter 105mm
		Bench Height 33m
	Drills	Halco 625H

| SECONDARY BREAKING | | Dropball on Ruston-Bucyrus 22-RB |

LOAD AND HAUL	**Trucks**	Aveling Barford
		Heathfield
	Loaders	Caterpillar 988B

CRUSHING PLANT	**Primary**	Sheepbridge 48 × 50 Double Impactor
	Secondary	Sheepbridge Single Impactor
	Tertiary	Mansfield No 4 Hammermill

TARMAC ROADSTONE (NORTH WEST) LTD (TARMAC QUARRY PRODUCTS LTD)

WATER SWALLOWS QUARRY

579

Buxton, Derbys SK17 7JB
Phone 0298 22395
OS Map 119 **OS Grid Reference** SK085750
Personnel R Hawksworth, Unit Manager

PRODUCT DATA	**Rock type**	Olivine Basalt
	Colour	Grey to green
	Grain	Fine
	Products	Crushed Stone
		Asphalt or Tarmacadam

DRILLING DATA		Hole Diameter 105mm
		Bench Height 10-14m
	Drills	Ingersoll-Rand CM340

| SECONDARY BREAKING | | Dropball on Smith 210 |

LOAD AND HAUL	**Trucks**	Terex
		Heathfield
	Loaders	Volvo BM 4600
		Caterpillar 926
		Caterpillar 980C

CRUSHING PLANT	**Primary**	Pegson Jaw
		Goodwin Barsby (Mobile)
	Secondary	(2)Hazemag APK60
	Tertiary	(2)Cone

TARMAC ROADSTONE (SCOTLAND) LTD (TARMAC QUARRY PRODUCTS LTD)

Head Office *134 Nithsdale Drive, Glasgow, Strathclyde G41 2PP*
Phone *0414 236611* **Fax** *0414 231034* **Telex** *779054*
Personnel *J Mawdsley, Chairman*
R J Harrison, Managing Director
D Warrender, Production Director
N Boyle, Production Manager
I McDonald, Production Manager
Vacant, Production Manager
Quarries *Blairhill, Clatchard Craig, Coatsgate, Craigenlow, Craighouse, Craignair, Craigpark, Craigs, Croy, Cunmont, Dunion, Ethiebeaton, Goat, High Craig, Kaimes, Kirkmabreck, Loanhead, Morefield, Morrinton, Murrayshall, Ravelrig, Shierglas*

TARMAC ROADSTONE (SCOTLAND) LTD (TARMAC QUARRY PRODUCTS LTD)

BLAIRHILL QUARRY 580

Black Ridge, Bathgate, Lothian EH48 3AG
Phone 05015 577
OS Map 65 **OS Grid Reference** NS888660
Personnel A Ray, Unit Manager

PRODUCT DATA	**Rock type**	Basalt
	Colour	Grey to black
	Grain	Fine
	Products	Crushed Stone
		Asphalt or Tarmacadam

| DRILLING DATA | | Contractor Ritchies Equipment |

NO FURTHER INFORMATION PROVIDED

TARMAC ROADSTONE (SCOTLAND) LTD (TARMAC QUARRY PRODUCTS LTD)

CLATCHARD CRAIG QUARRY 581

Newburgh, Cupar, Fife KY14 6JJ
Phone 0337 40226
OS Map 59 **OS Grid Reference** NO244178
Personnel D Scott, Unit Manager

PRODUCT DATA	**Rock type**	Basalt
	Colour	Not commercially significant
	Grain	Fine
	Products	Crushed Stone
		Asphalt or Tarmacadam

DRILLING DATA		Hole Diameter 105mm
		Bench Height 12m
	Drills	Halco 400C

| COMPRESSORS | | CompAir RO60-170 |

| SECONDARY BREAKING | | Dropball |

entry continues overleaf

| LOAD AND HAUL | **Trucks** | Volvo |
| | **Loaders** | Caterpillar 950 |

| CRUSHING PLANT | | No information provided |

TARMAC ROADSTONE (SCOTLAND) LTD (TARMAC QUARRY PRODUCTS LTD)

COATSGATE QUARRY 582

Beattock, Moffat, Dumfries & Galloway, DG10 9SL
Phone 06833 405
OS Map 78   **OS Grid Reference** NT064053
Personnel W Ireland, Unit Manager

PRODUCT DATA	**Rock type**	Silurian Greywacke
	Colour	Not commercially significant
	Products	Crushed Stone
		Asphalt or Tarmacadam

| DRILLING DATA | | Hole Diameter 110mm |
| | **Drills** | Halco 410C |

| COMPRESSORS | | CompAir |

NO FURTHER INFORMATION PROVIDED

TARMAC ROADSTONE (SCOTLAND) LTD (TARMAC QUARRY PRODUCTS LTD)

CRAIGENLOW QUARRY 583

Dunecht, Aberdeen, Grampian AB3 7ED
Phone 0330 3361
OS Map 38   **OS Grid Reference** NJ731093
Personnel W Henston, Unit Manager

PRODUCT DATA	**Rock type**	Hill of Fare Granite
	Colour	Not commercially significant
	Grain	Coarse
	Products	Crushed Stone
		Asphalt or Tarmacadam

DRILLING DATA		Hole Diameter 105mm
		Bench Height 11–23m
	Drills	Ingersoll-Rand CM351

| COMPRESSORS | | Ingersoll-Rand DXL700H |

| SECONDARY BREAKING | | Hydraulic Hammer |
| | | Dropball on Ruston-Bucyrus 38-RB |

LOAD AND HAUL	**Face shovel**	O&K
	Trucks	Terex
	Loaders	Caterpillar

| CRUSHING PLANT | | No information provided |

TARMAC ROADSTONE (SCOTLAND) LTD (TARMAC QUARRY PRODUCTS LTD)

CRAIGHOUSE QUARRY

584

Melrose, Roxburgh, Borders TD6 9DU
Phone 0896 822085
OS Map 73 **OS Grid Reference** NT600362
Personnel J Robertson, Unit Manager

PRODUCT DATA	**Rock type**	Basalt
	Colour	Grey
	Grain	Fine
	Products	Crushed Stone
		Asphalt or Tarmacadam
DRILLING DATA		Hole Diameter 105mm
		Bench Height 18m
	Drills	Ingersoll-Rand CM351
SECONDARY BREAKING		Dropball on Ruston-Bucyrus 22-RB
LOAD AND HAUL	**Trucks**	Terex
	Loaders	Caterpillar
CRUSHING PLANT	**Primary**	Lokomo 100
	Secondary	Allis Chalmers 1445
	Tertiary	Hazemag APK50
		Allis Chalmers 236

TARMAC ROADSTONE (SCOTLAND) LTD (TARMAC QUARRY PRODUCTS LTD)

CRAIGNAIR QUARRY

585

Dalbeattie, Dumfries & Galloway, DG5 4LZ
Phone 0556 610206
OS Map 84 **OS Grid Reference** NY820605
Personnel S K Ward, Unit Manager

PRODUCT DATA	**Rock type**	Griftell Granite
	Colour	Not commercially significant
	Grain	Coarse
	Products	Crushed Stone
		Asphalt or Tarmacadam
DRILLING DATA		Hole Diameter 105mm
		Bench Height 20m
	Drills	Holman Voltrak
COMPRESSORS		CompAir RO60-170
LOAD AND HAUL	**Trucks**	Lever
		Heathfield
	Loaders	Michigan
CRUSHING PLANT	**Primary**	Baxter 50 × 36 Jaw
	Secondary	Kue Ken 42 × 14 Jaw
	Tertiary	(2)Allis Chalmers Autocone

CRAIGPARK QUARRY

586

Wilkieston Rd, Ratho, Newbridge, Lothian EH28 8RT
Phone 0313 331363
OS Map 66   **OS Grid Reference** NT127706
Personnel W Greenock, Unit Manager

PRODUCT DATA	**Rock type**	Olivine Basalt
	Colour	Greenish black
	Grain	Fine
	Products	Crushed Stone
		Asphalt or Tarmacadam

DRILLING DATA		Hole Diameter 125mm
		Bench Height 18m
	Drills	Ingersoll-Rand LM600

| SECONDARY BREAKING | | Dropball on Ruston-Bucyrus 38-RB |

LOAD AND HAUL	**Trucks**	Terex R25
	Loaders	Terex
		Caterpillar 966C

CRUSHING PLANT	**Primary**	Goodwin Barsby 48 × 36 Jaw
	Secondary	Hazemag APK60
	Tertiary	None

CRAIGS QUARRY

587

Auchterderran, Cardenden, Fife KY5 0HE
Phone 0316 645555
OS Map 58   **OS Grid Reference** NT216965

QUARRY ON STAND

| ROCK DATA | **Rock type** | Basalt |

CROY QUARRY

588

Constarry Rd, Croy, Kilsyth, Strathclyde G65 9HY
Phone 0236 823274
OS Map 64   **OS Grid Reference** NS730758
Personnel R Fulton, Unit Manager

PRODUCT DATA	**Rock type**	Basalt
	Colour	Very dark grey
	Grain	Fine
	Products	Crushed Stone
		Asphalt or Tarmacadam

| DRILLING DATA | **Drills** | Holman Voltrak |

NO FURTHER INFORMATION PROVIDED

TARMAC ROADSTONE (SCOTLAND) LTD (TARMAC QUARRY PRODUCTS LTD)

CUNMONT QUARRY 589

Kingennie, Broughty Ferry, Dundee, Tayside DD5 3PX
Phone 0307 67374
OS Map 54   **OS Grid Reference** NO489370

QUARRY ON STAND

ROCK DATA	**Rock type**	Dolerite
	Colour	Bluish grey
	Grain	Coarse

TARMAC ROADSTONE (SCOTLAND) LTD (TARMAC QUARRY PRODUCTS LTD)

DUNION QUARRY 590

Jedburgh, Borders TD8 6SZ
Phone 0316 645555
OS Map 80   **OS Grid Reference** NT625191

QUARRY ON STAND

ROCK DATA	**Rock type**	Dolerite
	Colour	Pinkish blue
	Grain	Medium

TARMAC ROADSTONE (SCOTLAND) LTD (TARMAC QUARRY PRODUCTS LTD)

ETHIEBEATON QUARRY 591

Ethiebeaton, Broughty Ferry, Dundee, Tayside DD5 3RB
Phone 038253 4611
OS Map 54   **OS Grid Reference** NO478342
Personnel D Randal, Unit Manager

PRODUCT DATA	**Rock type**	Dolerite
	Colour	Greyish black
	Grain	Medium
	Product	Crushed Stone
DRILLING DATA		Hole Diameter 105mm
		Bench Height 18m
	Drills	Halco 400C
COMPRESSORS		Broomwade WR600
SECONDARY BREAKING		Dropball
LOAD AND HAUL	**Trucks**	Terex
	Loaders	Caterpillar
CRUSHING PLANT	**Primary**	Parker 32 × 42 Jaw
	Secondary	Allis Chalmers 836 Cone
	Tertiary	Allis Chalmers 436 Cone

TARMAC ROADSTONE (SCOTLAND) LTD (TARMAC QUARRY PRODUCTS LTD)

GOAT QUARRY

592

Aberdour, Burntisland, Fife KY3 0SB
Phone 0383 860517
OS Map 66   **OS Grid Reference** NT173866
Personnel L Petrie, Unit Manager

PRODUCT DATA	**Rock type**	Dolerite
	Colour	Greyish black
	Grain	Medium
	Product	Crushed Stone

DRILLING DATA		Hole Diameter 125mm
		Bench Height 26m
	Drills	Ingersoll-Rand LM600

| COMPRESSORS | | Ingersoll-Rand |

| SECONDARY BREAKING | | Dropball on Ruston-Bucyrus 30-RB |

LOAD AND HAUL	**Trucks**	Terex
	Loaders	Komatsu WA57
		Komatsu WA420

CRUSHING PLANT	**Primary**	Lokomo K200
	Secondary	Allis Chalmers
	Tertiary	None

TARMAC ROADSTONE (SCOTLAND) LTD (TARMAC QUARRY PRODUCTS LTD)

HIGH CRAIG QUARRY

593

Rannoch Road, Johnstone, Strathclyde PA5 0SP
Phone 0505 20344
OS Map 64   **OS Grid Reference** NS426614
Personnel G Anderson, Unit Manager

PRODUCT DATA	**Rock type**	Basalt
	Colour	Dark grey
	Grain	Fine
	Products	Crushed Stone
		Asphalt or Tarmacadam

DRILLING DATA		Hole Diameter 108mm
		Bench Height 20m
	Drills	Drillmark rig

| SECONDARY BREAKING | | Dropball on Ruston-Bucyrus 38-RB |

| LOAD AND HAUL | **Trucks** | Terex |
| | **Loaders** | Komatsu |

CRUSHING PLANT	**Primary**	Hadfield 54 × 42 Jaw
	Secondary	Allis Chalmers 1650
	Tertiary	(3)Allis Chalmers

TARMAC ROADSTONE (SCOTLAND) LTD (TARMAC QUARRY PRODUCTS LTD)

KAIMES QUARRY

594

Kirknewton, Lothian EH27 8EF
Phone 0316 645555
OS Map 66   **OS Grid Reference** NT130664

QUARRY ON STAND

ROCK DATA	**Rock type**	Dolerite
	Colour	Dark grey
	Grain	Medium

TARMAC ROADSTONE (SCOTLAND) LTD (TARMAC QUARRY PRODUCTS LTD)

KIRKMABRECK QUARRY

595

Kirkmabreck, Newton Stewart, Dumfries & Galloway, DG8 7DJ
Phone 0671 82217
OS Map 83   **OS Grid Reference** NY480570
Personnel L Harvey, Unit Manager

ROCK DATA	**Rock type**	Biotite Granodiorite
	Colour	Silver grey
	Grain	Fine
	Product	Crushed Stone

(Closing July 1990)

NO FURTHER INFORMATION PROVIDED

TARMAC ROADSTONE (SCOTLAND) LTD (TARMAC QUARRY PRODUCTS LTD)

LOANHEAD QUARRY

596

Beith, Strathclyde KA15 2JN
Phone 0505 52534
OS Map 63   **OS Grid Reference** NS364555
Personnel J McGregor, Unit Manager

PRODUCT DATA	**Rock type**	Basalt
	Colour	Dark grey
	Grain	Fine
	Product	Crushed Stone

DRILLING DATA		Hole Diameter 125mm
		Bench Height 18-27m
	Drills	Ingersoll-Rand CM350

SECONDARY BREAKING		Hydraulic Hammer
		Talisker on Caterpillar 966C

LOAD AND HAUL	**Trucks**	Terex
	Loaders	Komatsu WA420

CRUSHING PLANT		No information provided

TARMAC ROADSTONE (SCOTLAND) LTD (TARMAC QUARRY PRODUCTS LTD)

MOREFIELD QUARRY

597

Ullapool, Highland IV26 2TH
Phone 0854 2336
OS Map 19 **OS Grid Reference** NH135952
Personnel R Rollo, Unit Manager

PRODUCT DATA	**Rock type**	Cambro-Ordovician Durness Limestone
	Colour	Grey
	Products	Crushed Stone
		Asphalt or Tarmacadam
		Agricultural Lime
DRILLING DATA		Hole Diameter 105mm
		Bench Height 18m
		Contractor Ritchies Equipment
SECONDARY BREAKING		Dropball on Ruston-Bucyrus 22-RB
LOAD AND HAUL	**Face shovel**	Caterpillar 225
	Trucks	Terex
CRUSHING PLANT	**Primary**	Pegson 18 × 32 Jaw
	Secondary	Kennedy 38
	Tertiary	Parker 24 × 6 Granulator

TARMAC ROADSTONE (SCOTLAND) LTD (TARMAC QUARRY PRODUCTS LTD)

MORRINTON QUARRY

598

Dumfries, Dumfries & Galloway, DG2 0JW
Phone 0387 82206
OS Map 78 **OS Grid Reference** NX876813
Personnel W Shields, Unit Manager

PRODUCT DATA	**Rock type**	Silurian Greywacke
	Colour	Not commercially significant
	Products	Crushed Stone
		Asphalt or Tarmacadam
DRILLING DATA		Hole Diameter 110mm
		Bench Height 18–25m
	Drills	Holman Zoomtrak
COMPRESSORS		Holman
SECONDARY BREAKING		Dropball on Ruston-Bucyrus 22-RB
LOAD AND HAUL	**Trucks**	Terex
	Loaders	Volvo
CRUSHING PLANT		No information provided

TARMAC ROADSTONE (SCOTLAND) LTD (TARMAC QUARRY PRODUCTS LTD)

MURRAYSHALL QUARRY

599

Cambusbarron, Stirling, Central FK7 9QA
Phone 0786 51736
OS Map 57 **OS Grid Reference** NS770912
Personnel P Tweed, Unit Manager

PRODUCT DATA	**Rock type**	Basalt
	Colour	Medium grey
	Grain	Fine
	Products	Crushed Stone
		Asphalt or Tarmacadam
DRILLING DATA		Hole Diameter 105mm
		Bench Height 14 and 27m
	Drills	Halco 400C
SECONDARY BREAKING		Hydraulic Hammer
LOAD AND HAUL	**Trucks**	Caterpillar
	Loaders	Caterpillar 950
CRUSHING PLANT		No information provided

TARMAC ROADSTONE (SCOTLAND) LTD (TARMAC QUARRY PRODUCTS LTD)

RAVELRIG QUARRY

600

Kirknewton, Lothian EH27 8EF
Phone 0314 495523
OS Map 65 **OS Grid Reference** NT130665
Personnel A Dean, Unit Manager

PRODUCT DATA	**Rock type**	Basalt
	Colour	Greyish black
	Grain	Fine
	Product	Crushed Stone
DRILLING DATA		Hole Diameter 115mm
		Bench Height 18m
	Drills	Ingersoll-Rand CM351
COMPRESSORS		Ingersoll-Rand XHP750
SECONDARY BREAKING		Hydraulic Hammer
LOAD AND HAUL	**Face shovel**	Komatsu PC650
	Trucks	(2)Aveling Barford RXD25
		Euclid R25
	Loaders	Caterpillar 966C
		Komatsu WA420
CRUSHING PLANT	**Primary**	Kobelco 54 × 42 Jaw
	Secondary	Allis Chalmers 1445
	Tertiary	(2)Allis Chalmers Gyratory
		Kobelco Impactor

TARMAC ROADSTONE (SCOTLAND) LTD (TARMAC QUARRY PRODUCTS LTD)

SHIERGLAS QUARRY

601

Killiecrankie, Pitlochry, Tayside PH16 5LL
Phone 079681 325
OS Map 43 **OS Grid Reference** NO882640
Personnel G Stewart, Unit Manager

PRODUCT DATA	**Rock type**	Pre-Cambrian Dalradian Limestone
	Colour	Not commercially significant
	Products	Crushed Stone
		Asphalt or Tarmacadam
		Agricultural Lime
DRILLING DATA		Hole Diameter 125mm
		Bench Height 17–31m
		Contractor Ritchies Equipment
SECONDARY BREAKING		Dropball on Ruston-Bucyrus 22-RB
LOAD AND HAUL	**Trucks**	Caterpillar
	Loaders	Caterpillar
		Komatsu
CRUSHING PLANT		No information provided

TARMAC ROADSTONE (SOUTHERN) LTD (TARMAC QUARRY PRODUCTS LTD)
Head Office *2A Bath Road, Newbury, Berks RG13 1JJ*
Phone *0635 35755*
Personnel *M W Whittle, Managing Director*
Quarries *Swanworth*

TARMAC ROADSTONE (SOUTHERN) LTD (TARMAC QUARRY PRODUCTS LTD)

SWANWORTH QUARRY

602

Worth Matravers, Swanage, Dorset BE19 3LE
Phone 092943 213/214
OS Map 194 **OS Grid Reference** SY972785
Personnel G Hudson, Unit Manager

PRODUCT DATA	**Rock type**	Jurassic Portland Limestone
	Colour	Cream
	Grain	Fine
	Products	Crushed Stone
		Asphalt or Tarmacadam
		Industrial Limestone
		Agricultural Lime
DRILLING DATA		Hole Diameter 100mm
		Bench Height 15m
	Drills	Tamrock DHA800
LOAD AND HAUL	**Trucks**	Aveling Barford RD40
		Heathfield
	Loaders	Caterpillar 966C
		Caterpillar 980C

CRUSHING PLANT	**Primary**	Goodwin Barsby 42 × 24 Jaw
	Secondary	BJD Impactor
	Tertiary	None

TARMAC ROADSTONE (WESTERN) LTD (TARMAC QUARRY PRODUCTS LTD)

Head Office *Whitehall House, Whitehall Road, Halesowen, W Midlands B63 3LE*
Phone *0215 504797*
Personnel *R L Isaacs, Managing Director*
T Peters, Director
P Holloway, Production Manager
H P Ragdale, Production Manager
M Underwood, Production Manager
Quarries *Bayston Hill, Caldon Low, Callow Hill, Cefn Gawr, Creigiau, Croxden, Fenacre, Hailstone, Hindlow,*
Nuneaton, Rhyader (Carrigwynion), Shipham, Stowfield, Triscombe

TARMAC ROADSTONE (WESTERN) LTD (TARMAC QUARRY PRODUCTS LTD)

BAYSTON HILL QUARRY   **603**

Sharpstone Lane, Bayston Hill, Shrewsbury, Shrops SY3 0AW
Phone 0743 723311
OS Map 126   **OS Grid Reference** SJ493090
Personnel S Webb, Unit Manager

PRODUCT DATA	**Rock type**	Pre-Cambrian Greywacke
	Colour	Medium grey
	Grain	Angular
	Products	Crushed Stone
		Asphalt or Tarmacadam
		Ready-mixed Concrete
DRILLING DATA		Hole Diameter 135mm
		Bench Height 12m
	Drills	Bohler DTC122
SECONDARY BREAKING		Hydraulic Hammer - Furukawa
		Dropball on Ruston-Bucyrus 22-RB
LOAD AND HAUL	**Trucks**	Volvo
		Terex 2366/64
	Loaders	Volvo
CRUSHING PLANT	**Primary**	Kue Ken 36 × 42 Jaw
	Secondary	(4)Allis Chalmers Hydrocone
		(2)Barmac Duopactor

TARMAC ROADSTONE (WESTERN) LTD (TARMAC QUARRY PRODUCTS LTD)

CALDON LOW QUARRY   **604**

PO Box 1, Waterhouses, Stoke-on-Trent, Staffs ST10 3EW
Phone 0538 308282
OS Map 119   **OS Grid Reference** SK077488
Personnel R Tyrer, Unit Manager

entry continues overleaf

PRODUCT DATA	**Rock type**	Carboniferous Milldale Limestone
	Colour	Light creamy buff
	Products	Crushed Stone
		Asphalt or Tarmacadam
		Industrial Limestone
		Agricultural Lime
DRILLING DATA		Hole Diameter 125mm
		Bench Height 21m
		Contractor EMK Drilling & Blasting
SECONDARY BREAKING		Talisker
LOAD AND HAUL	**Trucks**	Aveling Barford
	Loaders	Caterpillar
		Michigan
CRUSHING PLANT	**Primary**	Hadfield 54 × 48 Jaw
	Secondary	(2)Mansfield No 4 Hammermill
		(2)Sheepbridge Impactor
	Tertiary	None

TARMAC ROADSTONE (WESTERN) LTD (TARMAC QUARRY PRODUCTS LTD)

CALLOW HILL QUARRY

605

Callow Lane, Minsterley, Shrewsbury, Shrops SY5 0DA
Phone 0743 791258
OS Map 126   **OS Grid Reference** SJ387050
Personnel D Tomlins, Unit Manager

PRODUCT DATA	**Rock type**	Dolerite
	Colour	Dark grey
	Grain	Medium
	Products	Crushed Stone
		Asphalt or Tarmacadam
DRILLING DATA		Hole Diameter 125mm
		Bench Height 10.5-13m
	Drills	Bohler DTC122
SECONDARY BREAKING		Dropball on Ruston-Bucyrus 22-RB
LOAD AND HAUL	**Trucks**	Aveling Barford WXL118
	Loaders	Bray
		JCB 418
CRUSHING PLANT	**Primary**	Pegson 18 × 32 Jaw
	Secondary	Ajax C
	Tertiary	Ajax D

TARMAC ROADSTONE (WESTERN) LTD (TARMAC QUARRY PRODUCTS LTD)

CEFN GAWR QUARRY

606

Tongwynlais, Nr Caerphilly, S Glam CF4 7YZ
Phone 0272 858261
OS Map 171   **OS Grid Reference** ST739829

QUARRY ON STAND

ROCK DATA	**Rock type**	Carboniferous Dolomitic Limestone
	Colour	Pinkish grey
	Grain	Coarse

TARMAC ROADSTONE (WESTERN) LTD (TARMAC QUARRY PRODUCTS LTD)

CREIGIAU QUARRY 607

Heol Plant-y-Gored, Cardiff, Mid Glam CF4 8NF
Phone 0222 892234
OS Map 170 **OS Grid Reference** ST086818
Personnel N Townsend, Unit Manager

PRODUCT DATA	**Rock type**	Carboniferous Dolomitic Limestone
	Colour	Pinkish grey
	Grain	Coarse
	Product	Crushed Stone
DRILLING DATA		Hole Diameter 110mm
		Bench Height 15m
		Contractor Rees Blasting
LOAD AND HAUL	**Loaders**	Faun F1410
CRUSHING PLANT		All crushing by Contractor

TARMAC ROADSTONE (WESTERN) LTD (TARMAC QUARRY PRODUCTS LTD)

CROXDEN QUARRY 608

Croxden Common, Freeway, Cheadle, Stoke-on-Trent, Staffs ST10 1RH
Phone 0538 722393
OS Map 128 **OS Grid Reference** SK001410
Personnel G Allen, Unit Manager

PRODUCT DATA	**Rock type**	Bunter Sandstone with Conglomerate
	Colour	Not commercially significant
	Product	Crushed Stone
DRILLING DATA		Hole Diameter 110mm
		Bench Height 10–12m
	Drills	Halco 625H
LOAD AND HAUL	**Face shovel**	Caterpillar 245
		Liebherr 961
	Trucks	Heathfield
	Loaders	Caterpillar 966C
		Caterpillar 988B
		(2)Volvo BM4500
CRUSHING PLANT	**Primary**	(2)Allis Chalmers 36″ Hydrocone
	Secondary	Kue Ken 36×8 Jaw
		Parker 48″ Jaw
		Parker 36″ Jaw
	Tertiary	None

TARMAC ROADSTONE (WESTERN) LTD (TARMAC QUARRY PRODUCTS LTD)

FENACRE QUARRY

609

Burlescombe, Tiverton, Devon EX16 7JD
Phone 0823 672206
OS Map 181   **OS Grid Reference** ST066178
Personnel J Proctor, Unit Manager

PRODUCT DATA	**Rock type**	Carboniferous Limestone
	Colour	Medium grey
	Products	Crushed Stone
		Asphalt or Tarmacadam
		Industrial Limestone
DRILLING DATA		Hole Diameter 105mm
		Bench Height 12.5m
		Contractor Ritchies Equipment
COMPRESSORS		CompAir
SECONDARY BREAKING		Hydraulic Hammer - Hitachi
LOAD AND HAUL	**Trucks**	Aveling Barford RD30
		(2)Heathfield
CRUSHING PLANT	**Primary**	Goodwin Barsby 42×30 Jaw (Double Toggle)
	Secondary	Magco Hammermill
	Tertiary	None

TARMAC ROADSTONE (WESTERN) LTD (TARMAC QUARRY PRODUCTS LTD)

HAILSTONE QUARRY

610

Portway Road, Warley, W Midlands B65 9DW
Phone 0215 504797
OS Map 139   **OS Grid Reference** SO968879

QUARRY ON STAND

ROCK DATA	**Rock type**	Dolerite
	Colour	Medium to dark grey
	Grain	Medium

TARMAC ROADSTONE (WESTERN) LTD (TARMAC QUARRY PRODUCTS LTD)

HINDLOW QUARRY

611

Wormhill, Buxton, Derbys SK17 9QD
Phone 0298 768444
OS Map 119   **OS Grid Reference** SK097680
Personnel J C M Millar, Unit Manager

PRODUCT DATA	**Rock type**	Carboniferous Bee Low Limestone
	Colour	Light grey
	Products	Crushed Stone
		Industrial Limestone
		Burnt Limestone
		Agricultural Lime

DRILLING DATA		Hole Diameter 105mm
		Bench Height 24–26m
	Drills	Drilltec 40K
		Holman Voltrak

SECONDARY BREAKING   Dropball on Ruston-Bucyrus 22-RB

LOAD AND HAUL	**Trucks**	Aveling Barford RD40
	Loaders	Komatsu WA500

CRUSHING PLANT	**Primary**	Hadfield Jaw (Double Toggle)
	Secondary	Mansfield No 4 Hammermill
	Tertiary	Mansfield No 3 Hammermill

TARMAC ROADSTONE (WESTERN) LTD (TARMAC QUARRY PRODUCTS LTD)

NUNEATON QUARRY (HARTSHILL QUARRY)   **612**

Hartshill, Nuneaton, Warwicks CV10 10HP
Phone 0203 393713
OS Map 140   **OS Grid Reference** SP335947
Personnel M J Burgess, Unit Manager

PRODUCT DATA	**Rock type**	Cambrian Hartshill Quartzitic Sandstone
	Colour	Grey
	Grain	Medium
	Products	Crushed Stone
		Asphalt or Tarmacadam

DRILLING DATA		Hole Diameter 105-130mm
		Bench Height 8.5-12m
		Contractor Blastrite
	Drills	Halco SPD43-22

COMPRESSORS   Atlas Copco XRH350S

SECONDARY BREAKING   Hydraulic Hammer - Krupp on Cat 219

LOAD AND HAUL	**Trucks**	(3)Aveling Barford RD40
		(3)Aveling Barford RBX20
		(3)Aveling Barford RBX30
		(3)Aveling Barford RBX40
	Loaders	Komatsu WA500
		(8)Caterpillar 966C
		Caterpillar 980C

CRUSHING PLANT	**Primary**	Pegson Mobile (Jaw)
	Secondary	Pegson Mobile (Jaw)

TARMAC ROADSTONE (WESTERN) LTD (TARMAC QUARRY PRODUCTS LTD)

RHYADER QUARRY (CARRIGWYNION QUARRY)   **613**

PO Box 1, Rhyader, Powys LD6 5OT
Phone 0215 504797
OS Map 147   **OS Grid Reference** SN972657

entry continues overleaf

QUARRY ON STAND

ROCK DATA	**Rock type**	Ordovician Greywacke
	Colour	Grey to black
	Grain	Angular

TARMAC ROADSTONE (WESTERN) LTD (TARMAC QUARRY PRODUCTS LTD)

SHIPHAM QUARRY

614

Shipham Gorge, Cheddar, Somerset BS27 3DQ
Phone 0934 744308
OS Map 182 **OS Grid Reference** ST445558
Personnel T Santina, Unit Manager

PRODUCT DATA	**Rock type**	Carboniferous Limestone
	Colour	Not commercially significant
	Product	Crushed Stone

DRILLING DATA		Hole Diameter 105mm
		Bench Height 12m
		Contractor Ritchies Equipment

LOAD AND HAUL	**Trucks**	Aveling Barford RD30
	Loaders	(2)Komatsu WA350
		Komatsu WA470
		Komatsu WA500
		Komatsu WA950

CRUSHING PLANT	**Primary**	Pegson 1100 × 800 Jaw (Mobile)
	Secondary	Pegson 900 Autocone (Mobile)
	Tertiary	(2)Pegson
		Barmac 9600 Duopactor

TARMAC ROADSTONE (WESTERN) LTD (TARMAC QUARRY PRODUCTS LTD)

STOWFIELD QUARRY

615

Scowles Pitch, Coleford, Glos GL16 8NR
Phone 0594 32066
OS Map 162 **OS Grid Reference** SO556107
Personnel D Barge, Unit Manager

PRODUCT DATA	**Rock type**	Carboniferous Dolomitic Limestone
	Colour	Reddish grey
	Grain	Coarse
	Products	Crushed Stone
		Asphalt or Tarmacadam

DRILLING DATA		Hole Diameter 110mm
		Bench Height 10m
	Drills	Bohler DTC122 DHD

| COMPRESSORS | | Atlas Copco XRH350 |

| SECONDARY BREAKING | | Hydraulic Hammer |
| | | Dropball on Ruston-Bucyrus 19-RB |

LOAD AND HAUL	**Trucks**	Aveling Barford RD30
		Heathfield H28
		Heathfield H30
		Volvo 5350
	Loaders	Michigan 275C
		Broyt X41
CRUSHING PLANT	**Primary**	Goodwin Barsby 48 × 36 Jaw
	Secondary	Hazemag APK60
	Tertiary	Hazemag APK60

TARMAC ROADSTONE (WESTERN) LTD (TARMAC QUARRY PRODUCTS LTD)

TRISCOMBE QUARRY   **616**

Bishops Lydeard, Taunton, Somerset TA4 3HE
Phone 09848 672
OS Map 193   **OS Grid Reference** ST163366
Personnel D Bidgood, Unit Manager

PRODUCT DATA	**Rock type**	Devonian Sandstone
	Colour	Dull red
	Grain	Medium
	Products	Crushed Stone
		Asphalt or Tarmacadam
DRILLING DATA		Hole Diameter 105mm
		Bench Height 21m
		Contractor Ritchies Equipment
SECONDARY BREAKING		Hydraulic Hammer - Montabert
LOAD AND HAUL	**Trucks**	Terex
		Volvo
	Loaders	Volvo
CRUSHING PLANT		No information provided

TAYSIDE REGION COUNCIL
Head Office *Tayside House, 28 Crichton Street, Dundee, Tayside DD1 3RE*
Phone *0382 23281*   **Fax** *0382 28354*
Personnel *J F White, Director of Roads*
Quarries *Boysack, Collace*

TAYSIDE REGION COUNCIL

BOYSACK QUARRY   **617**

Boysackmill, Arbroath, Tayside DD99 9ZZ
Phone 0241 2270
OS Map 54   **OS Grid Reference** NO628496
Personnel D Cowan, Unit Manager

PRODUCT DATA	**Rock type**	Devonian Lava (Basalt)
	Colour	Greyish black
	Grain	Fine
	Product	Crushed Stone

entry continues overleaf

DRILLING DATA		Hole Diameter 105mm Bench Height 30m Contractor Drilling & Blasting
SECONDARY BREAKING		Hydraulic Hammer
LOAD AND HAUL	**Trucks** **Loaders**	(3)Terex Caterpillar 930
CRUSHING PLANT	**Primary** **Secondary** **Tertiary**	Goodwin Barsby Jaw (3)Kennedy Gyratory None

TAYSIDE REGION COUNCIL

COLLACE QUARRY

618

Dunsinane Hill, Kinrossie, Perth, Tayside PH1 2EE
Phone 0281 5222
OS Map 53   **OS Grid Reference** NO207316
Personnel J Harley, Unit Manager

PRODUCT DATA	**Rock type** **Colour** **Grain** **Product**	Dolerite Black Fine Crushed Stone
DRILLING DATA		Hole Diameter 105mm Bench Height 15-25m Contractor Drilling & Blasting
SECONDARY BREAKING		Hydraulic Hammer
LOAD AND HAUL	**Trucks** **Loaders**	(3)Terex Caterpillar 950 Komatsu WA350
CRUSHING PLANT	**Primary** **Secondary** **Tertiary**	Goodwin Barsby Jaw (3)Kennedy Gyratory None

TETBURY STONE COMPANY LTD

Head Office Veizey's Quarry, Chavenage Lane, Tetbury, Glos GL8 8JT
Phone 0666 53455
Personnel D Binns, Director
Quarries Veizey's

TETBURY STONE COMPANY LTD

VEIZEY'S QUARRY

619

Chavenage Lane, Tetbury, Glos GL8 8JT
Phone 0666 53455
OS Map 162   **OS Grid Reference** ST881945
Personnel A E Wheat, Unit Manager

PRODUCT DATA	**Rock type**	Jurassic Oolitic Limestone
	Colour	Cream to buff
	Grain	Coarse
	Products	Dressed or Lump Stone
		Industrial Limestone
		Agricultural Lime

RIPPABLE STRATA

| LOAD AND HAUL | **Loaders** | Bray |

| CRUSHING PLANT | | None |

T S THOMAS AND SONS (LYDNEY) LTD
Head Office Albion Chambers, Hill Street, Lydney, Glos GL15 5HN
Phone 0594 842333
Personnel T S Thomas, Managing Director
        M L Thomas, Director
Quarries Dayhouse

T S THOMAS AND SONS (LYDNEY) LTD

DAYHOUSE QUARRY   **620**

Tidenham, Chepstow, Gwent NT6 7LH
Phone 0291 622416
OS Map 162   **OS Grid Reference** ST554955
Personnel T S Thomas, Managing Director and Unit Manager

PRODUCT DATA	**Rock type**	Carboniferous Limestone
	Colour	Brownish grey
	Grain	Medium
	Product	Crushed Stone

DRILLING DATA		Hole Diameter 105mm
		Bench Height 15m
		Contractor W C D Sleaman

| SECONDARY BREAKING | | Dropball on Ruston-Bucyrus 22-RB |

| LOAD AND HAUL | **Trucks** | (2)Aveling Barford RD30 |
| | **Loaders** | Caterpillar |

CRUSHING PLANT	**Primary**	Hazemag SAP5 Impactor
	Secondary	Hazemag APK60
	Tertiary	Parker Kubitizer

Wm THOMPSON AND SON (DUMBARTON) LTD
Head Office Birch Road, Dumbarton, Strathclyde G82 2RN
Phone 0389 62271
Personnel D Thompson, Director
Quarries Miltonhill, Riganagower

entry continues overleaf

Wm THOMPSON AND SON (DUMBARTON) LTD

MILTONHILL QUARRY

621

Milton, Dumbarton, Strathclyde G82 2UA
Phone 0389 62271
OS Map 64   **OS Grid Reference** NS435744
Personnel D Thompson, Director

PRODUCT DATA	**Rock type**	Dolerite
	Colour	Streaky brown
	Grain	Medium
	Product	Crushed Stone

NO INFORMATION PROVIDED

Wm THOMPSON AND SON (DUMBARTON) LTD

RIGANAGOWER QUARRY

622

Milton, Dumbarton, Strathclyde G82 2TX
Phone 0389 62271
OS Map 64   **OS Grid Reference** NS438754

QUARRY ON STAND

ROCK DATA	**Rock type**	Basalt
	Colour	Dark grey
	Grain	Medium

W & M THOMPSON (QUARRIES) LTD
Head Office *Princess Way, Low Prudhoe, Northumb NE42 6PL*
Phone 0661 32422/32423
Personnel W Thompson, Director
Quarries Bishop Middleham

W & M THOMPSON (QUARRIES) LTD

BISHOP MIDDLEHAM QUARRY

623

Ferryhill, Durham DL17 8NJ
Phone 0740 654120/128
OS Map 93   **OS Grid Reference** NY333323
Personnel J Purvis, Unit Manager

PRODUCT DATA	**Rock type**	Permian Magnesian Limestone
	Colour	Grey
	Grain	Coarse
	Products	Crushed Stone
		Agricultural Lime
RIPPABLE STRATA	**Ripper**	Caterpillar D11H
DRILLING DATA		Contractor Steavenson & Co (Blasting)
SECONDARY BREAKING		Dropball on Ruston-Bucyrus 22-RB
CRUSHING PLANT	**Primary**	Parker 42 × 20 Jaw
	Secondary	Parker Kubitizer

TILCON QUARRIES LTD
Corporate Head Office Low Lane, Lingerfield, Knaresborough, W Yorks HG5 9JN
Phone 0423 864041 **Fax** 0423 864555
Personnel B Howarth, Managing Director
Companies Tilcon (Midlands and South) Ltd, Tilcon (North Central) Ltd
Tilcon (Northern) Ltd, Tilcon (Scotland) Ltd

TILCON (MIDLAND AND SOUTH) LTD (TILCON QUARRIES LTD)
Head Office Cornett Ends Lane, Merriden, Coventry, Warwicks CV7 7LG
Phone 0242 221111
Personnel D W Hallsworth, Managing Director
D A Allen, Production Manager
D Dickenson, Operations Manager
P W Gerrard, Operations Manager
J K Winnard, Operations Manager
Quarries Ballidon, Clearwell, Gore, Kevin, Mancetter, Rogers

TILCON (MIDLAND AND SOUTH) LTD (TILCON QUARRIES LTD)

BALLIDON QUARRY   624
PO Box 6, Ashbourne, Derbys DE6 1GU
Phone 0335 25301
OS Map 119 **OS Grid Reference** SK200555
Personnel L Fradley, Unit Manager

PRODUCT DATA	**Rock type**	Carboniferous Bee Low Limestone
	Colour	Creamy grey
	Products	Crushed Stone
		Asphalt or Tarmacadam
		Ready-mixed Concrete
		Industrial Limestone
DRILLING DATA		Hole Diameter 125mm
		Bench Height 22m
	Drills	Halco 650H
SECONDARY BREAKING		Hydraulic Hammer - Montabert
LOAD AND HAUL	**Trucks**	Aveling Barford
	Loaders	Caterpillar 988B
		Komatsu
CRUSHING PLANT	**Primary**	Hazemag SAP5
	Secondary	Mansfield No 5 Hammermill
	Tertiary	Mansfield No 5 Hammermill

TILCON (MIDLAND AND SOUTH) LTD (TILCON QUARRIES LTD)

CLEARWELL QUARRY   625
Stowe Green, St Briavels, Lydney, Glos GL15 6QW
Phone 0594 32445
OS Map 162 **OS Grid Reference** SO566069
Personnel A Towell, Unit Manager

entry continues overleaf

PRODUCT DATA	**Rock type**	Carboniferous Limestone
	Colour	Light brown
	Products	Crushed Stone
		Industrial Limestone
		Agricultural Lime

DRILLING DATA		Hole Diameter 105mm
		Bench Height 15m
		Contractor Ritchies Equipment
	Drills	Halco 650H

| COMPRESSORS | | CompAir RO60-170 |

| SECONDARY BREAKING | | Dropball |

LOAD AND HAUL	**Trucks**	Aveling Barford
		Komatsu
	Loaders	Caterpillar

CRUSHING PLANT	**Primary**	Goodwin Barsby ST36 × 24 Jaw (Single Toggle)
		Parker Jaw (Mobile)
	Secondary	Mansfield No 3 Hammermill
	Tertiary	None

TILCON (MIDLAND AND SOUTH) LTD (TILCON QUARRIES LTD)

GORE QUARRY 626

Walton, Presteigne, Powys LD8 2DL
Phone 0544 230421
OS Map 148   **OS Grid Reference** SO258593
Personnel B Evans, Unit Manager

PRODUCT DATA	**Rock type**	Pre-Cambrian Sandstone
	Colour	Light grey, light green, red
	Grain	Medium to coarse
	Products	Crushed Stone
		Asphalt or Tarmacadam

DRILLING DATA		Hole Diameter 105mm
		Bench Height 20m
	Drills	Halco 410C

| SECONDARY BREAKING | | Dropball on Ruston-Bucyrus 30-RB |

LOAD AND HAUL	**Face shovel**	Akermans
	Trucks	(3)Aveling Barford RD017
	Loaders	(3)Caterpillar

CRUSHING PLANT	**Primary**	Goodwin Barsby 42 × 30 Jaw
		Goodwin Barsby 42 × 24 Jaw (Mobile)
	Secondary	Hazemag APK50
	Tertiary	Pegson 24S Gyrasphere
		Pegson 900 Autocone

TILCON (MIDLAND AND SOUTH) LTD (TILCON QUARRIES LTD)

KEVIN QUARRY

627

Ramshorn, Waterhouses, Staffs STIO 3BX
Phone 0538 702385
OS Map 119   **OS Grid Reference** SK085465
Personnel J Mitchell, Unit Manager

PRODUCT DATA	**Rock type**	Carboniferous Kevin Limestone
	Colour	Light creamy grey
	Products	Crushed Stone
		Asphalt or Tarmacadam
		Industrial Limestone
		Agricultural Lime
DRILLING DATA		Hole Diameter 110-125mm
		Bench Height 15-22m
		Contractor EMK Drilling & Blasting
SECONDARY BREAKING		Dropball on Ruston-Bucyrus 30-RB
		Dropball on Ruston-Bucyrus 38-RB
LOAD AND HAUL	**Trucks**	Aveling Barford
	Loaders	Komatsu WA450
CRUSHING PLANT	**Primary**	Goodwin Barsby 42×30 Jaw
	Secondary	Mansfield No 4 Hammermill
	Tertiary	None

TILCON (MIDLAND AND SOUTH) LTD (TILCON QUARRIES LTD)

MANCETTER QUARRY

628

Mancetter, Atherstone, Warwicks CV9 2RF
Phone 0827 712324
OS Map 140   **OS Grid Reference** SP308956
Personnel P A J Lee, Unit Manager

PRODUCT DATA	**Rock type**	Dolerite
	Colour	Light grey
	Grain	Medium
	Products	Crushed Stone
		Asphalt or Tarmacadam
DRILLING DATA		Hole Diameter 110mm
		Bench Height 21m
	Drills	Halco 410C
SECONDARY BREAKING		Hydraulic Hammer - Krupp
LOAD AND HAUL	**Face shovel**	Komatsu PC650
	Trucks	Heathfield
	Loaders	None
CRUSHING PLANT		No information provided

TILCON (MIDLAND AND SOUTH) LTD (TILCON QUARRIES LTD)

ROGERS QUARRY

629

Scowles, Coleford, Glos GLI6 8QT
Phone 0242 221111
OS Map 162 **OS Grid Reference** SO558111

QUARRY ON STAND

ROCK DATA		
	Rock type	Carboniferous Limestone
	Colour	Light brown, grey

TILCON (NORTH CENTRAL) LTD (TILCON QUARRIES LTD)
Head Office Harrogate House, Parliament Street, Harrogate, N Yorks HG1 2RF
Phone 0423 68092 **Telex** 57998
Personnel J Hague, Director
T Allen, Production Manager
G Mullinder, Production Manager
Quarries Abergele, Brodsworth, Giggleswick, Spring Lodge, Swinden, Trimm Rock

---

TILCON (NORTH CENTRAL) LTD (TILCON QUARRIES LTD)

ABERGELE QUARRY

630

St George, Abergele, Clwyd LL22 9BD
Phone 0745 833172
OS Map 116 **OS Grid Reference** SH969759
Personnel M Williams, Unit Manager

PRODUCT DATA	**Rock type**	Carboniferous Limestone
	Colour	Grey, fawn
	Products	Crushed Stone
		Ready-mixed Concrete
		Burnt Limestone
		Agricultural Lime
DRILLING DATA		Hole Diameter 105mm
		Bench Height 9-21m
		Contractor A Jones Rock Drillers
SECONDARY BREAKING		Hydraulic Hammer
		Dropball on Smith
LOAD AND HAUL	**Trucks**	Terex
	Loaders	Caterpillar
CRUSHING PLANT	**Primary**	Hadfield 36 × 24 Jaw
	Secondary	(2)Mansfield No 4 Hammermill
	Tertiary	(3)Christy & Norris Ball-mill

TILCON (NORTH CENTRAL) LTD (TILCON QUARRIES LTD)

BRODSWORTH QUARRY

631

Green Lane, Scawthorpe, Doncaster, S Yorks DN5 7UY
Phone 0302 722743
OS Map 111 **OS Grid Reference** SE528065
Personnel B Lunn, Unit Manager

PRODUCT DATA	**Rock type**	Permian Limestone
	Colour	Light brown
	Product	Crushed Stone

RIPPABLE STRATA

TILCON (NORTH CENTRAL) LTD (TILCON QUARRIES LTD)

GIGGLESWICK QUARRY

632

Giggleswick, Settle, N Yorks BD24 0DH
Phone 0729 22563
OS Map 98 **OS Grid Reference** SD809649
Personnel B Moore, Unit Manager

PRODUCT DATA	**Rock type**	Carboniferous Great Scar Limestone
	Colour	Grey
	Product	Crushed Stone

DRILLING DATA		Hole Diameter 105mm
		Bench Height 21m
	Drills	Halco 410C

| COMPRESSORS | | CompAir 650HE |

| SECONDARY BREAKING | | Dropball on Ruston-Bucyrus 22-RB |

LOAD AND HAUL	**Trucks**	Terex
	Loaders	Caterpillar
		Terex

CRUSHING PLANT	**Primary**	Goodwin Barsby 24 × 36 Jaw
	Secondary	Mansfield No 4 Hammermill
	Tertiary	None

TILCON (NORTH CENTRAL) LTD (TILCON QUARRIES LTD)

SPRING LODGE QUARRY

633

Womersley, Nr Doncaster, S Yorks DN6 9BB
Phone 0977 83120
OS Map 111 **OS Grid Reference** SE520202
Personnel B Whitehead, Unit Manager

PRODUCT DATA	**Rock type**	Permian Upper Magnesian Limestone
	Colour	Light brownish grey
	Grain	Coarse
	Products	Crushed Stone
		Ready-mixed Concrete
		Industrial Limestone

RIPPABLE STRATA

TILCON (NORTH CENTRAL) LTD (TILCON QUARRIES LTD)

SWINDEN QUARRY

634

Linton, Nr Skipton, N Yorks BD23 6BE
Phone 0756 752671
OS Map 98 **OS Grid Reference** SD980617
Personnel D Banks, Unit Manager

entry continues overleaf

PRODUCT DATA	**Rock type**	Carboniferous Reef Limestone
	Colour	Grey
	Products	Crushed Stone
		Industrial Limestone
		Burnt Limestone
		Agricultural Lime
DRILLING DATA		Hole Diameter 125mm
		Bench Height 24m
	Drills	Hausherr HB80
SECONDARY BREAKING		Dropball on Ruston-Bucyrus 38-RB
LOAD AND HAUL	**Face shovel**	Liebherr R974
	Trucks	Aveling Barford
	Loaders	Caterpillar
CRUSHING PLANT	**Primary**	Hazemag APK50 Impactor
		Hadfield Jaw
	Secondary	Mansfield No 3 Hammermill
		Mansfield No 4 Hammermill
		Mansfield No 5 Hammermill
	Tertiary	None

TILCON (NORTH CENTRAL) LTD (TILCON QUARRIES LTD)

TRIMM ROCK QUARRY

635

Cilcain, Mold, Clwyd CH7 5HR
Phone 0352 741353
OS Map 116   **OS Grid Reference** SJ191659
Personnel M Tetley, Unit Manager

PRODUCT DATA	**Rock type**	Carboniferous Limestone
	Colour	Light grey
	Products	Crushed Stone
		Industrial Limestone
DRILLING DATA		Hole Diameter 105mm
		Bench Height 15–25m
	Drills	Halco 410C
SECONDARY BREAKING		Dropball on Ruston-Bucyrus 22-RB
LOAD AND HAUL	**Trucks**	Euclid
		(2)Terex
	Loaders	Komatsu WA450
CRUSHING PLANT	**Primary**	BJD Jaw
	Secondary	Mansfield No 4 Hammermill
	Tertiary	(3)Christy & Norris Ball-mill

TILCON (NORTHERN) LTD (TILCON QUARRIES LTD)
Head Office PO Box 5, Fell Bank, Birtley, Chester-le-Street, Durham DH3 2ST
Phone 0914 109611
Personnel K Riley, Managing Director
            R Huddleston, Production Manager
Quarries Belford, Forcett, Harden, Marsden, Skipton Rock, Stainton

TILCON (NORTHERN) LTD (TILCON QUARRIES LTD)

BELFORD QUARRY

636

Belford, Northumb NE70 7DZ
Phone 0914 109611
OS Map 75   **OS Grid Reference** NU129341

QUARRY ON STAND

ROCK DATA	**Rock type**	Dolerite Whin Sill
	Colour	Very dark grey
	Grain	Fine

TILCON (NORTHERN) LTD (TILCON QUARRIES LTD)

FORCETT QUARRY

637

East Layton, Richmond, N Yorks DL11 7PH
Phone 0325 718291
OS Map 92   **OS Grid Reference** NZ155111
Personnel J Moses, Unit Manager

PRODUCT DATA	**Rock type**	Carboniferous Limestone
	Colour	Light brownish grey
	Products	Crushed Stone
		Asphalt or Tarmacadam
		Industrial Limestone
DRILLING DATA		Hole Diameter 125mm
		Bench Height 11–14m
	Drills	Atlas Copco ROC404
COMPRESSORS		Atlas Copco XRH350
SECONDARY BREAKING		Dropball on Ruston-Bucyrus 22-RB
LOAD AND HAUL	**Face shovel**	Ruston-Bucyrus
	Trucks	(4)Aveling Barford RD30
	Loaders	(3)Caterpillar 950
		Caterpillar 966C
CRUSHING PLANT	**Primary**	Sheepbridge 36 × 48 Jaw
	Secondary	Mansfield No 5 Hammermill
	Tertiary	None

TILCON (NORTHERN) LTD (TILCON QUARRIES LTD)

HARDEN QUARRY

638

Biddlestone, Harbottle, Morpeth, Northumb NE65 7DU
Phone 0669 30281
OS Map 81   **OS Grid Reference** NT959085
Personnel A Teasdale, Unit Manager

PRODUCT DATA	**Rock type**	Andesite
	Colour	Red
	Grain	Fine
	Product	Crushed Stone

entry continues overleaf

DRILLING DATA		Hole Diameter 110mm
		Bench Height 15m
	Drills	Halco 410C

SECONDARY BREAKING		Dropball on Ruston-Bucyrus 22-RB

LOAD AND HAUL	**Trucks**	Terex
	Loaders	Caterpillar 966C
		Caterpillar 980C

CRUSHING PLANT	**Primary**	Goodwin Barsby 48×30 Jaw
	Secondary	Mansfield No 4 Hammermill
	Tertiary	None

TILCON (NORTHERN) LTD (TILCON QUARRIES LTD)

MARSDEN QUARRY

639

Whitburn, Sunderland, Tyne and Wear SR6 7NG
Phone 0915 292441
OS Map 88 **OS Grid Reference** NZ405641
Personnel M Parkins, Unit Manager

PRODUCT DATA	**Rock type**	Permian Magnesian Limestone
	Colour	Buff, grey
	Grain	Coarse
	Products	Crushed Stone
		Asphalt or Tarmacadam
		Industrial Limestone
		Agricultural Lime

DRILLING DATA		Hole Diameter 105mm
		Bench Height 6–15m
	Drills	Halco 410C

COMPRESSORS		CompAir 700HE

SECONDARY BREAKING		Dropball on Ruston-Bucyrus 22-RB

LOAD AND HAUL	**Face shovel**	Ruston-Bucyrus
	Trucks	Aveling Barford
	Loaders	(3)Caterpillar 950
		Caterpillar 966C
		Komatsu WA450

CRUSHING PLANT	**Primary**	Sheepbridge Jaw
	Secondary	Mansfield No 4 Hammermill
	Tertiary	Mansfield No 5 Hammermill
		Bradley Pulverizer

TILCON (NORTHERN) LTD (TILCON QUARRIES LTD)

SKIPTON ROCK QUARRY

640

Harrogate Road, Skipton, N Yorks BD23 6BJ
Phone 0914 109611
OS Map 104 **OS Grid Reference** SE330538

QUARRY ON STAND

ROCK DATA **Rock type** Carboniferous Limestone
Colour Dark grey

TILCON (NORTHERN) LTD (TILCON QUARRIES LTD)

STAINTON QUARRY **641**

Stainton, Barrow-in-Furness, Cumbria CA13 0NN
Phone 0229 65197
OS Map 96   **OS Grid Reference** SD245730
Personnel A Burton, Unit Manager

PRODUCT DATA **Rock type** Carboniferous Limestone
Colour Drab pink
Products Crushed Stone
Agricultural Lime

DRILLING DATA Hole Diameter 105mm
Bench Height 16m
Drills Atlas Copco ROC404-04

COMPRESSORS CompAir RO160-70

SECONDARY BREAKING Dropball on Ruston-Bucyrus 22-RB

LOAD AND HAUL **Trucks** Terex
Loaders Caterpillar

CRUSHING PLANT **Primary** Parker 32 × 42 Jaw
Secondary Sheepbridge Impactor
Tertiary Christy & Norris Hammermill

TILCON (SCOTLAND) LTD (TILCON QUARRIES LTD)
Head Office 250 Alexandra Parade, Glasgow, Strathclyde G31 3AX
Phone 0415 541818
Personnel G Greenhalgh, Managing Director
J Jameson, Production Director
G Gray, Production Manager
J Hall, Production Manager
Quarries Banavie, Bannerbank, Cairneyhill, Cruicks, Dumbuckhill Furnace, Waulkmill

TILCON (SCOTLAND) LTD (TILCON QUARRIES LTD)

BANAVIE QUARRY **642**

Corpach, Fort William, Highland PH33 7LY
Phone 0397 7267
OS Map 41   **OS Grid Reference** NN112772
Personnel D Robertson, Unit Manager

PRODUCT DATA **Rock type** Granite
Colour Silver grey with black speckles
Grain Coarse
Products Crushed Stone
Asphalt or Tarmacadam

entry continues overleaf

DRILLING DATA		Hole Diameter 105mm
		Bench Height 20m
		Contractor Ritchies Equipment

SECONDARY BREAKING		Hydraulic Hammer

LOAD AND HAUL	**Trucks**	Aveling Barford
	Loaders	Caterpillar

CRUSHING PLANT		No information provided

TILCON (SCOTLAND) LTD (TILCON QUARRIES LTD)

BANNERBANK QUARRY
643

Stewarton Road, Doddside, Newton Mearns, Glasgow, Strathclyde G77 6QA
Phone 0416 398350
OS Map 64   **OS Grid Reference** NS495525
Personnel J Hadrow, Unit Manager

PRODUCT DATA	**Rock type**	Carboniferous Clyde Plateau Lavas (Basalt)
	Colour	Dark grey
	Grain	Fine
	Products	Crushed Stone
		Asphalt or Tarmacadam

DRILLING DATA		Hole Diameter 125mm
		Bench Height 27–30m
		Contractor Ritchies Equipment

SECONDARY BREAKING		Dropball on Ruston-Bucyrus 22-RB

LOAD AND HAUL	**Trucks**	None
	Loaders	Caterpillar

CRUSHING PLANT	**Primary**	Lokomo
	Secondary	Hazemag APK50

TILCON (SCOTLAND) LTD (TILCON QUARRIES LTD)

CAIRNEYHILL QUARRY
644

Caldercruix, Strathclyde ML6 8NX
Phone 0236 842351
OS Map 65   **OS Grid Reference** NS852665
Personnel A Rogers, Unit Manager

PRODUCT DATA	**Rock type**	Quartzitic Dolerite Sill
	Colour	Dark grey
	Grain	Medium
	Products	Crushed Stone
		Asphalt or Tarmacadam

DRILLING DATA		Hole Diameter 105mm
		Bench Height 15m
		Contractor Ritchies Equipment

SECONDARY BREAKING		Dropball on Ruston-Bucyrus 38-RB
		Talisker on Priestman 120

LOAD AND HAUL	**Trucks**	Aveling Barford
	Loaders	Caterpillar 980C
CRUSHING PLANT	**Primary**	Lokomo Mk120
	Secondary	Allis Chalmers 1650 Gyratory
	Tertiary	Allis Chalmers 945
		(2)Allis Chalmers 36″ Cone
		(2)Nordberg 36″ Gyradisc

TILCON (SCOTLAND) LTD (TILCON QUARRIES LTD)

CRUICKS QUARRY　　　　　　645

Inverkeithing, Fife KY11 1HH
Phone 0383 413241
OS Map 66　**OS Grid Reference** NT132817
Personnel B Giles, Unit Manager

PRODUCT DATA	**Rock type**	Dolerite Sill
	Colour	Dark grey
	Grain	Medium
	Products	Crushed Stone
		Asphalt or Tarmacadam
DRILLING DATA		Hole Diameter 105mm
		Bench Height 15m
		Contractor Ritchies Equipment
SECONDARY BREAKING		Hydraulic Hammer
		Dropball on Ruston-Bucyrus 30-RB
LOAD AND HAUL	**Trucks**	Aveling Barford
	Loaders	Caterpillar
CRUSHING PLANT		No information provided

TILCON (SCOTLAND) LTD (TILCON QUARRIES LTD)

DUMBUCKHILL QUARRY　　　646

Milton, Dumbarton, Strathclyde G82 2SE
Phone 0397 7267
OS Map 64　**OS Grid Reference** NS420747

QUARRY ON STAND

ROCK DATA	**Rock type**	Basalt
	Colour	Dark grey
	Grain	Fine

TILCON (SCOTLAND) LTD (TILCON QUARRIES LTD)

FURNACE QUARRY　　　　　647

Furnace, Inveraray, Strathclyde PA32 8XW
Phone 0499 5268/5284
OS Map 55　**OS Grid Reference** NN028003
Personnel A Campell, Unit Manager

entry continues overleaf

PRODUCT DATA	**Rock type**	Porphyritic Microgranite
	Grain	Fine
	Products	Crushed Stone
		Asphalt or Tarmacadam

DRILLING DATA		Hole Diameter 110mm
		Bench Height 15m
		Contractor Ritchies Equipment

| SECONDARY BREAKING | | Dropball on Ruston-Bucyrus 38-RB |

LOAD AND HAUL	**Trucks**	Caterpillar
	Loaders	Caterpillar 950
		Caterpillar 980C

| CRUSHING PLANT | | No information provided |

TILCON (SCOTLAND) LTD (TILCON QUARRIES LTD)

WAULKMILL QUARRY

648

Inverkeilor, Angus, Tayside DD1 4UT
Phone 02413 286
OS Map 54 **OS Grid Reference** NO632494
Personnel D Walker, Unit Manager

PRODUCT DATA	**Rock type**	Andesite
	Colour	Brownish grey
	Grain	Fine
	Products	Crushed Stone
		Asphalt or Tarmacadam

DRILLING DATA		Hole Diameter 105mm
		Bench Height 17–20m
		Contractor Ritchies Equipment

| SECONDARY BREAKING | | Dropball |

| LOAD AND HAUL | **Trucks** | Aveling Barford |
| | **Loaders** | Caterpillar |

| CRUSHING PLANT | | All crushing by Contractor |

TILLICOULTRY QUARRIES LTD

Head Office Craigfoot Quarry, Craigfoot, Tillicoultry, Fife FK13 6AZ
Phone 0259 50347
Personnel D Menzies, Managing Director
　　　　　 I Menzies, Director
Quarries Craigfoot

TILLICOULTRY QUARRIES LTD

CRAIGFOOT QUARRY

649

Craigfoot, Tillicoultry, Fife FK13 6AZ
Phone 0259 50347
OS Map 58 **OS Grid Reference** NJ913977
Personnel I Menzies, Director and Unit Manager

ROCK DATA	**Rock type**	Dolerite
	Colour	Greyish black
	Grain	Medium
	Product	Crushed Stone

DRILLING DATA		Hole Diameter 108mm
		Bench Height 26-28m
		Contractor Forth Quarry Services

| SECONDARY BREAKING | | Dropball |

| LOAD AND HAUL | **Loaders** | Komatsu |

NO FURTHER INFORMATION PROVIDED

TORRINGTON STONE LTD
Head Office *Torrington, Devon EX38 8JF*
Phone *0805 22438*
Personnel *B Setchell, Owner*
Quarries *Beam*

TORRINGTON STONE LTD
BEAM QUARRY

650

Torrington, Devon EX38 8JF
Phone 0805 22438
OS Map 180   **OS Grid Reference** SS485205
Personnel B Setchell, Owner and Unit Manager

PRODUCT DATA	**Rock type**	Triassic Bunter Sandstone
	Colour	Rustic brown with blue hue
	Grain	Medium
	Product	Dressed or Lump Stone

| DRILLING DATA | | Hand-held equipment |

| COMPRESSORS | | CompAir |

TRACTOR SHOVELS TAWSE LTD (EVERED HOLDINGS plc)
Head Office *Forth House, North Road, Inverkeithing, Fife KY11 1HG*
Phone *0383 412155*   **Fax** *0383 415242*   **Telex** *727767*
Personnel *J Murray, Managing Director*
Quarries *Achnagart, Ardchronie, Bower, Corrennie, Craigie Hill, Kemnay, Marybank, Tom's Forest, Tyrebagger*

TRACTOR SHOVELS TAWSE LTD (EVERED HOLDINGS plc)
ACHNAGART QUARRY

651

Achnagart, Glenshiel, Inverinate, Highland IV40 8HJ
Phone 059981 286
OS Map 25   **OS Grid Reference** NG963143
Personnel D Henderson, Unit Manager

| PRODUCT DATA | **Rock type** | Pre-Cambrian Moine Schist |
| | **Product** | Crushed Stone |

entry continues overleaf

DRILLING DATA		Hole Diameter 105mm
	Drills	(2)Atlas Copco ROC404-04

COMPRESSORS	Ingersoll-Rand

LOAD AND HAUL	No information

CRUSHING PLANT	**Primary**	Goodwin Barsby 36X24 Jaw (Single Toggle)
	Secondary	Goodwin Barsby 36×6 Granulator
		Pegson 2′ Gyrasphere

TRACTOR SHOVELS TAWSE LTD (EVERED HOLDINGS plc)

ARDCHRONIE QUARRY

652

Ardgay, Highland IV24 3DJ
Phone 021 711 1717
OS Map 21   **OS Grid Reference** NH617884

QUARRY ON STAND

ROCK DATA	**Rock type**	Pre-Cambrian Moine Schist

TRACTOR SHOVELS TAWSE LTD (EVERED HOLDINGS plc)

BOWER QUARRY

653

Gillock, Thurso, Caithness, Highland KW14 8JL
Phone 09558 6222
OS Map 12   **OS Grid Reference** ND202585
Personnel C Sutherland, Unit Manager

PRODUCT DATA	**Rock type**	Pre-Cambrian Moine Schist
	Products	Dressed or Lump Stone
		Crushed Stone

DRILLING DATA		Hole Diameter 105mm
		Bench Height 8-10m
		Contractor Rocklift

LOAD AND HAUL	**Trucks**	Aveling Barford
	Loaders	Caterpillar

CRUSHING PLANT	**Primary**	Pegson Jaw
	Secondary	Pegson Jaw
	Tertiary	None

TRACTOR SHOVELS TAWSE LTD (EVERED HOLDINGS plc)

CORRENNIE QUARRY

654

Tillyfourie, Alford, Inverurie, Grampian AB3 8DX
Phone 021 711 1717
OS Map 37   **OS Grid Reference** NJ646127

QUARRY ON STAND

ROCK DATA	**Rock type**	Gabbro

TRACTOR SHOVELS TAWSE LTD (EVERED HOLDINGS plc)

CRAIGIE HILL QUARRY
655

Graigie, Kilmarnock, Strathclyde KA1 5NA
Phone 056386 292
OS Map 70 **OS Grid Reference** NS423328
Personnel D Jones, Unit Manager

PRODUCT DATA	**Rock type**	Basalt
	Colour	Greyish black
	Grain	Fine
	Products	Dressed or Lump Stone
		Crushed Stone
DRILLING DATA		Hole Diameter 110mm
		Bench Height 10m
		Contractor Dobson Anfolite
SECONDARY BREAKING		Dropball on Ruston-Bucyrus 22-RB
LOAD AND HAUL	**Trucks**	Heathfield
	Loaders	Caterpillar 950
		Caterpillar 980C
		Fiatallis
CRUSHING PLANT	**Primary**	Pegson Jaw
	Secondary	(2)Pegson Jaw
	Tertiary	(2)Pegson Autocone

TRACTOR SHOVELS TAWSE LTD (EVERED HOLDINGS plc)

KEMNAY QUARRY
656

Kemnay, Inverurie, Grampian AB5 9PB
Phone 0467 43861
OS Map 38 **OS Grid Reference** NJ736169
Personnel A W Gibb, Unit Manager

PRODUCT DATA	**Rock type**	Hill of Fare Granite
	Colour	Light grey
	Grain	Coarse
	Product	Crushed Stone
DRILLING DATA		Hole Diameter 105mm
		Bench Height 16m
	Drills	Halco 410C
COMPRESSORS		CompAir 650HE
SECONDARY BREAKING		Dropball on Ruston-Bucyrus 22-RB
LOAD AND HAUL	**Face shovel**	Akermans
	Trucks	Aveling Barford
	Loaders	Caterpillar
CRUSHING PLANT	**Primary**	Kue Ken 40×27 Jaw (Double Toggle)
	Secondary	Kue Ken 30×20 Jaw
	Tertiary	Kennedy

MARYBANK QUARRY

657

Stornaway, Isle of Lewis, Western Isles, PA86 0DF
Phone 085 13227
OS Map 8 **OS Grid Reference** NB405330
Personnel D MacLeod, Unit Manager

PRODUCT DATA	**Rock type**	Pre-Cambrian Moine Schist
	Product	Crushed Stone
DRILLING DATA		Contractor Rocklift
		No further information
CRUSHING PLANT	**Primary**	Parker Jaw
	Secondary	Parker Jaw
	Tertiary	Allis Chalmers
		Lokomo

TOM'S FOREST QUARRY

658

Kintore, Inverurie, Grampian AB5 0YU
Phone 0467 42467
OS Map 38 **OS Grid Reference** NJ764168
Personnel A W Gibb, Unit Manager

PRODUCT DATA	**Rock type**	Hill of Fare Granite
	Colour	Not commercially significant
	Grain	Coarse
	Products	Crushed Stone
		Asphalt or Tarmacadam
		Ready-mixed Concrete
DRILLING DATA		Hole Diameter 110mm
		Bench Height 18.5m
	Drills	Halco 410C
COMPRESSORS		CompAir 650HE
SECONDARY BREAKING		Dropball on Ruston-Bucyrus 22-RB
LOAD AND HAUL	**Face shovel**	Akermans
	Trucks	(2)Aveling Barford RD30
	Loaders	(3)Caterpillar
CRUSHING PLANT	**Primary**	Kue Ken 40×36 Jaw
	Secondary	(2)Kue Ken 30×12 Jaw
	Tertiary	Allis Chalmers 3′ Cone
		Nordberg 3′ Cone

TYREBAGGER QUARRY

659

Blackburn, Aberdeen, Grampian AB5 0TT
Phone 021 711 1717
OS Map 38 **OS Grid Reference** NJ835123

QUARRY ON STAND

ROCK DATA

Rock type Gabbro
Colour Bluish grey
Grain Coarse

TREBARWITH QUARRIES LTD
Head Office Trebarwith Quarry, Trebarwith Road, Delabole, Camelford, Cornwall PL33 9DE
Phone 0840 213354
Personnel R Williams, Director
Quarries Trebarwith

TREBARWITH QUARRIES LTD

TREBARWITH QUARRY   660

Trebarwith Road, Delabole, Camelford, Cornwall PL33 9DE
Phone 0840 213354
OS Map 200   **OS Grid Reference** SX071858
Personnel R Williams, Director and Unit Manager

PRODUCT DATA

Rock type Upper Devonian Slate
Colour Blue to grey
Grain Fine
Product Dressed or Lump Stone

DRILLING DATA

Hole Diameter 38mm
Hand-held equipment

COMPRESSORS

Holman

TREDINNICK QUARRY LTD
Head Office Tredinnick Quarry, Tredinnick Hill, Grampound, Truro, Cornwall TR2 5TQ
Phone 0726 883487
Personnel A Mathews, Managing Director
       B Williams, Director
       P Williams, Director

TREDINNICK QUARRY LTD

TREDINNICK QUARRY   661

Tredinnick Hill, Grampound, Truro, Cornwall TR2 5TQ
Phone 0726 883487
OS Map 204   **OS Grid Reference** SW935428
Personnel P Williams, Director and Unit Manager

PRODUCT DATA

Rock type Devonian Grampound Grit
Colour Fawn, red
Grain Medium
Product Dressed or Lump Stone

DRILLING DATA

Hole Diameter 105mm
Bench Height 30m
Contractor Saxton Co (Deep Drilling)

SECONDARY BREAKING   Dropball   *entry continues overleaf*

LOAD AND HAUL	**Face shovel**	Priestman
	Trucks	None
	Loaders	Caterpillar 960
CRUSHING PLANT	**Primary**	Parker 24 × 36 Jaw (Mobile)
	Secondary	None
	Tertiary	None

P J TRESISE
Head Office *Trenoweth Quarry, Mabe Burnthouse, Penryn, Cornwall TR10 9HY*
Phone *0326 72546*
Personnel *P J Tresise, Director*
Quarries *Trenoweth*

P J TRESISE
TRENOWETH QUARRY 662
Mabe Burnthouse, Penryn, Cornwall TR10 9HY
Phone 0326 72546
OS Map 204   **OS Grid Reference** SW755333
Personnel P J Tresise, Director and Unit Manager

PRODUCT DATA	**Rock type**	Granite
	Colour	Silver grey
	Grain	Fine
	Product	Dressed or Lump Stone

RIPPABLE STRATA

TREVONE QUARRIES LTD
Head Office *Bosahan Quarry, Constantine, Falmouth, Cornwall TR11 5QP*
Phone *0326 40604*   **Fax** *0326 40403*   **Telex** *45665*
Personnel *M Cook, Director*
Quarries *Bosahan, Trevone*

TREVONE QUARRIES LTD
BOSAHAN QUARRY 663
Constantine, Falmouth, Cornwall TR11 5QP
Phone 0326 40604
OS Map 203   **OS Grid Reference** SW716281
Personnel P Morgan, General Manager and Unit Manager

PRODUCT DATA	**Rock type**	Carnmenellis Granite
	Colour	Grey
	Grain	Medium
	Product	Dressed or Lump Stone
DRILLING DATA		Hand-held equipment
COMPRESSORS		Atlas Copco
		Broomwade Static
LOAD AND HAUL	**Loaders**	Hyster Fork-lift

TREVONE QUARRIES LTD

TREVONE QUARRY

664

Longdowns, Falmouth, Cornwall TR11 5QD
Phone 0326 40604
OS Map 204   **OS Grid Reference** SW747324
Personnel P Morgan, General Manager and Unit Manager

PRODUCT DATA	**Rock type**	Carnmenellis Granite
	Colour	Grey
	Grain	Medium
	Product	Dressed or Lump Stone
DRILLING DATA		Hand-held equipment

WADDINGTON FELL QUARRIES LTD
Head Office Waddington Fell Quarry, Fell Road, Waddington, Clitheroe, Lancs BB7 3AA
Phone 02006 334
Personnel W J Brown, Director
Quarries Waddington Fell

WADDINGTON FELL QUARRIES LTD

WADDINGTON FELL QUARRY

665

Fell Road, Waddington, Clitheroe, Lancs BB7 3AA
Phone 02006 334
OS Map 103   **OS Grid Reference** SD718479
Personnel F Brown, Unit Manager

PRODUCT DATA	**Rock type**	Carboniferous Millstone Grit
	Colour	Buff, pink, yellow
	Grain	Fine
	Products	Dressed or Lump Stone
		Crushed Stone
DRILLING DATA		Hole Diameter 105mm
		Bench Height 10–15m
	Drills	Ingersoll-Rand CM350
COMPRESSORS		Ingersoll-Rand
SECONDARY BREAKING		Dropball on Smith 26
LOAD AND HAUL	**Trucks**	Aveling Barford
	Loaders	Caterpillar
CRUSHING PLANT	**Primary**	Kue Ken 120 Jaw
	Secondary	Kue Ken 95 Jaw
	Tertiary	None

JOHN WAINWRIGHT AND COMPANY LTD
Head Office Downside Quarry, Downside, Shepton Mallett, Somerset BA4 4JF
Phone 0749 2366
Personnel J Laff, Director
Quarries Moons Hill

entry continues overleaf

JOHN WAINWRIGHT AND COMPANY LTD

MOONS HILL QUARRY

666

Stoke St Michael, Bath, Avon BA3 5JU
Phone 0749 840274
OS Map 183 **OS Grid Reference** ST662462
Personnel P Barkwell, Unit Manager

PRODUCT DATA	**Rock type**	Andesite
	Colour	Brownish grey
	Grain	Fine
	Products	Crushed Stone
		Asphalt or Tarmacadam
		Pre-cast Concrete Products
DRILLING DATA		Hole Diameter 110mm
		Bench Height 12m
		Contractor W C D Sleeman
SECONDARY BREAKING		Dropball on Ruston-Bucyrus 22-RB
LOAD AND HAUL	**Trucks**	Euclid R25
	Loaders	Caterpillar
CRUSHING PLANT	**Primary**	Pegson Jaw
	Secondary	Pegson Gyratory
	Tertiary	Pegson Autocone

THOMAS W WARD ROADSTONE LTD (RMC ROADSTONE PRODUCTS LTD)
Head Office *Albion Works, Saville Street, Sheffield, S Yorks S4 7UL*
*Phone 0742 701900 **Fax** 0742 722744 **Telex** 547831*
***Personnel** B Whitworth, Managing Director*
T M Lenagh, General Manager
T Bentley, Production Manager
***Quarries** Goddards, Shining Bank*

THOMAS W WARD ROADSTONE LTD (RMC ROADSTONE PRODUCTS LTD)

GODDARDS QUARRY

667

Stoney Middleton, Derbys S30 1TR
Phone 0433 30343
OS Map 119 **OS Grid Reference** SK225756
Personnel K Wiles, Unit Manager

PRODUCT DATA	**Rock type**	Carboniferous Monsal Dale Limestone
	Colour	Grey
	Products	Dressed or Lump Stone
		Crushed Stone
		Industrial Limestone
DRILLING DATA		Hole Diameter 125mm
		Bench Height 13-17m
	Drills	Hausherr HBM60
SECONDARY BREAKING		Dropball on Ruston-Bucyrus 22-RB
LOAD AND HAUL	**Trucks**	Aveling Barford
		Terex
	Loaders	(2)Michigan

CRUSHING PLANT	**Primary**	Pegson 42 × 48 Jaw
	Secondary	Parker 105 Kubitizer
	Tertiary	Mansfield No 3 Hammermill

THOMAS W WARD ROADSTONE LTD (RMC ROADSTONE PRODUCTS LTD)

SHINING BANK QUARRY

668

Alport, Bakewell, Derbys DE4 1LE
Phone 062 986366
OS Map 119   **OS Grid Reference** SK228650
Personnel J Shaw, Unit Manager

PRODUCT DATA	**Rock type**	Carboniferous Monsal Dale Limestone
	Colour	Dark blue, bluish grey
	Products	Dressed or Lump Stone
		Crushed Stone
		Asphalt or Tarmacadam
		Industrial Limestone
DRILLING DATA		Hole Diameter 125mm
		Bench Height 15-20m
	Drills	Hausher HBM60
SECONDARY BREAKING		Dropball on Ruston-Bucyrus 22-RB
LOAD AND HAUL	**Trucks**	Aveling Barford
	Loaders	Caterpillar
CRUSHING PLANT	**Primary**	Kue Ken 48″ Jaw
	Secondary	Hazemag APK60
	Tertiary	Hazemag APK30

WATERMILL AGGREGATES LTD
Head Office *Concraigs Quarry, Rathen, Fraserburgh, Grampian AB4 4ED*
Phone 0346 28226
Personnel I F Corbett, Managing Director
Quarries *Concraigs*

WATERMILL AGGREGATES LTD

CONCRAIGS QUARRY

669

Rathen, Fraserburgh, Grampian AB4 4ED
Phone 0346 28226
OS Map 30   **OS Grid Reference** NK017611
Personnel I F Corbett, Managing Director and Unit Manager

PRODUCT DATA	**Rock type**	Pre-Cambrian Dalradian Limestone
	Product	Crushed Stone
DRILLING DATA		Contractor

NO FURTHER INFORMATION PROVIDED

W CLIFFORD WATTS LTD
Head Office *118–122 Scarborough Road, Bridlington, Humberside YO16 5NX*
Phone *0262 675383* *Fax* *0262 604629*
Personnel *R W Watts, Director*
Quarries *Brantingham*

W CLIFFORD WATTS LTD

BRANTINGHAM QUARRY 670

Brough, Humberside HU15 1QC
Phone 0262 675383
OS Map 106 **OS Grid Reference** SE945300
Personnel J S Warcup, General Manager

PRODUCT DATA	**Rock type**	Jurassic Limestone
	Colour	Not commercially significant
	Grain	Fine
	Product	Crushed Stone
DRILLING DATA		Hole Diameter 108mm
		Bench Height 17m
	Drills	(2)Holman Voltrak
COMPRESSORS		CompAir 650HE
LOAD AND HAUL	**Trucks**	Volvo
	Loaders	Caterpillar
CRUSHING PLANT	**Primary**	Goodwin Barsby 36×24 Jaw (Mobile)

THE WEELAND SAND COMPANY
Head Office *High Street, South Milford, Leeds, W Yorks LS25 5AA*
Phone *0532 682337*
Personnel *J Connors, Director*
Quarries *Bank Top*

THE WEELAND SAND COMPANY

BANK TOP QUARRY 671

Blackhill Road, Arthington, Leeds, W Yorks LS21 1P2
Phone 0532 842235
OS Map 104 **OS Grid Reference** SE268436
Personnel K Fletcher, Unit Manager

PRODUCT DATA	**Rock type**	Carboniferous Millstone Grit
	Colour	Not commercially significant
	Grain	Medium
	Products	Dressed or Lump Stone
		Crushed Stone
DRILLING DATA		Hole Diameter 105mm
		Bench Height 7-15m
	Drills	Halco 300
COMPRESSORS		Ingersoll-Rand

LOAD AND HAUL	**Trucks**	Haulomatic
	Loaders	Komatsu WA350
CRUSHING PLANT	**Primary**	Goodwin Barsby Jaw
	Secondary	None
	Tertiary	None

WESTERN ROADSTONE LTD (RMC ROADSTONE PRODUCTS LTD)
Head Office *London Road, Wick, Bristol, Avon BS15 5SJ*
Phone *027582 4237*   **Fax** *027582 2272*
Personnel *P J Owen, Chairman*
        G R C Smith, Divisional Director
        M J McDowall, Director
        D O Glasson, Director of Quarries
        K Denning, Production Manager
Quarries *Backwell, Quartzite, Wick, Wickwar*

WESTERN ROADSTONE LTD (RMC ROADSTONE PRODUCTS LTD)

BACKWELL QUARRY   **672**

Backwell, Nr Bristol, Avon BS19 3DE
Phone 027583 2041
OS Map 172   **OS Grid Reference** ST493678
Personnel A Day, Unit Manager

PRODUCT DATA	**Rock type**	Carboniferous Limestone
	Colour	Light grey, buff
	Products	Crushed Stone
		Ready-mixed Concrete
DRILLING DATA		Hole Diameter 110mm
		Bench Height 15m
		Contractor Drill Quip UK
COMPRESSORS		CompAir
SECONDARY BREAKING		Hydraulic Hammer on Ford H44
LOAD AND HAUL	**Trucks**	(2)Aveling Barford RD40
	Loaders	Caterpillar 980C
CRUSHING PLANT		No information provided

WESTERN ROADSTONE LTD (RMC ROADSTONE PRODUCTS LTD)

QUARTZITE QUARRY   **673**

c/o Wickwar Quarry, The Downs Road, Wickwar, Wotton-under-Edge, Glos GL12 8NF
Phone 0454 294513
OS Map 172   **OS Grid Reference** ST690898
Personnel A Millard, Unit Manager

PRODUCT DATA	**Rock type**	Carboniferous Cromhall Sandstone
	Colour	Pink
	Grain	Medium
	Product	Crushed Stone

entry continues overleaf

DRILLING DATA		Hole Diameter 105mm
		Bench Height 12m
		Contractor Drill Quip UK

SECONDARY BREAKING		Dropball

LOAD AND HAUL	**Trucks**	Foden
	Loaders	Caterpillar 966C

CRUSHING PLANT		No information provided

WESTERN ROADSTONE LTD (RMC ROADSTONE PRODUCTS LTD)

WICK QUARRY
674

London Road, Wick, Bristol, Avon BS15 5SJ
Phone 027582 4237
OS Map 172   **OS Grid Reference** ST711732
Personnel R Piper, Unit Manager

PRODUCT DATA	**Rock type**	Carboniferous Limestone
	Colour	Light grey
	Products	Crushed Stone
		Asphalt or Tarmacadam
		Pre-cast Concrete Products

DRILLING DATA		Hole Diameter 110–165mm
		Bench Height 12-15m
		Contractor Drill Quip UK

SECONDARY BREAKING		Hydraulic Hammer - Furukawa

LOAD AND HAUL	**Face shovel**	Caterpillar 980C
	Trucks	(2)Aveling Barford RD40

CRUSHING PLANT	**Primary**	Parker 32 × 42 Jaw (Single Toggle)
	Secondary	Mansfield No 5 Hammermill
	Tertiary	Mansfield No 3 Hammermill

WESTERN ROADSTONE LTD (RMC ROADSTONE PRODUCTS LTD)

WICKWAR QUARRY
675

The Downs Road, Wickwar, Wotton-under-Edge, Glos GL12 8NF
Phone 0454 294225
OS Map 172   **OS Grid Reference** ST720896
Personnel M Markloves, Unit Manager

PRODUCT DATA	**Rock type**	Carboniferous Limestone
	Colour	Light grey
	Products	Crushed Stone
		Asphalt or Tarmacadam
		Ready-mixed Concrete

DRILLING DATA		Hole Diameter 105mm
		Bench Height 12m
		Contractor Drill Quip UK

SECONDARY BREAKING		Hydraulic Hammer - Furukawa

LOAD AND HAUL	**Trucks**	Caterpillar 769C
	Loaders	Caterpillar 950
		Caterpillar 966D
		Caterpillar 980C
		Komatsu WA450
		Komatsu WA500
CRUSHING PLANT	**Primary**	Nordberg Bergeaud Jaw
	Secondary	Hazemag APK50
		SBM Wagoner
	Tertiary	Babbitless

THOMAS WILLIAMS (EUXTON) LTD
Head Office Springfield, Wigan Road, Euxton, Chorley, Lancs PR7 6LB
Phone 02572 63437
Personnel T Williams, Director
Quarries Whittle Hill

THOMAS WILLIAMS (EUXTON) LTD
WHITTLE HILL QUARRY   **676**
Whittle-le-Woods, Chorley, Lancs PR6 0LB
Phone 02572 64167
OS Map 102 **OS Grid Reference** SD585220
Personnel T Williams, Director and Unit Manager

PRODUCT DATA	**Rock type**	Carboniferous Millstone Grit
	Colour	Brownish buff
	Grain	Coarse
	Product	Dressed or Lump Stone
DRILLING DATA		Hand-held equipment

GEORGE WIMPEY plc
Corporate Head Office Hammersmith Grove, Hammersmith, London W6 7EN
Phone 081 748 2000 **Fax** 081 748 0076 **Telex** 25666
Personnel Sir C Chetwood, Chairman and Chief Executive
        J A Dwyer, Group Managing Director Contracting and Minerals
        T S Ross, Managing Director Minerals
Companies Wimpey Asphalt Ltd, Wimpey Hobbs Ltd

WIMPEY ASPHALT LTD (GEORGE WIMPEY plc)
Head Office Chiswick Green House, Acton Lane, London W4 5HF
Phone 081 995 7733 **Fax** 081 995 6746
Personnel J A Dwyer, Chairman
        A E Panter, Managing Director
Companies Wimpey Asphalt Ltd (Scotland)

WIMPEY ASPHALT LTD (SCOTLAND) (WIMPEY ASPHALT LTD)
Head Office Barnton Grove, Edinburgh, Lothian EH4 6BT
Phone 0313 391721 **Fax** 0313 395803
Personnel A Windmill, Regional Director
Quarries Boards, Borthwick, Daviot, Friarton, Hillwood, Orrock

entry continues overleaf

WIMPEY ASPHALT LTD (SCOTLAND) (WIMPEY ASPHALT LTD)

BOARDS QUARRY

677

Denny, Central FK6 6RA
Phone 0324 823555
OS Map 57   **OS Grid Reference** NS800854
Personnel J McCulloch, Unit Manager

PRODUCT DATA	**Rock type**	Quartzitic Dolerite Sill
	Colour	Medium grey
	Grain	Medium
	Products	Crushed Stone
		Asphalt or Tarmacadam

DRILLING DATA		Hole Diameter 105mm
		Bench Height 20m
		Contractor Ritchies Equipment

| SECONDARY BREAKING | | Hydraulic Hammer |

| LOAD AND HAUL | **Trucks** | Volvo |
| | **Loaders** | Caterpillar |

CRUSHING PLANT	**Primary**	Kue Ken 160 Jaw
	Secondary	Allis Chalmers 1850
	Tertiary	Nordberg 1350 Autocone
		Nordberg 4' Cone (Shorthead)

WIMPEY ASPHALT LTD (SCOTLAND) (WIMPEY ASPHALT LTD)

BORTHWICK QUARRY

678

Duns, Borders TD11 3NR
Phone 0361 83717
OS Map 67   **OS Grid Reference** NT775543
Personnel T Y Whiteman, Unit Manager

PRODUCT DATA	**Rock type**	Olivine Dolerite
	Colour	Dark grey, green
	Grain	Medium
	Products	Crushed Stone
		Asphalt or Tarmacadam

DRILLING DATA		Hole Diameter 105mm
		Bench Height 20m
		Contractor Ritchies Equipment

| SECONDARY BREAKING | | Hydraulic Hammer |

| LOAD AND HAUL | **Trucks** | Volvo |
| | **Loaders** | Volvo |

CRUSHING PLANT	**Primary**	Parker 24 × 36 Jaw
	Secondary	Baxter ST32 × 22 Jaw (Single Toggle)
	Tertiary	Allis Chalmers 43" Gyratory
		Pegson 36" Cone
		Pegson 24" Cone

WIMPEY ASPHALT LTD (SCOTLAND) (WIMPEY ASPHALT LTD)

DAVIOT QUARRY   **679**

Daviot, Highland IV1 2XD
Phone 0463 85210
OS Map 27   **OS Grid Reference** NH714396
Personnel J Conway, Unit Manager

PRODUCT DATA	**Rock type**	Pre-Cambrian Moine Schist
	Colour	Pinkish grey
	Products	Crushed Stone
		Asphalt or Tarmacadam
DRILLING DATA		Hole Diameter 105mm
		Bench Height 15-25m
		Contractor Ritchies Equipment
SECONDARY BREAKING		Hydraulic Hammer
LOAD AND HAUL	**Trucks**	Volvo
	Loaders	Volvo
CRUSHING PLANT	**Primary**	Kue Ken 120 Jaw
	Secondary	Kue Ken 106 Jaw
	Tertiary	Allis Chalmers 736 Cone
		Allis Chalmers 436 Cone

WIMPEY ASPHALT LTD (SCOTLAND) (WIMPEY ASPHALT LTD)

FRIARTON QUARRY   **680**

Gleneagles Road, Perth, Tayside PH2 0AW
Phone 0738 24131
OS Map 58   **OS Grid Reference** NO118213
Personnel W Gowrie, Unit Manager

PRODUCT DATA	**Rock type**	Basalt
	Colour	Dark grey
	Grain	Fine
	Products	Crushed Stone
		Asphalt or Tarmacadam
DRILLING DATA		Hole Diameter 105mm
		Bench Height 12-20m
		Contractor Ritchies Equipment
LOAD AND HAUL		Contractor
CRUSHING PLANT	**Primary**	Parker 32 × 42 Jaw
	Secondary	Allis Chalmers 1136 Cone
	Tertiary	Allis Chalmers 436 Cone
		Allis Chalmers 236 Cone

WIMPEY ASPHALT LTD (SCOTLAND) (WIMPEY ASPHALT LTD)

HILLWOOD QUARRY   **681**

Ratho, Lothian EH28 8LU
Phone 0313 331202   **Fax** 0313 331553
OS Map 66   **OS Grid Reference** NT135716
Personnel S Fullerton, Unit Manager

entry continues overleaf

PRODUCT DATA	**Rock type**	Dolerite
	Colour	Dark grey
	Grain	Medium
	Products	Crushed Stone
		Asphalt or Tarmacadam

DRILLING DATA		Hole Diameter 105mm
		Bench Height 18–20m
		Contractor Ritchies Equipment

| SECONDARY BREAKING | | Hydraulic Hammer |
| | | Dropball on Ruston-Bucyrus 38-RB |

| LOAD AND HAUL | | Contractor |

CRUSHING PLANT	**Primary**	Pegson 44 × 48 Jaw
	Secondary	Allis Chalmers 1858 Gyratory
	Tertiary	Nordberg Cone
		Allis Chalmers 436 Cone
		(2)Allis Chalmers 336 Cone

WIMPEY ASPHALT LTD (SCOTLAND) (WIMPEY ASPHALT LTD)

ORROCK QUARRY

682

Burntisland, Fife KY2 5XD
Phone 0592 872661
OS Map 66 **OS Grid Reference** NT217887
Personnel J McKenna, Unit Manager

PRODUCT DATA	**Rock type**	Basalt
	Colour	Very dark grey
	Grain	Fine
	Products	Crushed Stone
		Asphalt or Tarmacadam

| DRILLING DATA | | Contractor Ritchies Equipment |
| | | No further information provided |

CRUSHING PLANT	**Primary**	Pegson 36 × 46 Jaw
	Secondary	Allis Chalmers 1136 Gyratory
	Tertiary	Allis Chalmers 736 Cone
		Allis Chalmers 336 Cone
		Allis Chalmers 236 Cone

WIMPEY HOBBS LTD (GEORGE WIMPEY plc)

Head Office 4 High Street, Nailsea, Bristol, Avon BS19 1BW
Phone 0272 852281
Personnel A E Panter, Chairman
*   J Ellis, Director and General Manager*
*   M Rushbey, Commercial Director*
Quarries Blaengwynlais, Cornelly, Darlton, Flax Bourton, Halecombe, Harnhill, Hendy, Leapers Wood,
*   Minffordd, Pant, Pant (Halkyn), Penwyllt, Stancombe, Torcoed*

WIMPEY HOBBS LTD (GEORGE WIMPEY plc)

BLAENGWYNLAIS QUARRY

683

Caerphilly, Mid Glam CF8 1NG
Phone 0272 852281
OS Map 170   **OS Grid Reference** ST146841

QUARRY ON STAND

ROCK DATA		
	Rock type	Carboniferous Limestone
	Colour	Pinkish grey
	Grain	Medium

WIMPEY HOBBS LTD (GEORGE WIMPEY plc)

CORNELLY QUARRY

684

Pyle, Mid Glam CF33 4RD
Phone 0656 740413
OS Map 170   **OS Grid Reference** SS811813
Personnel M Llewellyn, Unit Manager

PRODUCT DATA		
	Rock type	Carboniferous Limestone
	Colour	Brownish grey
	Grain	Medium
	Products	Crushed Stone
		Ready-mixed Concrete
		Industrial Limestone
DRILLING DATA		Hole Diameter 125mm
		Bench Height 16–21m
		Contractor Ritchies Equipment
SECONDARY BREAKING		Dropball on Ruston-Bucyrus 22-RB
LOAD AND HAUL	**Trucks**	Terex
	Loaders	No information
CRUSHING PLANT	**Primary**	BJD 69 Impactor
	Secondary	BJD 42 Swinghammer
		BJD 66 Swinghammer

WIMPEY HOBBS LTD (GEORGE WIMPEY plc)

DARLTON QUARRY

685

Stoney Middleton, Nr Sheffield, S Yorks S30 1TR
Phone 0433 30681
OS Map 119   **OS Grid Reference** SK219755
Personnel T Elliot, Unit Manager

PRODUCT DATA		
	Rock type	Carboniferous Monsal Dale Limestone
	Colour	Light creamy grey
	Products	Crushed Stone
		Asphalt or Tarmacadam
DRILLING DATA		Hole Diameter 105mm
		Bench Height 18m
	Drills	Halco 400C

entry continues overleaf

SECONDARY BREAKING		Dropball on Ruston-Bucyrus 22-RB
LOAD AND HAUL	**Trucks**	Aveling Barford
	Loaders	Caterpillar
CRUSHING PLANT	**Primary**	Sheepbridge Impactor
	Secondary	(2)Mansfield No 6 Hammermill
	Tertiary	None

WIMPEY HOBBS LTD (GEORGE WIMPEY plc)

FLAX BOURTON QUARRY

686

Flax Bourton, Bristol, Avon BS19 3QB
Phone 0275 832031
OS Map 172 **OS Grid Reference** ST504685
Personnel P Oddy, Unit Manager

PRODUCT DATA	**Rock type**	Carboniferous Limestone
	Colour	Not commercially significant
	Products	Crushed Stone
		Asphalt or Tarmacadam
		Ready-mixed Concrete
		Pre-cast Concrete Products
DRILLING DATA		Hole Diameter 105mm
		Bench Height 12m
		Contractor Ritchies Equipment
SECONDARY BREAKING		Hydraulic Hammer - Montabert on JCB
LOAD AND HAUL	**Trucks**	Terex
	Loaders	Volvo
		Caterpillar 966D
CRUSHING PLANT	**Primary**	Kue Ken 32 x 42 Jaw
		Pegson Jaw
	Secondary	Pegson 20B Gyratory
	Tertiary	(2)Edgar Allan KB4
		(3)Edgar Allan KB3

WIMPEY HOBBS LTD (GEORGE WIMPEY plc)

HALECOMBE QUARRY

687

Leigh upon Mendip, Mews, Frome, Somerset BA3 5QG
Phone 0373 812731
OS Map 183 **OS Grid Reference** ST702475
Personnel R Handley, Unit Manager

PRODUCT DATA	**Rock type**	Carboniferous Limestone
	Colour	Not commercially significant
	Products	Crushed Stone
		Asphalt or Tarmacadam
DRILLING DATA		Hole Diameter 105mm
		Bench Height 16–19m
		Contractor Ritchies Equipment

SECONDARY BREAKING		Dropball
LOAD AND HAUL	**Face shovel**	O&K
	Trucks	(3)Terex R35
		Terex 33-07
	Loaders	Volvo
		Caterpillar 966D
CRUSHING PLANT		No information provided

WIMPEY HOBBS LTD (GEORGE WIMPEY plc)

HARNHILL QUARRY

688

Alveston, Avon BS12 3AS
Phone 0454 612670
OS Map 172 **OS Grid Reference** ST599879
Personnel K Farrar, Unit Manager

PRODUCT DATA	**Rock type**	Carboniferous Limestone
	Colour	Not commercially significant
	Products	Crushed Stone
		Asphalt or Tarmacadam
		Ready-mixed Concrete
DRILLING DATA		Hole Diameter 110mm
		Bench Height 12m
		Contractor Ritchies Equipment
SECONDARY BREAKING		Hydraulic Hammer on JCB
LOAD AND HAUL	**Face shovel**	O&K
	Trucks	Terex
	Loaders	Caterpillar
CRUSHING PLANT	**Primary**	Parker 36 × 10 Jaw
	Secondary	Acme 30 × 10 Jaw
	Tertiary	Ajax Impactor
		Pegson 36X6 Granulator

WIMPEY HOBBS LTD (GEORGE WIMPEY plc)

HENDY QUARRY

689

Pontyclun, Mid Glam CF3 7YU
Phone 0443 227552 **Fax** 0443 228618
OS Map 170 **OS Grid Reference** ST053813
Personnel M Davies, Unit Manager

PRODUCT DATA	**Rock type**	Carboniferous Limestone
	Colour	Medium brownish grey
	Grain	Medium
	Product	Crushed Stone
DRILLING DATA		Hole Diameter 105mm
		Bench Height 11m
		Contractor Ritchies Equipment

entry continues overleaf

LOAD AND HAUL	**Face shovel**	O&K
	Trucks	Terex
	Loaders	Caterpillar
		Terex

CRUSHING PLANT	**Primary**	Hazemag APK4 Hammermill
	Secondary	GEC Pensylvania Hammermill
	Tertiary	Symons Autocone

WIMPEY HOBBS LTD (GEORGE WIMPEY plc)

LEAPERS WOOD QUARRY 690

Kellet Road, Carnforth, Lancs LA6 1HJ
Phone 0524 732135 **Fax** 0524 736035
OS Map 97 **OS Grid Reference** SD514695
Personnel P Webster, Unit Manager

PRODUCT DATA	**Rock type**	Carboniferous Limestone
	Colour	Creamy grey
	Products	Crushed Stone
		Asphalt or Tarmacadam
		Agricultural Lime

DRILLING DATA		Hole Diameter 110mm
		Bench Height 15–17m
	Drills	Atlas Copco ROC404-04

| COMPRESSORS | | Atlas Copco XRH350 |

| SECONDARY BREAKING | | Dropball on Ruston-Bucyrus 22-RB |

| LOAD AND HAUL | **Trucks** | Aveling Barford |
| | **Loaders** | (4)Caterpillar |

CRUSHING PLANT	**Primary**	BJD Impactor
	Secondary	BJD Impactor
	Tertiary	(2)Mansfield No 3 Hammermill

WIMPEY HOBBS LTD (GEORGE WIMPEY plc)

MINFFORDD QUARRY 691

Penrhyndeudraeth, Nr Porthmadog, Gwynedd LL48 6HP
Phone 0766 770212
OS Map 124 **OS Grid Reference** SH593390
Personnel B Sweeney, Unit Manager

PRODUCT DATA	**Rock type**	Basalt
	Colour	Medium greyish green
	Grain	Fine
	Products	Crushed Stone
		Asphalt or Tarmacadam
		Ready-mixed Concrete

DRILLING DATA		Hole Diameter 110mm
		Bench Height 18–20m
		Contractor Rock Drillers

SECONDARY BREAKING		Hydraulic Hammer
		Dropball

LOAD AND HAUL	**Face shovel**	Akermans
	Trucks	Terex
		Volvo
	Loaders	Caterpillar 966D
		Caterpillar 968

CRUSHING PLANT	**Primary**	Kue Ken 36×42 Jaw
	Secondary	Nordberg 4.25' Cone
	Tertiary	Hazemag APK160
		Allis Chalmers 36" Gyratory

WIMPEY HOBBS LTD (GEORGE WIMPEY plc)

PANT QUARRY

692

St Brides Major, Bridgend, Mid Glam CF32 0SU
Phone 0656 56948
OS Map 170   **OS Grid Reference** SS895760
Personnel J Jones, Unit Manager

PRODUCT DATA	**Rock type**	Carboniferous Limestone
	Colour	Medium brownish grey
	Grain	Medium
	Products	Crushed Stone
		Asphalt or Tarmacadam

DRILLING DATA		Hole Diameter 105mm
		Bench Height 16–18m
		Contractor Ritchies Equipment

SECONDARY BREAKING		Dropball on Ruston-Bucyrus 22-RB

LOAD AND HAUL	**Face shovel**	Poclain
	Trucks	Euclid
	Loaders	Caterpillar
		Volvo

CRUSHING PLANT	**Primary**	Parker 24×36 Jaw
	Secondary	Parker Kubitizer

WIMPEY HOBBS LTD (GEORGE WIMPEY plc)

PANT (HALKYN) QUARRY

693

Halkyn, Holywell, Clwyd CH8 8BP
Phone 0352 780441   **Fax** 0352 781207
OS Map 116   **OS Grid Reference** SJ203703
Personnel A Edwards, Unit Manager

PRODUCT DATA	**Rock type**	Carboniferous Limestone
	Colour	Creamy grey
	Grain	Medium
	Products	Crushed Stone
		Asphalt or Tarmacadam

entry continues overleaf

DRILLING DATA		Hole Diameter 105mm Bench Height 20m Contractor A Jones Rock Drillers
SECONDARY BREAKING		Dropball on Ruston-Bucyrus 22-RB
LOAD AND HAUL	**Face shovel**	Akermans
	Trucks	Aveling Barford RD40
		Volvo
	Loaders	Caterpillar
CRUSHING PLANT	**Primary**	BJD 55 Impactor
	Secondary	BJD 44 Impactor
	Tertiary	Mansfield No 4 Hammermill

WIMPEY HOBBS LTD (GEORGE WIMPEY plc)

PENWYLLT QUARRY

694

Abercrave, Powys SA9 1GE
Phone 0637 30681
OS Map 170 **OS Grid Reference** SN856158
Personnel W Burton, Unit Manager

PRODUCT DATA	**Rock type**	Carboniferous Limestone
	Colour	Not commercially significant
	Product	Crushed Stone
DRILLING DATA		Hole Diameter 125mm Bench Height 35m Contractor Ritchies Equipment
SECONDARY BREAKING		Dropball
LOAD AND HAUL	**Trucks**	(2)Terex
	Loaders	Caterpillar
		Demag
CRUSHING PLANT	**Primary**	BJD 45 Impactor
	Secondary	BJD 30 Impactor

WIMPEY HOBBS LTD (GEORGE WIMPEY plc)

STANCOMBE QUARRY

695

Backwell, Bristol, Avon BS19 3DE
Phone 0272 852281
OS Map 172 **OS Grid Reference** ST503684

QUARRY ON STAND

ROCK DATA	**Rock type**	Carboniferous Limestone
	Colour	Light buffish grey
	Grain	Medium

WIMPEY HOBBS LTD (GEORGE WIMPEY plc)

TORCOED QUARRY

696

Porthyrhyd, Carmarthen, Dyfed SA32 8PY
Phone 0267 86261
OS Map 159 **OS Grid Reference** SN482136
Personnel T Jenkins, Unit Manager

PRODUCT DATA	**Rock type**	Carboniferous Limestone
	Colour	Not commercially significant
	Grain	Medium
	Products	Crushed Stone
		Asphalt or Tarmacadam
DRILLING DATA		Hole Diameter 125mm
		Bench Height 18m
		Contractor Ritchies Equipment
SECONDARY BREAKING		Dropball on Ruston-Bucyrus 22-RB
LOAD AND HAUL	**Trucks**	Terex
	Loaders	Michigan
CRUSHING PLANT	**Primary**	BJD 55 Mk3 Impactor
	Secondary	BJD Floating Hammer

WINCILATE LTD

Head Office *Aberllefeni Slate Quarries Ltd, Machynlleth, Powys SY20 9RU*
Phone *065473 602* **Fax** *065473 418*
Personnel *J F Lloyd, Managing Director*
W S Simmonds, Director
Quarries *Aberllefenni*

WINCILATE LTD

ABERLLEFENI QUARRY

697

Machynlleth, Powys SY20 9RU
Phone 065473 602
OS Map 124 **OS Grid Reference** SH768100
Personnel J F Lloyd, Director and Unit Manager

PRODUCT DATA	**Rock type**	Ordovician Slate
	Colour	Bluish grey
	Grain	Fine
	Product	Dressed or Lump Stone
DRILLING DATA		Hand-held equipment

WINDY HILL NATURAL STONE COMPANY

Head Office *Woodside House, 8A Barnard Street, Staindrop, Darlington, Durham DL2 3ND*
Phone *0833 60367*
Personnel *F Jackson, Director*
L Wilkinson, Director
Quarries *Windy Hill*

entry continues overleaf

WINDY HILL NATURAL STONE COMPANY

WINDY HILL QUARRY

698

Marwood, Barnard Castle, Durham DL2 3ND
Phone 0835 60367
OS Map 92   **OS Grid Reference** NY964223
Personnel F Jackson, Director and Unit Manager

PRODUCT DATA	**Rock type**	Carboniferous Sandstone
	Colour	Greyish brown
	Grain	Medium
	Product	Dressed or Lump Stone

DRILLING DATA		Hand-held equipment

WOTTON ROADSTONE LTD (RMC ROADSTONE PRODUCTS LTD)
Head Office *Wenvoe Quarry, Walston, Wenvoe, Cardiff, S Glam CF5 6XE*
Phone *0222 593378*   **Fax** *0222 595840*
Personnel *M J McDowall, Managing Director*
D Glasson, Divisional Director
R S Millard, General Manager
D Matysua, Production Manager
Quarries *Blaenyfan, Gellihalog, Gilfach, Wenvoe*

WOTTON ROADSTONE LTD (RMC ROADSTONE PRODUCTS LTD)

BLAENYFAN QUARRY

699

Pontantwn, Kidwelly, Dyfed SA17 5LA
Phone 0269 860204
OS Map 159   **OS Grid Reference** SN456114
Personnel H Isles, Unit Manager

PRODUCT DATA	**Rock type**	Carboniferous Limestone
	Colour	Grey
	Grain	Medium
	Products	Crushed Stone
		Asphalt or Tarmacadam
		Ready-mixed Concrete
		Pre-cast Concrete Products
		Agricultural Lime

DRILLING DATA		Hole Diameter 110mm
		Bench Height 11–16m
	Drills	Halco 410C

SECONDARY BREAKING		Dropball on Ruston-Bucyrus 22-RB

LOAD AND HAUL	**Trucks**	Aveling Barford RD30
		Aveling Barford RD40
	Loaders	Caterpillar 966C

CRUSHING PLANT	**Primary**	BJD Impactor
	Secondary	SBM Impactor

WOTTON ROADSTONE LTD (RMC ROADSTONE PRODUCTS LTD)

GELLIHALOG QUARRY

700

Ludchurch, Narbeth, Dyfed SA67 8LE
Phone 083483 504
OS Map 158   **OS Grid Reference** SN159107
Personnel R Bevinton, Unit Manager

PRODUCT DATA	**Rock type**	Carboniferous Limestone
	Colour	Dark grey
	Grain	Medium
	Product	Crushed Stone

DRILLING DATA — Hole Diameter 105mm
Bench Height 15–17m
Contractor A Jones

SECONDARY BREAKING — Dropball

NO FURTHER INFORMATION PROVIDED

WOTTON ROADSTONE LTD (RMC ROADSTONE PRODUCTS LTD)

GILFACH QUARRY

701

The Rhyddings, Neath, W Glam SA10 8AD
Phone 0639 643751
OS Map 170   **OS Grid Reference** SS752999
Personnel K Lee, Unit Manager

PRODUCT DATA	**Rock type**	Carboniferous Pennant Sandstone
	Colour	Olive green, silver grey
	Grain	Medium to coarse
	Products	Crushed Stone
		Asphalt or Tarmacadam
		Ready-mixed Concrete

DRILLING DATA — Hole Diameter 105mm
Bench Height 30m
Contractor Drill Quip UK

SECONDARY BREAKING — Dropball on Ruston-Bucyrus 22-RB

LOAD AND HAUL	**Trucks**	Aveling Barford RD30
		Terex R35B
	Loaders	Caterpillar 966C
		Komatsu WA450

CRUSHING PLANT	**Primary**	Kue Ken 120 Jaw
	Secondary	Kue Ken 106 Jaw
	Tertiary	Kue Ken 3000 Cone
		Allis Chalmers Hydrcone

WOTTON ROADSTONE LTD (RMC ROADSTONE PRODUCTS LTD)

WENVOE QUARRY

702

Walston, Wenvoe, Cardiff, S Glam CF5 6XE
Phone 0222 593378
OS Map 171   **OS Grid Reference** ST133742
Personnel L Fisher, Unit Manager

entry continues overleaf

PRODUCT DATA	**Rock type**	Carboniferous Limestone
	Colour	Light grey
	Grain	Medium
	Products	Crushed Stone
		Asphalt or Tarmacadam
		Ready-mixed Concrete
DRILLING DATA		Hole Diameter 105mm
		Bench Height 18m
	Drills	Halco 425P
COMPRESSORS		Atlas Copco XR350DD
SECONDARY BREAKING		Dropball on Ruston-Bucyrus 22-RB
LOAD AND HAUL	**Trucks**	Aveling Barford RD30
		Terex R35T
	Loaders	Caterpillar 966C
		Komatsu WA450
CRUSHING PLANT	**Primary**	Pegson 44 × 48 Jaw
	Secondary	Hazemag APK60
	Tertiary	BJD 44 Impactor

FOSTER YEOMAN LTD

Corporate Head Office Marsden House, Marsden Bigot, Frome, Somerset BA11 5DU
Phone 0373 52002 **Fax** 0373 84501
Personnel Mrs Yeoman, Chairman
        D Tidmarsh, Managing Director
        G Smart, Director
        J Yeoman, Director
Companies Yeoman (Morvern) Ltd   — see separate entry
Quarries Dulcote, Merehead

FOSTER YEOMAN LTD

DULCOTE QUARRY 703

Wells, Somerset BA5 3PY
Phone 0749 72691
OS Map 183 **OS Grid Reference** ST567444
Personnel C Roberts, Unit Manager

PRODUCT DATA	**Rock type**	Carboniferous Limestone
	Colour	Not commercially significant
	Grain	Fine
	Product	Crushed Stone
DRILLING DATA		Hole Diameter 150mm
		Bench Height 15m
	Drills	Halco 625H
LOAD AND HAUL	**Trucks**	Caterpillar
	Loaders	Caterpillar

CRUSHING PLANT   No information provided

FOSTER YEOMAN LTD

MEREHEAD QUARRY   **704**

Torr Works, East Cranmore, Shepton Mallet, Somerset BA4 4SQ
Phone 0373 51001   **Telex** 449561
OS Map 183   **OS Grid Reference** ST695439
Personnel R Neale, Unit Manager

PRODUCT DATA	**Rock type**	Carboniferous Limestone
	Colour	Light grey
	Grain	Fine to medium
	Products	Crushed Stone
		Asphalt or Tarmacadam
		Industrial Limestone
		Agricultural Lime

DRILLING DATA		Hole Diameter 195mm
		Bench Height 18m
	Drills	Ingersoll-Rand DM40
		Ingersoll-Rand DM50

SECONDARY BREAKING		(4)Hydraulic Hammer

LOAD AND HAUL	**Trucks**	(2)Caterpillar 777
	Loaders	Caterpillar

CRUSHING PLANT	**Primary**	(2)Fuller 54″ Gyratory (Mobile)
	Secondary	GEC Pensyvania Hammermill
	Tertiary	(2)Nordberg 5.5′ Cone

YEOMAN (MORVERN) LTD (FOSTER YEOMAN LTD)
Head Office Glensanda, Morvern by Oban, Highland PA34 5EM
Phone 0631 73415
Personnel D Yeoman, Director
Quarries Glensanda

YEOMAN (MORVERN) LTD (FOSTER YEOMAN LTD)

GLENSANDA QUARRY   **705**

Morvern by Oban, Highland PA34 5EM
Phone 0631 73415
OS Map 49   **OS Grid Reference** NM815475
Personnel E Gilroy, Site Manager

PRODUCT DATA	**Rock type**	Biotite Granite
	Colour	Red
	Grain	Medium
	Product	Crushed Stone

entry continues overleaf

DRILLING DATA	**Drills**	(5)Ingersoll-Rand CM350
		Ingersoll-Rand LM500
		Halco 450H
		Tamrock DHA1000
COMPRESSORS		(8)Ingersoll-Rand DXL600P
LOAD AND HAUL	**Face shovel**	Demag H185
	Trucks	Caterpillar
	Loaders	Caterpillar
CRUSHING PLANT	**Primary**	Nordberg 54–75 Gyratory
	Secondary	Nordberg 7′ Cone (Standard)
	Tertiary	Nordberg 7′ Gyratory

Index 1

Quarry Companies
by company name

Entries in this index refer to the page numbers
of the main Directory

Geotechnical and Geological Engineering

Editor, **D G Toll**, School of Engineering and Applied Science, University of Durham

Co Editor, **Prof J M Kemeny**, Dept., of Mining and Geological Engineering, University of Arizona, Tucson, Arizona

Geotechnical and Geological Engineering is a new quarterly international journal. Formerly International Journal of Mining and Geological Engineering, the journal's title has been changed to reflect a widening of its scope.

With a new Editor, Co-Editor, *Geotechnical and Geological Engineering* will emphasize the practical aspects of geotechnical engineering and engineering geology. Papers on theoretical and experimental advances in soil and rock mechanics will also be welcomed. Particular priority will be given to the following areas:

* Case histories describing ground engineering projects, including contractual and logistical aspects.
* Novel geotechnical construction techniques
* Pollution and environmental problems
* Tropical soil and rock engineering
* Ground investigation; engineering geological and hydrological appraisals
* Computer-aided geotechnical and geological engineering including the application of information technology

The high technical and publication standards of *International Journal of Mining and Geological Engineering* will be maintained by *Geotechnical and Geological Engineering*.

Subscription Information

ISSN: 0960-3182 Published quarterly Volume 9 will published in 1991
European Community: £120
USA/Canada: $215
Rest of World: £132

For more information about this journal and others published by us, please contact:
The Journals Promotion Dept., Chapman and Hall, 2-6 Boundary Row, London SE1 8HN *(Telephone 071 865 0066)*

Construction Equipment

Experience. The difference

A S QUARRIES LTD   3
43-53 Trafalgar Road
Greenwich
London SE10 9TT
Phone 081 858 5161
Fax 081 293 4103

ABBEY QUARRIES LTD   3
Abbey Quarry
Quarry Lane
Linby
Hucknall
Notts NG15 8GA
Phone 0602 630760

ABER STONE LTD   4
Aberstrecht Quarry
Moelfre
Anglesey
Gwynedd LL72 8NN
Phone 024888 644

ACRESFORD SAND AND GRAVEL LTD   4
High Street
Syston
Leicester
Leics LE7 8GS
Phone 0533 609666

AGGETTS LTD   5
Knowle Quarry
North Road
Okehampton
Devon EX20 1RQ
Phone 0837 52609

D G AITKENS   5
Fornham Park
Fornham St Martin
Bury St Edmunds
Suffolk IP28 6TT
Phone 028484 432

ANTRON HILL GRANITE LTD   6
Meadowbank
Lestraynes Road
Rame
Penryn
Cornwall TR10 9EL
Phone 0209 860672

ARC LTD (HANSON plc)   6
The Ridge
Chipping Sodbury
Bristol
Avon BS17 6AY
Phone 0454 316000
Fax 0454 325161
Telex 449353

ARC CENTRAL LTD (ARC LTD)   6
Ashby Road East
Shepshed
Loughborough
Leics LE12 9BU
Phone 0509 503161
Fax 0509 504120

ARC NORTHERN LTD (ARC LTD)   11
Clifford House
Wetherby Business Park
York Road
Wetherby
W Yorks LS22 4NS
Phone 0937 61977
Fax 0937 61610

ARC POWELL DUFFRYN LTD   19
  (ARC LTD)
Canal Road
Cwmbach
Aberdare
Mid Glam CF44 0AG
Phone 0685 884444
Fax 0685 884444 Ext 227

ARC SOUTHERN LTD (ARC LTD)   24
Stoneleigh House
Frome
Somerset BA11 2HB
Phone 0373 63211
Fax 0373 65843

ARC SOUTH WEST LTD (ARC LTD)   30
Grace Road
Marsh Barton
Trading Estate
Exeter
Devon EX2 8PU
Phone 0392 215353
Fax 0392 51849

C D ASHBRIDGE   33
Pigeon Cote Farm
Malton Road
York
N Yorks YO3 9TD
Phone 0904 425301

J & J ASHCROFT LTD   *34*
Little Quarries
Whittle-le-Woods
Chorley
Lancs PR6 7QR
Phone 02572 62137

ASKERNISH QUARRY COMPANY LTD   *34*
Askernish Quarry
Hillside Garage
Lochboisdale
South Uist
Western Isles PA81 5TS
Phone 08784 278

D S ATKINSON AND PARTNERS   *35*
10 Royal Crescent
Whitby
N Yorks YO21 1RT
Phone 0947 810936

R E ATKINSON   *35*
42 Garden Road
Brighouse
Halifax
W Yorks HD6 2ES
Phone 0274 832379

ATTWOODS plc   *36*
The Pickeridge
Stoke Common Road
Fulmer
Bucks SL3 6HA
Phone 02816 2700
Fax 02816 2464

ALLEN BAILEY   *36*
Hainworth Shaw Quarry
Winfield
Shaw Lane
Hainworth Shaw
Keighley
W Yorks BD21 5QR
Phone 0535 604340

BAIRD AND STEVENSON   *36*
  (QUARRYMASTERS) LTD
Locharbriggs Quarry
Locharbriggs
Dumfries
Dumfries & Galloway DG1 1QS
Phone 0387 710237

S J BAKER   *37*
Trevillet Quarry
Tintagel
Cornwall PL34 0HL
Phone 0840 770659

BALFOUR BEATTY LTD   *37*
7 Mayday Road
Thornton Heath
Croydon
Surrey CR4 7XA
Phone 081 684 6922
Fax 081 684 6597
Telex 264042

BARDON QUARRIES LTD   *37*
Bardon Quarry
Bardon Hill
Coalville
Leics LE6 2TL
Phone 0530 510066
Fax 0530 35780
Telex 341017

BARHAM BROS LTD   *38*
Station Quarry
Charlton Mackrell
Somerton
Somerset TA11 6AG
Phone 045822 3249

BARLAND QUARRY LTD   *38*
Barland Quarry
Pennard Road
Bishopston
Swansea
W Glam SA3 3JE
Phone 044128 2284/2434

BARNEY'S QUARRY DIVISION LTD   *39*
Woodeaton
Oxford
Oxon OX3 9TP
Phone 0865 52646

W & J BARR AND SONS LTD   *40*
Tongland Quarry
Heathfield
Ayr
Strathclyde KA8 9SX
Phone 0292 281311

BATH AND PORTLAND STONE LTD 41
(ARC LTD)
Moor Park House
Moor Green
Corsham
Wilts SN13 9SE
Phone 0225 810456
Fax 0225 811234

BATH STONE CO LTD 42
Midford Lane
Limpley Stoke
Bath
Avon BA2 5SA
Phone 022122 3792

BECKSTONE QUARRY LTD 42
Burford Quarry
Burford Road
Brize Norton
Oxon OX8 8NN
Phone 0993 842391

THE BELLISTON QUARRY 43
COMPANY LTD
Sconnie Park
Leven
Fife KY8 4TD
Phone 0333 26841

BEN BENNETT JNR LTD 43
Lisle Road
Rotherham
S Yorks S60 2RL
Phone 0709 382251

PETER BENNIE LTD 44
Oxwich Close
Brackmills Ind Estate
Northampton
Northants NN4 0BH
Phone 0604 765252

BERWYN SLATE QUARRY LTD 46
Clogau Quarry
Horse Shoe Pass
Llangollen
Clwyd LL20 8BS
Phone 0792 851468

BILLOWN LIME QUARRIES LTD 46
Balladoole Quarry
Malew
Ballasalla
Isle of Man
Phone 0624 822527

BINNS BROS 47
Southowram
Halifax
W Yorks HX3 9QH
Phone 0422 50700

D A BIRD LTD 47
Pury End Quarry
Bugbrooke
Northampton
Northants NN7 3PH
Phone 0604 830455

BLUE CIRCLE INDUSTRIES plc 48
Eccleston Square
London SW1V 1PX
Phone 071 828 3456
Fax 071 245 8169
Telex 849939

BLUE CIRCLE INDUSTRIES plc 48
(CEMENT WORKS)
12th Floor
Churchill Plaza
Churchill Way
Basingstoke
Hants RG21 1QU
Phone 0256 844500
Fax 0256 470608

BODFARI QUARRIES LTD 54
Midland Bank Chambers
Holywell
Clwyd CH8 7TH
Phone 0352 712581

BORDER HARDCORE AND 54
ROCKERY STONE CO LTD
Middletown Quarry
Middletown
Welshpool
Powys SY21 8DJ
Phone 093874 253

BORDERS REGION COUNCIL 55
Council Offices
Newtown St Boswells
Melrose
Borders TD6 0SA
Phone 0835 23301
Fax 0835 22145

BOSTON SPA MACADAMS LTD		*55*
Inglebank Quarry		
Tadcaster Road		
Boston Spa		
Wetherby		
W Yorks LS23 6BZ		
Phone 0937 843493		

BOTHEL LIMESTONE AND   *56*
  BRICK COMPANY
Moota Quarry
Cockermouth
Cumbria CA13 0QE
Phone 0965 202234

H V BOWEN AND SONS LTD   *56*
Tan-y-Foel Quarry
Cefncoch
Llanfair Caereinion
Welshpool
Powys SY21 0AN
Phone 0691 830642

BOYNE BAY LIME COMPANY LTD   *57*
54 Park Road
Aberdeen
Grampian AB2 1PA
Phone 0224 632281

BREEDON plc   *57*
Breedon Hill Quarry
Breedon on the Hill
Derby
Leics DE7 1AP
Phone 0332 862254
Fax 0332 863149

R A BRIGGS AND COMPANY LTD   *58*
Woodside Quarry
Ring Road
Leeds
W Yorks LS16 6QG
Phone 0532 582214

BRIGHOUSE BRICK, TILE AND   *58*
  STONE CO LTD
Tower Farm
Hollins Lane
Steeton
Keighley
W Yorks BD20 6LY
Phone 0535 654864

BRITISH GYPSUM LTD   *59*
Ruddington Hall
Ruddington
Notts NG11 6LX
Phone 0602 844844
Fax 0602 846644
Telex 377292

BRITISH RAIL   *60*
Meldon Quarry
Okehampton
Devon EX20 4LT
Phone 0837 2749

BRITISH STEEL plc   *60*
20 Inch Mill
Brenda Road
Hartlepool
Cleveland TS25 2EG
Phone 0429 266611
Fax 0429 266611
Telex 341561

BRITISH STEEL plc   *60*
  (QUARRY PRODUCTS)
Hardendale Quarry
Shap
Penrith
Cumbria CA10 3QG
Phone 09316 647

BRODEN CONSTRUCTION (NE) LTD   *61*
Industrial Estate
New Herrington
Houghton-le-Spring
Tyne and Wear DH4 7BG
Phone Not available

BRONTE STONE COMPANY   *61*
Coldedge
Withens
Halifax
W Yorks HX3 5LL
Phone 0422 240909

BROWN BROS (LONGRIDGE) LTD   *62*
Altham Quarry
Whinney Hill Road
Accrington
Lancs BB5 6NR
Phone 0254 871049

BRUCE PLANT LTD 63
Spring Works
High Street
Inverbervie by Montrose
Grampian DD10 0RH
Phone 0561 61265

BULLIMORES SAND AND 64
GRAVEL LTD
South Witham
Grantham
Lincs NG33 5QE
Phone 057283 393/627

BURLINGTON SLATE LTD 65
Cavendish House
Kirkby-in-Furness
Cumbria LA17 7UN
Phone 022989 661/666
Fax 022989 466
Telex 65157

NEIL BUTCHER 68
Park Nook Quarry
Bannister Lane
Skelbrooke
Doncaster
S Yorks DN6 8LT
Phone 0302 727924

BUTTERLY AGGREGATES LTD 68
(RMC GROUP plc)
17-21 West Parade
Lincoln
Lincs LN1 1NP
Phone 0522 523391
Fax 0522 511230

CASTLE CEMENT (KETTON) LTD 69
Priestgate
Peterborough
Cambs PE1 1DF
Phone 0733 310505

CASTLE HILL QUARRY 70
COMPANY LTD
Castle Hill Quarry
Cannington
Bridgwater
Somerset TA5 2QD
Phone 0278 652280

CAST VALE QUARRIES 71
(HARGREAVES QUARRIES LTD)
Vale Road Quarry
Mansfield Woodhouse
Mansfield
Notts NG19 8DP
Phone 0623 24570

CAT CASTLE QUARRIES LTD 72
Cat Castle Quarry
Deepdale
Barnard Castle
Durham DL12 9DQ
Phone 0833 50681

CENTRAL REGION COUNCIL 72
Roads & Transport Dept
Viewforth
Stirling
Central FK8 2ET
Phone 0786 73111
Fax 0786 50802

D E CHANCE LTD 73
Bableigh Wood Quarry
2 Acre Road
Horns Cross
Bideford
Devon EX39 5EB
Phone 02375 279

CHARTER CONSOLIDATED plc 73
40 Holborn Viaduct
London EC1P 1A
Phone 071 353 1545
Fax 071 583 2950
Telex 929582

CHIC-GRIT LTD 73
Unit 12
Industrial Estate
Norwich Road
Watton
Norfolk IP25 6DR
Phone 0953 881642

CLARK CONTRACTORS 74
Grove Farm
Welders Lane
Chalfont St Peter
Bucks SL9 8TU
Phone 0247 3200

CLIPSHAM QUARRY COMPANY 74
Clipsham Hall
Clipsham
Leics LE15 7QS
Phone 0780 410204

CLOBURN QUARRY COMPANY LTD 74
Pettinain Quarry
Pettinain
Lanark
Strathclyde ML11 8SR
Phone 0555 4898

D R COLLYER 75
Ridgeway Quarry
Crich Lane
Ridgeway
Belper
Derbys DE4 5BQ
Phone 077 3852400

CORNISH ROADSTONES 75
  (BRIGHOUSE BRICK, TILE AND
  STONE CO LTD)
Tower Farm
Hollins Lane
Steeton
Keighley
W Yorks BD20 6LY
Phone 0535 654864

CORNISH RUSTIC STONE LTD 76
Bowithic Quarry
Trewarmet
Tintagel
Cornwall PL34 0AN
Phone 0836 350082

COTSWOLD WALLING AND MASONRY 76
The Old School
Naunton
Cheltenham
Glos GL54 3BA
Phone 04515 775

W S CROSSLEY (YORK STONE) LTD 77
Norcliffe Lane
Southowram
Halifax
W Yorks HX3 8PL
Phone 0422 44263

CULTS LIME LTD 77
Skelpie Mine
Limehills
Pitlessie
Cupar
Fife KY7 7TF
Phone 0334 52548

CUMBRIA STONE QUARRIES LTD 78
Silver Street
Crosby Ravensworth
Penrith
Cumbria CA10 3JA
Phone 09315 227
Fax 09315 367

J CURTIS AND SONS LTD 80
Thrupp Lane
Radley
Abingdon
Oxon OX14 3NQ
Phone 0235 24545

DALGETY AGRICULTURE LTD 80
Parkmore Quarry
Parkmore
Dufftown
Keith
Grampian AB5 4DL
Phone 0340 20200

DARRINGTON QUARRIES LTD 81
Darrington Quarry
Darrington Leys
Cridling Stubbs
Knottingley
N Yorks WF11 0AH
Phone 0977 82368/82369

N DAVID 83
134 Victoria Rise
London SW4 0NW
Phone 081 720 8764

R DAVIDSON 83
Beacon Lodge Quarry
Long Lane
Southowram
Halifax
W Yorks HX3 9UD
Phone 0422 63674

DAVIDSON AND MURISON LTD The Garage Maryculter Aberdeen Grampian AB1 0BT Phone 0224 732391	*84*	DUNLOSSIT TRUSTEES LTD Ballygrant Quarry Ballygrant Isle of Islay Strathclyde PA45 7QL Phone 049684 232/652	*88*

DAVIDSON AND MURISON LTD   *84*
The Garage
Maryculter
Aberdeen
Grampian AB1 0BT
Phone 0224 732391

MANSELL DAVIES AND SON LTD   *85*
Station Yard
Crymmych
Dyfed SA35 0BZ
Phone 023973 631

DEVON COUNTY COUNCIL   *85*
Civic Centre
North Walk
Barnstable
Devon EX31 1ED
Phone 0271 47067
Fax 0271 47067

DEVON QUARRY AND   *86*
   TARPAVING COMPANY LTD
Blackaller Quarry
Drewsteignton
Exeter
Devon EX6 6RA
Phone 0647 21226

DRINKWATER SABEY LTD   *86*
   (ATTWOODS plc)
The Pickeridge
Stoke Common Road
Fulmer
Bucks SL3 6HA
Phone 02816 2700
Fax 02816 2464

DUMFRIES AND GALLOWAY   *87*
   REGION COUNCIL
Council Offices
Dumfries
Dumfries & Galloway DG2 2DD
Phone 0387 53141

DUNHOUSE QUARRY COMPANY LTD   *87*
7 South Church Road
Bishop Auckland
Durham DL14 7LB
Phone 0388 602322
Fax 0388 603832

DUNLOSSIT TRUSTEES LTD   *88*
Ballygrant Quarry
Ballygrant
Isle of Islay
Strathclyde PA45 7QL
Phone 049684 232/652

A M & K O EARL   *89*
Sycamore Quarry
Windmill Road
Kerridge
Macclesfield
Cheshire SK10 5AZ
Phone 0625 72125

ECC CALCIUM CARBONATES LTD   *89*
   (ENGLISH CHINA CLAYS LTD)
Wilton Road
Quidhampton
Salisbury
Wilts SP2 9AD
Phone 0722 743444
Fax 0722 744900
Telex 477054

ECC QUARRIES LTD   *90*
   (ENGLISH CHINA CLAYS LTD)
Greystones
Huntcote Road
Croft
Leics LE9 6GS
Phone 0455 285200
Fax 0455 283837

   ECC QUARRIES LTD   *90*
      Midlands and North Area
   Croft Quarry
   Croftworks
   Croft
   Leics LE9 6GS
   Phone 0455 282601

   ECC QUARRIES LTD   *93*
      Southern Area
   Rockbeare Hill Quarry
   Exeter
   Devon EX5 2HB
   Phone 0404 822494

   ECC QUARRIES LTD   *95*
      South West Area
   Moorcroft Quarry
   Billacombe
   Plymouth
   Devon PL9 8AJ
   Phone 0752 402661

EDENHALL CONCRETE PRODUCTS 98
Old Mill Quarry
Beith
Strathclyde KA15 1HY
Phone 05055 2721

R W T EDWORTHY 98
(HEATHFIELD MINERALS) LTD
Hayne Quarry
Johnsland
Bow
Crediton
Devon EX17 7HG
Phone 03633 283

ELGIN PRECAST CONCRETE LTD 99
(ROBERTSONS CONTRACTING)
Edgar Road
Elgin
Grampian IV30 3YQ
Phone 0343 549358/549634

R J ENGLAND 99
22-25 Brympton Way
Lynx W Trading Estate
Yeovil
Somerset BA20 2HP
Phone 0935 78173

ENGLISH CHINA CLAYS GROUP plc 100
John Keay House
St Austell
Cornwall PL25 4DJ
Phone 0726 74482
Fax 0726 623019
Telex 45526

    ENGLISH CHINA CLAYS LTD 100
      (ENGLISH CHINA CLAYS GROUP plc)
    Northernhay House East
    Northernhay Place
    Exeter
    Devon EX4 3QP
    Phone 0392 52231
    Fax 0392 412132
    Telex 42838

ESKETT QUARRIES LTD 100
   (EVERED HOLDINGS plc)
Stoneleigh
Park End Road
Workington
Cumbria CA14 4DG
Phone 0900 62271

EVERED HOLDINGS plc 103
6th Floor
Radcliffe House
Blenheim Court
Lode Lane
Solihull
W Midlands B91 2AA
Phone 021 711 1717

EVERED QUARRY PRODUCTS 103
   (ENGLAND) LTD (EVERED
   HOLDINGS plc)
Evered House
Burnwick Terrace
Newport Road
Stafford
Staffs ST16 1BB
W Midlands B91 2AA
Phone 0785 57500

FARMINGTON STONE LTD 104
37 Roslyn Road
Redland
Bristol
Avon BS6 6NJ
Phone 0272 247336

G FARRAR (QUARRIES) LTD 105
Bradford Street
Keighley
W Yorks BD21 3EB
Phone 0535 602344

FENSTONE QUARRIES LTD 105
Market Bridge
Bielby
York
W Yorks YO4 4JR
Phone 0759 318666

FFESTINIOG SLATE QUARRY LTD 106
Blaenau Ffestiniog
Gwynedd LL41 3NB
Phone 0766 830204/830218/830664
Fax 0766 831105

FILKINS QUARRIES LTD 107
Brook House
Cricklade
Wilts SN6 6DD
Phone 0793 750150

W E FORD AND SON (QUARRIES)   *107*
Fords Quarry
Bacup Road
Burnley
Lancs BB11 3RL
Phone 0282 56815

FOREST OF DEAN STONE FIRMS LTD   *108*
Bixland Stone Works
Parkend
Lydney
Glos GL15 4JS
Phone 0594 562304
Fax 0594 564184

GALA QUARRY COMPANY   *108*
Abbey Place
Melrose
Borders TD6 9LQ
Phone 089682 2595

J & A GARDNER AND COMPANY LTD   *109*
288 Clyde Street
Glasgow
Strathclyde G1 4JS
Phone 041 221 7845

D & J GARRICK QUARRIES LTD   *109*
Vatseter Quarry
Gott
Shetland ZE2 9SB
Phone 059584 279

GASKELL BROS (W M & C) LTD   *110*
Bryn Road
Ashton-in-Makerfield
Wigan
Lancs WN4 8AH
Phone 0942 725722
Fax 0942 271189

DAVID GEDDES (QUARRIES) LTD   *110*
Swirlburn
Colliston-by-Arbroath
Tayside DD11 3SH
Phone 024189 266
Fax 024189 445

T W GIBBS   *111*
Horn Park Quarry
Broadwindsor
Beaminster
Dorset DT8 3PT
Phone 0308 68419

DENNIS GILLSON AND SON   *111*
   (HAWORTH) LTD
Naylor Hill Quarries
Blackmore Road
Keighley
W Yorks BD22 9SW
Phone 0535 43317

F GILMAN LTD   *111*
Snowdrop Lane
Haverfordwest
Dyfed SA61 1ET
Phone 0437 765226
Fax 0437 760810

GLENDINNING GROUP   *113*
Glentor
Balland Lane
Ashburton
Newton Abbot
Devon TQ13 7LF
Phone 0364 52601

GLOUCESTER SAND AND GRAVEL   *114*
   COMPANY LTD
Overbury Estate Office
Overbury
Tewkesbury
Glos GL20 7NS
Phone 038689 217

GOETRE LTD   *114*
Station Yard
Abermule
Montgomery
Powys SY15 6NL
Phone 068686 667

GRAMPIAN REGION COUNCIL   *115*
Director of Works
Woodhill House
Ashgrove Road West
Aberdeen
Grampian AB9 2LU
Phone 0224 682222
Fax 0224 697445

GREAVES WELSH SLATE CO LTD   *117*
Llechwedd Slate Mines
Blaenau Ffestiniog
Gwynedd LL41 3NB
Phone 0766 830522
Fax 0766 831064

D GREENWOOD 118
Soil Hill Quarry
Coal Lane
Causeway Foot
Halifax
W Yorks HX2 9PG
Phone 0274 832384

GREGORY QUARRIES LTD 119
184 Nottingham Road
Mansfield
Notts NG18 5AP
Phone 0623 23092

GRINSHILL STONE QUARRIES LTD 120
    (ECC QUARRIES LTD)
Grinshill Quarry
Clive
Shrewsbury
Shrops SY4 5PU
Phone 093928 523

GRYPHON QUARRIES LTD 120
Old Mill Works
Pontllanfraith
Gwent NP2 2AH
Phone 0495 227331

GWYNDY QUARRIES LTD 121
Gwyndy Quarry
Llandrygarn
Llanerchymedd
Anglesey
Gwynedd LL71 7AS
Phone 0407 720236

ALEXANDER HALL AND SON 122
    (BUILDERS) LTD
Granitehill
Northfield
Aberdeen
Grampian AB9 2AW
Phone 0224 693155

JOHN HANCOCK AND SONS 122
    (BATH) LTD
1A Sommer Dale Ave
Odd Down
Bath
Avon BA2 2PG
Phone 0225 833428

HANSON plc 122
1 Grosvenor Place
London SW1X 7JH
Phone 071 245 1245
Fax 071 245 3455
Telex 917698

HARDROCK LTD 123
Stoney Brow Quarry
Roby Mill
Upholland
Wigan
Lancs WN8 0QE
Phone 0695 622950

HARGREAVES QUARRIES LTD 123
    (CHARTER CONSOLIDATED plc)
PO Box 1
Crossgate Lane
Pickering
N Yorks YO18 7ER
Phone 0751 72231
Fax 0751 76380

C F HARRIS LTD 126
3 High Street
South Milford
Leeds
N Yorks LS25 5AA
Phone 0977 682337

HARRIS QUARRIES 127
    (STONECRAFT YORKSHIRE)
61 Lockwood Avenue
South Anston
Sheffield
S Yorks S31 7GQ
Phone 0742 872478

W J HAYSOM AND SON 127
St Aldhelm's Quarry
Worth Matravers
Swanage
Dorset BH19 3HL
Phone 092943 217

HEARSON QUARRY (SWIMBRIDGE) 128
Hearson Quarry
Swimbridge
Barnstable
Devon EX32 0QH
Phone 0271 830055

HIGHLAND LIME COMPANY   *128*
Torlundy Quarry
Torlundy
Fort William
Highland PH33 6SW
Phone 0397 2227

HIGHLAND REGION COUNCIL   *129*
Regional Buildings
Glenurquhart Road
Inverness
Highland IV3 5NX
Phone 0463 234121
Fax 0463 223201

SETH HILL AND SON LTD   *129*
The Quarries
Bonvilston
Cardiff
S Glam CF5 6TQ
Phone 04468 207

HILLHOUSE QUARRY COMPANY   *130*
Hillhouse Quarry
Troon
Strathclyde KA10 7HX
Phone 0292 313311
Fax 0292 314640

R HINCHCLIFFE AND SON LTD   *130*
   (MARSHALS HALIFAX plc)
Appleton Quarry
Shepley
Huddersfield
W Yorks HD8 8BB
Phone 0484 606390

HORN CRAG FARM LTD   *131*
Horn Crag Quarry
Barclay Bank Chambers
Kirkgate
Silsden
Keighley
W Yorks BD20 0AJ
Phone 0535 55442

HORTON QUARRIES LTD   *131*
Edgehill
Banbury
Oxon OX15 6HS
Phone 029 587238

ROGER HUGHES AND COMPANY   *132*
Plas Gwilym Quarry
Llysfaen Road
Old Colwyn
Colwyn Bay
Clwyd LL29 9HE
Phone 0492 515255

W H HUMPHREYS   *132*
Sunnyside
Penygroes
Caernarfon
Gwynedd LL54 6RN
Phone 0286 880502/881028

HUNTSMANS QUARRIES LTD   *132*
   (COTSWOLD PLANT COMPANY)
The Old School
Naunton
Cheltenham
Hereford & Worcs GL54 3AE
Phone 0451 5555

HUNTSMANS QUARRIES   *133*
   (PERTON) LTD
Perton Quarry
Stoke Edith
Hereford
Hereford & Worcs HR1 4HW
Phone 0432 79258

ICI (CHEMICALS AND POLYMERS) LTD *134*
PO Box 14
The Heath
Runcorn
Cheshire SK17 8TH
Phone 0928 514444
Telex 629655

ICI (CHEMICALS AND   *134*
   POLYMERS) LTD
   MOND DIVISION
PO Box 3
Buxton
Derbys SK17 8TH
Phone 0298 768444

ISLE OF MAN GOVERNMENT   *134*
   DEPT OF HIGHWAYS, PORTS
   AND PROPERTIES
Douglas Borough Council
Town Hall
Douglas
Isle of Man
Phone 0624 842387

J A JACKSON (PRESTON) LTD   *135*
Green Lane
Lightfoot Green
Preston
Lancs PR4 0AP
Phone 0772 861230

BILL JARVIS AND SON LTD   *136*
Barnshill Yard
Otley Road
Baildon
W Yorks BD17 7JF
Phone 0274 586235

DONALD A JOHNSON   *137*
  (NORTH UIST) LTD
Cnoc nan Locha
Sollas
North Uist
Western Isles PA82 5BU
Phone 08766 281

JOHNSON WELLFIELD QUARRIES LTD   *137*
Crosland Moor Quarry
Crosland Hill
Huddersfield
W Yorks HD4 7AB
Phone 0484 652311

JOHNSTON ROADSTONES LTD   *138*
Leaton Quarry
Leaton
Wellington
Shrops TF6 5HB
Phone 0952 86351
Fax 0952 86413

R A JONAS AND SONS   *139*
Tredennick Downs Quarry
St Issey
Wadebridge
Cornwall PL27 7QZ
Phone 0841 540332

D GWYNDOF JONES   *139*
Vronlog Quarry
Llanllyfni
Caernafon
Gwynedd LL54 6RT
Phone 0286 880574

D JONES (FFYNON PLANT   *140*
  HIRE COMPANY LTD)
Mynydd y Garreg Quarry
Crymmych
Dyfed SA41 3QS
Phone 0239 73223

K W & H E KEATES   *140*
31 Easington Road
Worth Matravers
Swanage
Dorset BH19 3LF
Phone 092943 207

KEY QUARRIES LTD   *141*
Quickburn Quarry
Salters Gate
Tow Law
Bishop Auckland
Durham DL13 4JN
Phone 0388 730500

KINTYRE FARMERS   *141*
Glengyle
Glebe Street
Campbeltown
Strathclyde PA28 6LS
Phone 0586 52602

KIRKSTONE GREEN SLATE   *142*
  QUARRIES LTD
Skelwith Bridge
Ambleside
Cumbria LA22 9NN
Phone 05394 33296
Fax 05394 34006
Telex 65106

C KNIVETON LTD   *142*
Turkeyland Quarry
Ballasalla
Isle of Man
Phone 0624 823594

LADYCROSS STONE COMPANY   *143*
72 Churchill Close
Shotley Bridge
Consett
Durham DH8 0EU
Phone 043473 302

JOHN LAING (CIVIL) LTD   *143*
The Marlowes
Hemel Hempstead
Herts HP2 4TP
Phone 0442 65566

LAING STONEMASONRY   *143*
  (JOHN LAING (CIVIL) LTD)
Dalston Road
Carlisle
Cumbria CA2 5NR
Phone 0228 21401

LANDELLE LTD   *144*
Kirby Clough Farm
Kettleshulme
Stockport
Greater Manchester SK12 5AY
Phone 0663 46871

LANDERS QUARRIES   *144*
Landers Quarry
Kingston Road
Langton Matravers
Swanage
Dorset BH19 3JP
Phone 0929 43205

LAWER BROS   *144*
Chywoon Quarry
Longdowns
Penryn
Cornwall TR10 9AF
Phone 0209 860520

G LAWRIE   *145*
64 Berelands Road
Prestwick
Strathclyde KA9 1ER
Phone 0292 79226

W E LEACH (SHIPLEY) LTD   *145*
Westside Mills
Ripley Road
Bradford
W Yorks BD18 1BD
Phone 0274 72700

LEITH'S TRANSPORT (ABERDEEN) LTD *146*
Linksfield Road
Aberdeen
Grampian AB2 1RW
Phone 0224 484198

LEWIS LAND SERVICES LTD   *147*
New Buildings
Newmarket
Stornoway
Isle of Lewis
Western Isles PA86 0ED
Phone 0851 3232

LIME KILN HILL QUARRY COMPANY   *147*
Lime Kiln Hill Quarry
Mells
Frome
Somerset BA11 3PH
Phone 0373 812225

LINCOLN STONE (1986) LTD   *148*
32 Lord Street
Gainsborough
Lincs DN21 2DB
Phone 0427 617464

G LINDLEY AND SONS   *148*
Sovereign Quarry
Shepley
Huddersfield
W Yorks HD8 8BH
Phone 0484 606203

LOCAL STONE COMPANY LTD   *148*
Bognor Common Quarry
The Grove
Little Bognor
Fittleworth
W Sussex RH20 1JY
Phone 0798 42888

LONGCLIFFE QUARRIES LTD   *149*
Brassington
Derbys DE4 4BZ
Phone 062985 284
Telex 377039

LOTHIAN REGION COUNCIL   *149*
c/o Department of Highways
19 Market Street
Edinburgh
Lothian EH1 1BL
Phone 031 2299292
Fax 031 2298600

D & P LOVELL QUARRIES   *150*
2A Cranbourne Road
Swanage
Dorset
BH19 1EA
Phone 0929 422657

J LOVIE AND SON LTD   *150*
Cowbog
New Pitsligo
Fraserburgh
Grampian AB4 4PR
Phone 07717 272

R MACASKILL (CONTRACTORS) LTD   *151*
Ardhasaig Quarry
Harris
Western Isles PA85 3AJ
Phone 0859 2066

MACCLESFIELD STONE   *151*
  QUARRIES LTD
Bridge Quarry
Windmill Lane
Kerridge
Macclesfield
Cheshire SK10 5AZ
Phone 0625 73208

A MACE   *152*
16 James Street
Oakworth
Keighley
W Yorks BD22 7PE
Phone 0535 43492

JOHN MACLENNAN   *152*
129 Craigston
Castlebay
Isle of Barra
Western Isles PA80 5XS
Phone 08714 508

K MACRAE AND COMPANY LTD   *152*
Borrowston Quarry
Thrumster
Wick
Highland KW1 5TX
Phone 0955 85383

MANDALE MINING CO LTD   *153*
Sheldon
Bakewell
Derbys DE4 1QS
Phone 062981 3676

MARLOW STONE (CHILLATON)LTD   *153*
The Old Chapel
Chillaton
Devon PL16 0HU
Phone 082286 206

MARSHALLS HALIFAX plc   *154*
Brier Lodge
Southowram
Halifax
W Yorks HX3 9SY
Phone 0422 57155
Fax 0422 67093

MARSHALLS STONE COMPANY LTD   *156*
Spikers Hill Quarry
West Ayton
Scarborough
N Yorks YO13 9LB
Phone 0723 863175

MARSHINGTON STONE COMPANY LTD   *156*
Shire Hill Quarry
Sheffield Road
Glossop
Derbys SK13 9PU
Phone 0663 64211

SIR A McALPINE plc   *157*
Upton
South Wirral
Merseyside L66 7ND
Phone 051 3394141
Fax 051 3394748
Telex 627185

SIR A McALPINE QUARRY   *157*
  PRODUCTS LTD
   (SIR A McALPINE plc)
Borras Airfield
Holt Road
Wrexham
Clwyd LL13 9SE
Phone 0978 351921
Fax 0978 366445
Telex 617056

SIR A McALPINE QUARRY   *160*
  PRODUCTS (SCOTLAND) LTD
   (SIR A McALPINE plc)
Ashley Drive
Bothwell
Strathclyde G71 8BS
Phone 0698 810404

SIR A McALPINE SLATE   *162*
  PRODUCTS LTD
   (SIR A McALPINE plc)
Penrhyn Quarry
Bethesda
Bangor
Gwynedd LL57 4YG
Phone 0248 600656
Fax 0248 601171
Telex 61513

McLAREN ROADSTONE LTD   *163*
  (RMC ROADSTONE PRODUCTS LTD)
Cragmill Quarry
Belford
Northumb NE70 7EZ
Phone 0668 213 866
Fax 06683 844

P G MEDWELL AND SON   *164*
Holleywell Quarry
The Quarries
Clipsham
Oakham
Leics LE15 7SQ
Phone 078081 730

B MERRICK   *164*
Northview
Swallow House Lane
Hayfield
Stockport
Cheshire SK12 5HB
Phone 0663 43565

MILL HILL QUARRY (TAVISTOCK) LTD   *165*
Longford Quarry
Moorshop
Tavistock
Devon PL19 8NP
Phone 0822 612483/612786

W B MILLER (QUARRY MASTERS) LTD   *165*
Kennoway
Leven
Fife KY8 5SG
Phone 0333 350306

MILLMEAD QUARRY PRODUCTS LTD   *166*
Scabba Wood Quarry
Cadeby Road
Sprotborough
Doncaster
S Yorks DN5 7SY
Phone 0302 310018

MINE TRAIN QUARRY   *167*
39 Upton Gardens
Upton-upon-Severn
Worcester
Hereford & Worcs WR8 0NU
Phone 06846 2532

MONE BROS (EXCAVATIONS) LTD   *167*
Albert Road
Morley
Leeds
W Yorks LS27 8RU
Phone 0532 523636

MONTAGUE ESTATE LTD   *168*
Barn Close House
Itchen Abbas
Winchester
Hants SO21 1BJ
Phone 096278 464

MORAY STONE CUTTERS   *168*
Spynie Quarry
Birnie
Elgin
Grampian IV30 3SW
Phone 034386 244

D H MORGAN AND SONS   *169*
Victoria
Roch
Haverfordwest
Dyfed SA62 6JU
Phone 0437 710259

MORRIS AND PERRY   *169*
(GURNEY SLADE QUARRIES)
Gurney Slade Quarry
Bath
Avon BA3 4TE
Phone 0749 840441

MORRISON CONSTRUCTION LTD   *170*
Shandwick House
Chapel Street
Tain
Highland IV19 1JF
Phone 0542 22273

JOHN MOWLEM AND COMPANY plc   *170*
Foundation House
Eastern Road
Bracknell
Berks RG12 2UZ
Phone 0344 426826
Fax 0344 485779
Telex 847476

MULCAIR CONTRACTING COMPANY   *171*
Cibyn Industrial Estate
Caernarfon
Gwynedd LL55 2BB
Phone 0286 4928

MULTI-AGG LTD   *171*
Shellingford Quarry
Stanford Road
Stanford in the Vale
Faringdon
Oxon SN7 8NE
Phone 03677 8949/8986

NANHORON GRANITE QUARRY LTD   *172*
Nanhoron Quarry
Pwllheli
Gwynedd LL53 8PR
Phone 075883 221

W NANKIVELL AND SONS LTD *172*
Tor Down Quarry
St Breward
Bodmin
Cornwall PL30 4LZ
Phone 0208 850216

NANT NEWYDD QUARRY *173*
Nant Newydd Quarry
Brynteg
Anglesey
Gwynedd LL4 8RE
Phone 0248 852644

NASH ROCKS STONE AND *173*
  LIME COMPANY
P O Box 1
Kington
Hereford & Worcs HR5 3LQ
Phone 0544 230711
Fax 0544 231406

NATURAL STONE PRODUCTS LTD *174*
  (TARMAC BUILDING MATERIALS LTD)
Bellmoor
Retford
Notts DN22 8SG
Phone 0777 708771

NORTH WEST AGGREGATES LTD *176*
  (RMC ROADSTONE PRODUCTS LTD)
North West House
Spring Street
Widnes
Cheshire WA8 0NG
Phone 051 423 6699
Fax 051 495 1667
Telex 628065

ARCHIBALD NOTT AND SONS LTD *178*
Tinto House
Station Road
South Molton
Devon EX36 3LL
Phone 07695 2763

L OATES *179*
Squire Hill Quarry
Southowram
Halifax
W Yorks HX3 9SX
Phone 0484 715407

OGDEN ROADSTONE LTD *179*
  (EVERED HOLDINGS plc)
St Clair Road
Otley
W Yorks LS21 1HX
Phone 0943 464531
Fax 0943 466044
Telex 51189

J OLDHAM AND COMPANY *180*
  (STONEMASONS) LTD
Tearne House
Hollington
Uttoxeter
Staffs ST10 4HR
Phone 088926 353/354/355
Fax 088926 212

ORKNEY BUILDERS *181*
Heddle Hill Quarry
Finstown
Orkney KW17 2JW
Phone 085676 511

ORKNEY ISLAND COUNCIL *182*
Council Offices
Kirkwall
Orkney KW15 1NY
Phone 0856 3535

OTS HOLDINGS *182*
Happylands Quarry
Broadway
Hereford & Worcs WR12 7WD
Phone 0386 853582

HARRY PARKER AND SON *183*
Chellow Grange Quarry
Haworth Road
Bradford
W Yorks BD9 6NL
Phone 0274 42439

PATERSON OF GREENOAK HILL LTD *183*
Gartsherrie
Coatbridge
Strathclyde ML5 2EU
Phone 0236 33351/33356

E J PATERSON *184*
Woodside Quarry
Woodside
Auchlee
Longside
Grampian AB4 7UA
Phone 077982 314

PAWSON BROS LTD   *184*
Britannia Quarry
Howley Park
Morley
W Yorks LS27 0SW
Phone 0532 530464

J D PEARCE QUARRY COMPANY   *184*
Johnstone House
Church Street
Keinton Mandeville
Somerton
Somerset TA11 6ER
Phone 045822 3326

PENHILL QUARRY AND   *185*
   HAULAGE COMPANY LTD
Jubilee House
Kilhampton
Bude
Cornwall EX23 9RS
Phone 028882 326

PENRYN GRANITE LTD   *185*
   (CHARTER CONSOLIDATED plc)
15-17 Fairmantle St
Truro
Cornwall TR1 2EH
Phone 0872 75626

PETERBOROUGH QUARRIES LTD   *188*
   (EVERED HOLDINGS plc)
West End
Maxey
Peterborough
Cambs PE6 9HA
Phone 0778 342355

PHILPOTS QUARRIES LTD   *189*
West Hoathly
East Grinstead
West Sussex RH19 4PS
Phone 0342 810428

PERCY PICKARD (SALES) LTD   *189*
Fagley Quarries
Fagley Lane
Eccleshill
Bradford
W Yorks BD2 3NT
Phone 0274 637307
Fax 0274 626146

PILKINGTON QUARRIES LTD   *190*
The Dolomite Quarry
Warmsworth
Doncaster
S Yorks DN14 9RG
Phone 0302 853092/853354
Fax 0302 310259

PIONEER AGGREGATES (UK) LTD   *191*
59-60 Northolt Road
South Harrow
Middx HA2 0EY
Phone 081 423 3066
Fax 081 423 3845
Telex 928153

I C PIPER   *196*
4 Caradon View
Minions
Liskeard
Cornwall PL14 5LL
Phone 0579 62016

PREMIER LIME AND   *196*
   STONE COMPANY
Creeton Quarry
Little Bytham
Grantham
Lincs NG33 4QC
Phone 078081 202

P & H RADLEY LTD   *196*
Stone Quarry
Windy Ridge
Cartworth Moor
Holmfirth
W Yorks HD7 1QS
Phone 0484 683536

RATTEE AND KETT LTD   *197*
   (JOHN MOWLEM AND
   COMPANY plc)
Purbeck Road
Cambridge
Cambs CB2 2PF
Phone 0223 248061
Fax 0223 410284

RBS BROOKLYNS LTD   *197*
   (TARMAC BUILDING
   MATERIALS LTD)
Shaft Road
Combe Down
Bath
Avon BA2 7HP
Phone 0225 835650
Fax 0225 835950

GEORGE M READ 198
Wilderness Quarry
Mitcheldean
Glos GL17 0DS
Phone 0452 830395

REALSTONE LTD 198
Wingerworth
Chesterfield
Derbys S42 6RG
Phone 0246 270244
Fax 0246 201060
Telex 547779

REDLAND AGGREGATES LTD 200
Bradgate House
Groby
Nr Leicester
Leics LE6 0FA
Phone 0530 242151
Fax 0530 245396
Telex 34481

JAMES REID AND COMPANY (1947) 204
Lugton Lime Works
Lugton
Kilmarnock
Strathclyde KA3 4EB
Phone 050585 435

RHUDDLAN STONE LTD 205
  (C & M PARRY PLANT HIRE)
Rhuddlan Bach Quarry
Brynteg
Anglesey
Gwynedd LL78 8QA
Phone 0248 852282

RISKEND QUARRYING COMPANY 205
Riskend Quarry
Tak Ma Doon Road
Kilsyth
Glasgow
Strathclyde G65 9JY
Phone 0236 821486

RMC GROUP plc 206
RMC House
Coldharbour Lane
Thorpe
Egham
Surrey TW20 8TD
Phone 0932 568833
Fax 0932 568933

RMC DIMENSIONAL STONE LTD 206
  (RMC GROUP plc)
Ann Twyford Quarry
Birchover
Matlock
Derbys DE4 2BN
Phone 0629 88287

RMC INDUSTRIAL MINERALS LTD 207
  (RMC GROUP plc)
Hindlow
Buxton
Derbys SK17 0EL
Phone 0298 71155

RMC ROADSTONE PRODUCTS LTD 208
  (RMC GROUP plc)
RMC House
Church Lane
Bromsgrove
Worcs B61 8RA
Phone 0527 575777
Fax 0527 575263

ROBINSON ROCK 208
Holmescales Quarry
Old Hutton
Kendal
Cumbria LA8 0NB
Phone 0539 32255/33018

ROSELAND AGGREGATES LTD 209
Lean Quarry
Horningtops
Liskeard
Cornwall PL14 3QD
Phone 0579 42342

RTZ CORPORATION plc 209
6 St James's Square
London SW17 4LD
Phone 071 930 2399
Fax 071 930 3249
Telex 24639

RTZ MINING AND EXPLORATION LTD 209
  DELABOLE SLATE DIVISION
Pengelly House
Delabole
Cornwall PL33 9AZ
Phone 0840 212242
Fax 0840 212948
Telex 45157

RUGBY PORTLAND CEMENT plc   *210*
Crown House
Rugby
Warwicks CV21 2DT
Phone 0788 542111
Fax 0788 540256
Telex 31523

A SANDERS LTD   *212*
Tuckingmill Quarry
Bow
Crediton
Devon EX17 6HN
Phone 03633 315

ALEXANDER SANDISON AND   *212*
   SONS LTD
Northside
Baltasound
Unst
Shetland ZE2 9DS
Phone 095781 444

SANTIME LTD   *213*
Pilkington Quarry
Makinson Lane
Horwich
Bolton
Greater Manchester BL6 6NA
Phone 0204 693679

SAXONMOOR LTD   *213*
Hillhouse Edge Quarry
Cartworth Moor
Holmfirth
Huddersfield
W Yorks HD7 1RL
Phone 0484 683239

SCOTTISH AGGREGATES LTD   *214*
   (RMC GROUP plc)
6 Union Street
Bridge of Allan
Stirling
Central FK9 4NT
Phone 0786 834055

SCOTTISH NATURAL STONES LTD   *215*
   (BALFOUR BEATTY LTD)
Edinburgh Road
Springhill
Shotts
Strathclyde ML7 5DT
Phone 0501 23248

SCOTTS (BAMPTON)LTD   *216*
Kersdown Quarry
Tiverton Road Works
Bampton
Tiverton
Devon EX16 9DX
Phone 0398 31466

THOMAS SCOURFIELD AND SONS   *217*
Kiln Quarry
Carew Cheriton
Tenby
Dyfed SA70 8SR
Phone 0646 651326

SHAP GRANITE LTD   *218*
   (RMC ROADSTONE
   PRODUCTS LTD)
Shap
Penrith
Cumbria CA10 3QQ
Phone 09316 787
Fax 09316 775

R SHEPHERD   *218*
3 Raise Hamlet
Ward Way
Alston
Cumbria CA9 3AS
Phone 0498 81288

SHERBURN STONE COMPANY LTD   *219*
15 Front Street
Sherburn Hill
Durham DH6 1PA
Phone 0913 720636

SHETLAND ISLES COUNCIL   *221*
Council Offices
Grantfield
Lerwick
Shetland ZE1 0NT
Phone 0595 3535
Fax 0595 3535 Ext 223

SHIPLEY BANK QUARRIES   *222*
Rose Cottage
Lartington
Barnard Castle
Durham DL12 8BP
Phone 0833 50529

SINGLETON BIRCH LTD *222*
Melton Ross Quarry
Melton Ross
Brigg
Lincs DN38 6AE
Phone 0652 688386

SMITH AND SONS *223*
    (BLETCHINGTON) LTD
Enslow
Bletchington
Oxon OX5 3AY
Phone 086983 281

SPAR SUPPLY COMPANY *224*
Slieau Chiarn House
Braaid
Marown
Isle of Man
Phone 0624 851756/851904

SPEYSIDE SAND AND GRAVEL *224*
Rothes Glen Quarry
Rothes
Grampian AB3 9CB
Phone 03403 700

C W SPRONSTON (LIME) LTD *224*
Afonwen Quarry
Riverview
Townfield
Frodsham
Cheshire WA6 7RL
Phone 0928 32250

STAFFORDSHIRE STONE *225*
    (HOLLINGTON) LTD
Institute Road
Kings Heath
Birmingham B14 7EG
Phone 021 444 1087

S D & M STAINSBY *225*
Parkside
The Avenue
High Shincliffe
Durham DH1 2PT
Phone 0913 867098

STANCLIFFE STONE COMPANY *226*
Cressbrook Mill
Cressbrook
Buxton
Derbys SK17 8SY
Phone 0298 871228

STEETLEY plc *226*
PO Box 53
Brownsover Road
Rugby
Warwicks CV21 2UT
Phone 0788 535621
Fax 0788 535361
Telex 311677

STEETLEY QUARRY *226*
    PRODUCTS LTD
    (STEETLEY plc)
Coventry Point
Market Way
Coventry
Warwicks CY1 1EA
Phone 0203 633888
Fax 0203 26788

STEETLEY QUARRIES DIVISION *226*
    (STEETLEY QUARRY
    PRODUCTS LTD)
Kiveton Lane
Kiveton Park
Sheffield
S Yorks S31 8NN
Phone 0909 770581
Fax 0909 773628
Telex 54276

STEETLEY QUARRIES DIVISION *226*
    (STEETLEY QUARRY
    PRODUCTS LTD),
    Southern Region
Gibbet Lane
Shawell
Lutterworth
Leics LE17 6AA
Phone 0788 74841
Fax 0788 536226

STEETLEY QUARRIES DIVISION *227*
    (STEETLEY QUARRY
    PRODUCTS LTD),
    Southern Region, Eastern Area
Gibbet Lane
Shawell
Lutterworth
Leics LE17 6AA
Phone 0788 74841
Fax 0788 536226

STEETLEY QUARRIES DIVISION   *228*
  (STEETLEY QUARRY
  PRODUCTS LTD),
  Southern Region, Western Area
Heol Goch
Pentyrch
Taffs Well
Cardiff
S Glam CF4 8XB
Phone 0222 810549
Fax 0222 810747

STEETLEY QUARRIES DIVISION   *231*
  (STEETLEY QUARRY
  PRODUCTS LTD),
  Northern Region
Northern Region House
Copgrove
Harrogate
N Yorks HG3 3TB
Phone 0423 340781
Fax 0423 340750

GR STEIN REFRACTORIES LTD   *236*
Manuel Works
Linlithgow
Lothian EH49 6LH
Phone 0506 843333

J H L STEPHENS   *236*
Callywith Quarry
Penquite Lane
Bodmin
Cornwall PL31 2AZ
Phone 0208 72029

STOKE HALL QUARRY   *237*
  (STONE SALES) LTD
Grindleford
Sheffield
S Yorks S30 1HW
Phone 0433 30313

STONEGRAVE AGGREGATES LTD   *237*
Aycliffe Quarry
Aycliffe
Durham DL5 6NB
Phone 0325 313129

STOWE HILL QUARRIES   *238*
Stowe Hill Quarry
Stowe
St Briavels
Lydney
Glos GL15 6QH
Phone 0594 530100

STOWEY STONE LTD   *238*
2 Market Place
Radstock
Bath
Avon BA3 2AW
Phone 0761 33167

STRATHCLYDE REGION COUNCIL   *238*
Department of Roads
Hamilton
Strathclyde ML3 0AL
Phone 0698 454444
Fax 0698 454294

SULLOM QUARRIES   *239*
Gulberwick
Lerwick
Shetland ZE2 9EX
Phone 0595 2401

A & D SUTHERLAND   *240*
69 Princes Street
Thurso
Highland KW14 7DJ
Phone 0847 63224

J SUTTLE   *240*
Swanage Quarry
Off Panorama Road
Swanage
Dorset BH19 2QS
Phone 0929 423576

SUTTON LIMES LTD   *241*
Sutton Quarry
Ripon
N Yorks HG4 3JZ
Phone 0765 3841

M & L SYMONDS   *241*
Pilsamoor Quarry
Egloskerry
Launceston
Cornwall PL15 8SH
Phone 0566 86362

TADCASTER BUILDING   *242*
  LIMESTONE LTD
Highmoor Quarry
Warren Lane
Toulson
Tadcaster
N Yorks LS24 9NU
Phone 0937 833956

TARMAC plc   242
Hilton Hall
Essington
Wolverhampton
W Midlands WV11 2BQ
Phone 0902 307407
Fax 0902 307408
Telex 338544

    TARMAC BUILDING   242
      MATERIALS LTD
      (TARMAC plc)
    Millfield Road
    Ettingshall
    Wolverhampton
    W Midlands WV4 6JP
    Phone 0902 353522
    Fax 0902 353920
    Telex 339825

    TARMAC QUARRY   242
      PRODUCTS LTD
      (TARMAC plc)
    Millfield Road
    Ettingshall
    Wolverhampton
    W Midlands WV4 6JP
    Phone 0902 353522
    Fax 0902 353920
    Telex 339825

    TARMAC ROADSTONE   243
      (EASTERN) LTD (TARMAC
      QUARRY PRODUCTS LTD)
    John Hadfield House
    Dale Road
    Matlock
    Derbys DE4 3PL
    Phone 0629 580300
    Fax 0629 580204

    TARMAC ROADSTONE   249
      (EAST MIDLANDS) LTD (TARMAC
      QUARRY PRODUCTS LTD)
    John Hadfield House
    Dale Road
    Matlock
    Derbys DE4 3PL
    Phone 0629 580300

    TARMAC ROADSTONE   250
      (NORTH WEST) LTD (TARMAC
      QUARRY PRODUCTS LTD)
    Moorcroft
    Lismore Road
    Buxton
    Derbys SK17 9AP
    Phone 0298 25441
    Fax 0298 72413

    TARMAC ROADSTONE   257
      (SCOTLAND) LTD (TARMAC
      QUARRY PRODUCTS LTD)
    134 Nithsdale Drive
    Glasgow
    Strathclyde G41 2PP
    Phone 0414 236611
    Fax 0414 231034
    Telex 779054

    TARMAC ROADSTONE   266
      (SOUTHERN) LTD (TARMAC
      QUARRY PRODUCTS LTD)
    2A Bath Road
    Newbury
    Berks RG13 1JJ
    Phone 0635 35755

    TARMAC ROADSTONE   267
      (WESTERN) LTD (TARMAC
      QUARRY PRODUCTS LTD)
    Whitehall House
    Whitehall Road
    Halesowen
    W Midlands B63 3LE
    Phone 0215 504797

TAYSIDE REGION COUNCIL   273
Tayside House
28 Crichton Street
Dundee
Tayside DD1 3RE
Phone 0382 23281
Fax 0382 28354

TETBURY STONE COMPANY LTD   274
Veizey's Quarry
Chavenage Lane
Tetbury
Glos GL8 8JT
Phone 0666 53455

T S THOMAS AND SONS (LYDNEY) LTD *275*
Albion Chambers
Hill Street
Lydney
Glos GL15 5HN
Phone 0594 842333

Wm THOMPSON AND SON *275*
(DUMBARTON) LTD
Birch Road
Dumbarton
Strathclyde G82 2RN
Phone 0389 62271

W & M THOMPSON (QUARRIES) LTD *276*
Princess Way
Low Prudhoe
Northumb NE42 6PL
Phone 0661 32422/32433

TILCON QUARRIES LTD *277*
Low Lane
Lingerfield
Knaresborough
W Yorks HG5 9JN
Phone 0423 864041
Fax 0423 864555

TILCON (MIDLANDS AND *277*
SOUTH) LTD (TILCON
QUARRIES LTD)
Cornett Ends Lane
Merriden
Coventry
Warwicks CV7 7LG
Phone 0242 221111

TILCON (NORTH CENTRAL) *280*
LTD (TILCON QUARRIES LTD)
Harrogate House
Parliament Street
Harrogate
N Yorks HG1 2RF
Phone 0423 68092
Telex 57998

TILCON (NORTHERN) LTD *282*
(TILCON QUARRIES LTD)
PO Box 5
Fell Bank
Birtley
Chester-le-Street
Durham DH3 2ST
Phone 0914 109611

TILCON (SCOTLAND) LTD *285*
(TILCON QUARRIES LTD)
250 Alexandra Parade
Glasgow
Strathclyde G31 3AX
Phone 0415 541818

TILLICOULTRY QUARRIES LTD *288*
Craigfoot Quarry
Craigfoot
Tillicoultry
Fife FK13 6AZ
Phone 0259 50347

TORRINGTON STONE LTD *289*
Torrington
Devon EX38 8JF
Phone 0805 22438

TRACTOR SHOVELS TAWSE LTD *289*
(EVERED HOLDINGS plc)
Forth House
North Road
Inverkeithing
Fife KY11 1HG
Phone 0383 412155
Fax 0383 415242
Telex 727767

TREBARWITH QUARRIES LTD *293*
Trebarwith Quarry
Trebarwith Road
Delabole
Camelford
Cornwall PL33 9DE
Phone 0840 213354

TREDINNICK QUARRY LTD *293*
Tredinnick Quarry
Tredinnick Hill
Grampound
Truro
Cornwall TR2 5TQ
Phone 0726 883487

P J TRESISE *294*
Trenoweth Quarry
Mabe Burnthouse
Penryn
Cornwall TR10 9HY
Phone 0326 72546

TREVONE QUARRIES LTD   *294*
Bosahan Quarry
Constantine
Falmouth
Cornwall TR11 5QP
Phone 0326 40604
Fax 0326 40403
Telex 45665

WADDINGTON FELL QUARRIES LTD   *295*
Waddington Fell Quarry
Fell Road
Waddington
Clitheroe
Lancs BB7 3AA
Phone 02006 334

JOHN WAINWRIGHT AND   *295*
    COMPANY LTD
Downside Quarry
Downside
Shepton Mallett
Somerset BA4 4JF
Phone 0749 2366

THOMAS W WARD ROADSTONE LTD   *296*
    (RMC ROADSTONE PRODUCTS LTD)
Albion Works
Saville Street
Sheffield
S Yorks S4 7UL
Phone 0742 701900
Fax 0742 722744
Telex 547831

WATERMILL AGGREGATES LTD   *297*
Concraigs Quarry
Rathen
Fraserburgh
Grampian AB4 4ED
Phone 0346 28226

W CLIFFORD WATTS LTD   *298*
118-122 Scarborough Road
Bridlington
Humberside YO16 5NX
Phone 0262 675383
Fax 0262 604629

THE WEELAND SAND COMPANY   *298*
High Street
South Milford
Nr Leeds
W Yorks LS25 5AA
Phone 0532 682337

WESTERN ROADSTONE LTD   *299*
    (RMC ROADSTONE PRODUCTS LTD)
London Road
Wick
Bristol
Avon BS15 5SJ
Phone 027582 4237
Fax 027582 2272

THOMAS WILLIAMS (EUXTON) LTD   *301*
Springfield
Wigan Road
Euxton
Chorley
Lancs PR7 6LB
Phone 02572 63437

GEORGE WIMPEY plc   *301*
Hammersmith Grove
Hammersmith
London W6 7EN
Phone 081 748 2000
Fax 081 748 0076
Telex 25666

WIMPEY ASPHALT LTD   *301*
    (GEORGE WIMPEY plc)
Chiswick Green House
Acton Lane
London W4 5HF
Phone 081 995 7733
Fax 081 995 6746

WIMPEY ASPHALT LTD   *301*
    (SCOTLAND) (WIMPEY
    ASPHALT LTD)
Barnton Grove
Edinburgh
Lothian EH4 6BT
Phone 0313 391721
Fax 0313 395803

WIMPEY HOBBS LTD   *304*
    (GEORGE WIMPEY plc)
4 High Street
Nailsea
Bristol
Avon BS19 1BW
Phone 0272 852281

WINCILATE LTD   *311*
Aberllefeni Slate Quarries Ltd
Machynlleth
Powys SY20 9RU
Phone 065473 602
Fax 065473 418

WINDY HILL NATURAL
   STONE COMPANY   *311*
Woodside House
8A Barnard Street
Staindrop
Darlington
Durham DL2 3ND
Phone 0833 60367

WOTTON ROADSTONE LTD   *312*
   (RMC ROADSTONE PRODUCTS LTD)
Wenvoe Quarry
Walston
Wenvoe
Cardiff
S Glam CF5 6XE
Phone 0222 593378
Fax 0222 595840

FOSTER YEOMAN LTD   *314*
Marsden House
Marsden Bigot
Frome
Somerset BA11 5DU
Phone 0373 52002
Fax 0373 84501

YEOMAN (MORVERN) LTD   *315*
   (FOSTER YEOMAN LTD)
Glensanda
Morvern by Oban
Highland PA34 5EM
Phone 0631 73415

Index 2

Quarries

by quarry name

Entries in this index refer to the
unique quarry numbers (1 – 705)
in the main Directory

Top of our class...

24 x 15 Kue Ken

20ft x 6ft Niagara D14 Screen

The reports which we get from our clients are good.

They know that our workmanship is to a very high standard when rebuilding crushers, screens and associated plant. Our reputation for second-hand machinery is well known within the trade at home and to our overseas clients.

If there is equipment you require, either 'as is', rebuilt, or purpose built to suit your particular requirements, contact Port & Quarry Supplies who will be pleased to give you information on specifications and prices.

If you would like a copy of our next Plant List please contact us at the address below.

4ft Nordberg Standard

Illustrated above left
Hazemag APK40

42 x 12 Kue Ken
Mobile Secondary

Old Failand Quarry, Clevedon Road, Failand, Bristol BS8 3TU. Telephone (0272) 392567

Port & Quarry Supplies

Abbey	Nottinghamshire	3
Aberduna	Clwyd	433
Abergele	Clwyd	630
Aberllefeni	Powys	697
Aberstrecht	Gwynedd	4
Aberthaw	South Glamorgan	99
Achnagart	Highland	651
Admiralty	Dorset	2
Afonwen	Cheshire	511
Airdriehill	Strathclyde	361
Alkerton	Oxfordshire	90
Allington	Kent	47
Alltgoch	Dyfed	249
Altham	Lancashire	131
Ammanford	Dyfed	519
Ancaster	Lincolnshire	267
Ann Twyford	Derbyshire	468
Antron Hill	Cornwall	8
Apex	West Yorkshire	306
Appleton	West Yorkshire	293
Appley Bridge	Lancashire	21
Arcow	North Yorkshire	567
Ardchronie	Highland	652
Ardhasaig	Western Isles	340
Ardley	Oxfordshire	507
Ardownie	Tayside	246
Ashwood Dale	Derbyshire	563
Askernish	Western Isles	68
Auchinleck	Strathclyde	18
Avon Dassett	Warwickshire	222
Aycliffe	Durham	539
Bableigh Wood	Devon	157
Back Lane	Lancashire	203
Backwell	Avon	672
Balladoole	Isle of Man	96
Ballidon	Derbyshire	624
Ballygrant	Strathclyde	197
Balmedie	Grampian	257
Balmullo	Fife	372
Banavie	Highland	642
Bangley	Lothian	362
Bank Top	West Yorkshire	671
Bankend	Cumbria	168
Bankfield	Lancashire	568
Bannerbank	Strathclyde	643
Bantycock	Nottinghamshire	123
Bardon	Leicestershire	75
Barland	West Glamorgan	77
Barlockhart	Dumfries and Galloway	79
Barnsdale Bar	West Yorkshire	191
Barnshaw	West Yorkshire	192
Barrasford	Northumberland	551
Barrington	Cambridgeshire	476
Barton	North Yorkshire	499
Barton Wood	Devon	189
Batts Combe	Somerset	48
Baycliff	Cumbria	137

Bayston Hill	Shropshire	603
Beacon Lodge	West Yorkshire	185
Beam	Devon	650
Bearrah Tor	Cornwall	443
Beer	Devon	58
Belford	Northumberland	636
Belliston	Fife	88
Big Pits	Leicestershire	135
Birkhams	Cumbria	169
Bishop Middleham	Durham	623
Bixhead	Gloucestershire	241
Black	North Yorkshire	277
Blackaller	Devon	190
Blackhill	West Yorkshire	376
Black Hill	Cornwall	419
Black Hills	Grampian	133
Blackmoor	South Yorkshire	285
Black Pasture	Northumberland	487
Blaengwynlais	Mid Glamorgan	683
Blaenyfan	Dyfed	699
Blairhill	Lothian	580
Blaxter	Northumberland	392
Blencowe	Cumbria	223
Blodwell	Shropshire	10
Blue Bank	North Yorkshire	69
Bluehill	Grampian	258
Blynlee	Borders	242
Boards	Central	677
Bognor Common	West Sussex	335
Bold Venture	Lancashire	569
Bolehill	Derbyshire	449
Bolsover Moor	Derbyshire	552
Bolton Hill	Dyfed	250
Bolton Woods	West Yorkshire	429
Bonawe	Strathclyde	243
Boreland Fell	Dumfries and Galloway	193
Borrow Pit	Cheshire	245
Borrowston	Highland	344
Borthwick	Borders	678
Bosahan	Cornwall	663
Boughton	Northamptonshire	91
Bower	Highland	653
Bowithic	Cornwall	163
Bowyers	Dorset	1
Boyne Bay	Grampian	120
Boysack	Tayside	617
Braich Ddu	Gwynedd	262
Brandy Crag	Cumbria	138
Brantingham	Humberside	670
Brathay	Cumbria	139
Bray Valley	Devon	402
Breedon Hill	Leicestershire	121
Bridge	Cheshire	341
Brierlow	Derbyshire	470
Brindister	Shetland	545
Brittania	West Yorkshire	415
Broadlaw	Lothian	363

Broad Oak	Cumbria	170
Brockhill	Gloucestershire	164
Brodsworth	South Yorkshire	631
Broughton Moor	Lancashire	140
Bryn Engan	Gwynedd	384
Buckton Vale	Greater Manchester	354
Buddon Wood	Leicestershire	456
Builth Wells	Powys	36
Burford	Oxfordshire	87
Burley Hill	Clwyd	114
Burlington	Cumbria	141
Burnthouse	Cornwall	9
Bursting Stone	Cumbria	142
Cadeby	South Yorkshire	524
Caer Glaw	Gwynedd	19
Cairds Hill	Grampian	383
Cairneyhill	Strathclyde	644
Cairngryffe	Strathclyde	542
Cairnryan	Dumfries and Galloway	80
Caldon Low	Staffordshire	604
Calliburn	Strathclyde	319
Callow	Gwent	184
Callow Hill	Shropshire	605
Callow Rock	Somerset	209
Callywith	Cornwall	537
Camp Hill	Northamptonshire	98
Carew Quarries	Dyfed	494
Carnsew	Cornwall	420
Carrigwynion	Powys	613
Castle-an-Dinas	Cornwall	421
Castle Hill	Somerset	153
Cat Castle	Durham	155
Cauldon	Staffordshire	100
Cefn Gawr	South Glamorgan	606
Cerrigyrwyn	Dyfed	251
Charnage	Wiltshire	59
Charnwood	Leicestershire	11
Chellow Grange	West Yorkshire	412
Chillaton	Devon	346
Chinley Moor	Greater Manchester	369
Chinnor	Oxfordshire	477
Chipping Sodbury	Avon	49
Chywoon	Cornwall	326
Clashach	Grampian	379
Clatchard Craig	Fife	581
Clearwell	Gloucestershire	625
Cleat and Lower Grean	Western Isles	343
Clee Hill	Shropshire	12
Cliffe Hill	Leicestershire	564
Clipsham	Lincolnshire	135
Clock Face	West Yorkshire	347
Clogau	Clwyd	95
Coatsgate	Dumfries and Galloway	582
Coldstones	North Yorkshire	434
Colemans	Somerset	210
Collace	Tayside	618
Collyweston	Lincolnshire	136

Concraigs	Grampian	669
Cool Scar	North Yorkshire	224
Copp Grag	Northumberland	127
Corby	Northamptonshire	515
Corncockle	Dumfries and Galloway	194
Cornelly	Mid Glamorgan	684
Corrennie	Grampian	654
Corsehill	Dumfries and Galloway	195
Cotswold Hill	Gloucestershire	165
Cottonhill	Grampian	339
Cowdor and Hall Dale	Derbyshire	553
Cowltick	Northamptonshire	515
Coygen	Dyfed	252
Cragmill	Northumberland	366
Craig-yr-Hesg	Mid Glamorgan	37
Craigenlow	Grampian	583
Craigfoot	Fife	649
Craighouse	Borders	584
Craigie Hill	Strathclyde	655
Craiglash	Grampian	259
Craignair	Dumfries and Galloway	585
Craigpark	Lothian	586
Craigs	Fife	587
Creeton	Lincolnshire	444
Creigiau	Mid Glamorgan	607
Criggion	Shropshire	13
Crime Rig	Durham	500
Croft	Leicestershire	204
Crogarry Beag	Western Isles	308
Cromhall	Gloucestershire	50
Cromwell	West Yorkshire	348
Crosland Moor	West Yorkshire	310
Crossleys	Lincolnshire	425
Croxden	Staffordshire	608
Croy	Strathclyde	588
Cruicks	Fife	645
Cunmont	Tayside	589
Curister	Orkney	410
Cwmleyshon	Mid Glamorgan	38
Cwt-y-Bugail	Gwynedd	235
Daglingworth	Gloucestershire	51
Dairy	Cornwall	422
Darlton	South Yorkshire	685
Darney	Northumberland	393
Darrington	North Yorkshire	178
Daviot	Highland	679
Dayhouse	Gwent	620
Dead Friars	Durham	513
Dean	Cornwall	457
Deep Lane	West Yorkshire	430
Delabole	Cornwall	474
De Lank	Cornwall	469
Denbigh	Clwyd	570
Dene	Derbyshire	554
Dinas	Dyfed	355
Diphwys Casson	Gwynedd	263
Divethill	Northumberland	367

Doddington Hill	Northumberland	394
Dolomite (The)	South Yorkshire	432
Dolyhir and Strinds	Powys	390
Doulting	Somerset	83
Dove Holes	Derbyshire	471
Dowlow	Derbyshire	525
Downs	Dorset	338
Drybrook	Gloucestershire	52
Dry Rigg	North Yorkshire	458
Dukes	Derbyshire	450
Dulcote	Somerset	703
Dumbuckhill	Strathclyde	646
Dumfries	Dumfries and Galloway	20
Dunald Mill	Lancashire	571
Dunbar Northwest	Lothian	101
Dunduff	Strathclyde	413
Dunhouse	Durham	196
Dunion	Borders	590
Dunmore	Central	488
Duntilland	Strathclyde	543
East	Lancashire	21
Eastgate	Durham	102
East Grinstead	Wiltshire	200
Edstone	Borders	116
Edwin Richards	West Midlands	14
Eldon Hill	Derbyshire	399
Ellel	Lancashire	304
Elterwater	Cumbria	143
Eskett	Cumbria	225
Ethiebeaton	Tayside	591
Fagley	West Yorkshire	431
Farmington	Gloucestershire	231
Fenacre	Devon	609
Filkins	Oxfordshire	239
Fish Hill	Hereford and Worcester	508
Flaunden	Hertfordshire	159
Flax Bourton	Avon	686
Fletcher Bank	Greater Manchester	349
Fly Delph	West Yorkshire	342
Fly Flatts	West Yorkshire	129
Force Garth	Durham	278
Forcett	North Yorkshire	637
Fords	Lancashire	240
Forest Wood	Mid Glamorgan	435
Foxcliffe	North Yorkshire	179
Friarton	Tayside	680
Friendly	West Yorkshire	70
Furnace	Strathclyde	647
Gatelawbridge	Dumfries and Galloway	489
Gebdykes	North Yorkshire	526
Gedloch	Grampian	220
Gelligaer	Mid Glamorgan	436
Gellihalog	Dyfed	700
Ghyll Scaur	Cumbria	226
Giggleswick	North Yorkshire	632
Gilfach	West Glamorgan	701
Glebe	Lincolnshire	268

Glen	South Yorkshire	350
Glensanda	Highland	705
Glogue	Dyfed	188
Goat	Fife	592
Goddards	Derbyshire	667
Gloddfa Ganol	Gwynedd	236
Goldmire	Cumbria	22
Gore	Powys	626
Graig	Clwyd	459
Grange Mill	Derbyshire	89
Grange Top	Lincolnshire	151
Granite Mountain	Isle of Man	509
Great Blakenham	Suffolk	103
Great Gate	Staffordshire	512
Great Ponton	Lincolnshire	516
Greenbrey	Grampian	274
Greetham	Lincolnshire	426
Greetwell	Lincolnshire	149
Greys Pit	Kent	478
Greystone	Cornwall	212
Griff	Warwickshire	437
Grinshill	Shropshire	270
Groby	Leicestershire	15
Grove	Mid Glamorgan	438
Guiting	Gloucestershire	53
Gurney Slade	Avon	382
Gwyndy	Gwynedd	273
Hafod	Gwent	460
Hailstone	West Midlands	610
Hainworth Shaw	West Yorkshire	72
Halecombe	Somerset	687
Halkyn	Clwyd	400
Hallyards	Strathclyde	327
Ham Hill	Somerset	221
Ham Hill	Somerset	378
Hantergantick	Cornwall	395
Happylands	Hereford and Worcester	411
Harden	Northumberland	638
Hardendale	Cumbria	126
Harden Moor	West Yorkshire	307
Harnhill	Avon	688
Hartshill	Warwickshire	612
Harrycroft	South Yorkshire	527
Harthorpe East	Durham	535
Hart	Cleveland	501
Hartley	Cumbria	279
Hayes Wood Mine	Avon	86
Hayfield	Greater Manchester	324
Hayne	Devon	219
Hazelbank	Borders	337
Hearson	Devon	288
Heddle Hill	Orkney	409
Heights	Durham	230
Hendre	Clwyd	356
Hendy	Mid Glamorgan	689
Hengae	Gwynedd	385
Hertbury	Cornwall	417

Hessilhead	Strathclyde	463
High Craig	Strathclyde	593
High Force	Durham	280
Highmoor	Lancashire	227
Highmoor	North Yorkshire	550
High Nick	Northumberland	396
Hillend	Strathclyde	485
Hillhead	Derbyshire	572
Hillhouse	Strathclyde	292
Hillhouse Edge	West Yorkshire	484
Hillwood	Lothian	681
Hindlow	Derbyshire	611
Hingston Down	Cornwall	60
Holborough	Kent	104
Holleywell	Leicestershire	368
Holme Hall	South Yorkshire	555
Holme Park	Lancashire	205
Holmescales	Cumbria	472
Honister	Cumbria	357
Honley Wood	West Yorkshire	311
Hope	South Yorkshire	105
Hopton Wood	Derbyshire	565
Horn Crag	West Yorkshire	294
Horn Park	Dorset	247
Hornsleasow	Hereford and Worcester	298
Horton	Warwickshire	295
Horton	North Yorkshire	573
Houghton	Tyne and Wear	23
Hovingham	North Yorkshire	66
Howick	Northumberland	556
Huncote	Leicestershire	5
Hunters Hill	West Yorkshire	130
Huntsmans	Gloucestershire	299
Hutchbank	Lancashire	228
Ifton	Gwent	39
Independent	Dorset	2
Ingham	Suffolk	7
Ingleton	Lancashire	24
Ivonbrook	Derbyshire	405
Jackdaw Crag	North Yorkshire	180
Jamestone	Lancashire	305
Judkins	Warwickshire	16
Kaimes	Lothian	594
Keates	Dorset	317
Kelhead	Dumfries and Galloway	25
Kemnay	Grampian	656
Kendal Fell	Cumbria	574
Kensworth	Bedfordshire	479
Kersdown	Devon	493
Kessel Downs	Cornwall	423
Ketton	Lincolnshire	151
Kevin	Staffordshire	627
Kilbarchan	Strathclyde	486
Kilmond Wood	Durham	281
Kiln	Dyfed	495
Kingoodie	Fife	490
Kirkby	Cumbria	144

Kirkmabreck	Dumfries and Galloway	595
Kirton Lindsey	Lincolnshire	150
Knowle	Devon	6
Ladycross	Northumberland	322
Landers	Dorset	325
Laneast	Cornwall	475
Langside	Fife	373
Lazonby Fell	Cumbria	171
Lea	Shropshire	206
Lean	Cornwall	473
Leapers Wood	Lancashire	690
Leaton	Shropshire	312
Leeming	Lancashire	132
Leinthall	Hereford and Worcester	313
Leipsic	Cumbria	498
Lic	Western Isles	331
Lillishall	Shropshire	207
Lime Kiln Hill	Somerset	332
Linhay Hill	Devon	254
Lithalun	Mid Glamorgan	40
Little	Lancashire	67
Livox	Gwent	41
Llanddulas	Clwyd	26
Llechedd Slate Mine	Gwynedd	264
Llynclys	Shropshire	520
Loanhead	Strathclyde	596
Locharbriggs	Dumfries and Galloway	73
Long Bredy	Dorset	61
Longcliffe	Derbyshire	336
Longford	Devon	370
Long Lane	West Yorkshire	181
Longside	Grampian	134
Longwood	Avon	439
Lowthorpe	Humberside	529
Luxulyan	Cornwall	213
Machen	Gwent	42
Maenofferen	Gwynedd	265
Mancetter	Warwickshire	628
Mansfield	Nottinghamshire	269
Marsden	Tyne and Wear	639
Marybank	Western Isles	657
Masons	Suffolk	106
Meldon	Devon	125
Melton	Humberside	107
Melton Ross	Lincolnshire	506
Merehead	Somerset	704
Merivale	Devon	397
Middle Mill	Dyfed	253
Middle Peak	Derbyshire	557
Middlesgate	Humberside	480
Middleton Mine	Lothian	464
Middletown	Powys	115
Mill Hill	Devon	371
Miltonhill	Strathclyde	621
Minera	Clwyd	575
Mine Train	Gloucestershire	375
Minffordd	Gwynedd	691

Monks Park Mine	Wiltshire	84
Montcliffe	Lancashire	27
Moons Hill	Avon	666
Moorcroft	Devon	214
Moota	Cumbria	118
Mootlaw	Northumberland	558
Morefield	Highland	597
Morrinton	Dumfries and Galloway	598
Moss Rigg	Cumbria	145
Mount Pleasant	Norfolk	158
Mount Pleasant	Avon	447
Mountsorrel	Leicestershire	456
Murrayshall	Central	599
Mynydd y Garreg	Dyfed	316
Nanhoron	Gwynedd	387
Nantlle	Gwynedd	237
Nant Newydd	Gwynedd	389
Naylor Hill Quarries	West Yorkshire	248
Newbiggin	Fife	491
Newbridge	North Yorkshire	282
New England	Devon	215
Newthorpe	West Yorkshire	182
Norman	Cambridgeshire	108
Northfield	Central	156
Northfleet	Kent	109
North Lasts	Grampian	329
North Mains	Grampian	186
Northowram Hills	West Yorkshire	232
Nuneaton	Warwickshire	612
Oakeley	Gwynedd	238
Offham	Kent	54
Old Cliffe Hill	Leicestershire	566
Old Mill	Strathclyde	218
Old Quarrington	Durham	559
Once-a-Week	Derbyshire	345
Orrock	Fife	682
Palmers	Derbyshire	514
Pant	Mid Glamorgan	692
Pant (Halkyn)	Clwyd	693
Pantyffynnon	South Glamorgan	291
Pant-y-Pwll Dwr	Clwyd	400
Parish	Derbyshire	358
Parkhead	Cumbria	28
Park Nook	South Yorkshire	148
Parkmore	Grampian	177
Peckfield	West Yorkshire	406
Peddles Lane	Somerset	416
Pencaemawr	Mid Glamorgan	436
Penderyn	Mid Glamorgan	43
Penhow	Gwent	44
Penlee	Cornwall	424
Penmaenmawr	Gwynedd	29
Penrhyn	Gwynedd	365
Penryn	Cornwall	420
Penstrowed	Powys	256
Penwyllt	Powys	694
Perton	Hereford and Worcester	300

Pets	Cumbria	320
Pettinain	Strathclyde	160
Philpots	West Sussex	428
Pickerings	Cumbria	172
Pigsden	Cornwall	418
Pilkington	Greater Manchester	483
Pilsamoor	Cornwall	549
Pitcaple	Grampian	260
Pitsford	Northamptonshire	92
Plaistow	Devon	403
Plas Gwilym	Clwyd	296
Plymstock	Devon	110
Poortown	Isle of Man	302
Portland	Dorset	62
Potgate	North Yorkshire	528
Prince of Wales	Cornwall	451
Prudham	Northumberland	128
Pury End	Northamptonshire	98
Quartzite	Gloucestershire	673
Queensgate	Humberside	201
Quickburn	Durham	318
Quidhampton	Wiltshire	202
Raisby	Durham	560
Ravelrig	Lothian	600
Rawdon	West Yorkshire	328
Raynes	Clwyd	401
Red Hole	Staffordshire	512
Redmire	North Yorkshire	561
Rhuddlan Bach	Gwynedd	466
Rhyader	Powys	613
Rhyndaston	Dyfed	381
Ribblesdale Works	Lancashire	152
Ridgeway	Derbyshire	161
Riganagower	Strathclyde	622
Risca	Gwent	45
Riskend	Strathclyde	467
Rock Cottage	North Yorkshire	377
Rogers	Gloucestershire	629
Rooks	Cumbria	173
Ross of Mull	Strathclyde	492
Rothes Glen	Grampian	510
Round 'O'	Lancashire	30
Ruston Parva	Humberside	529
Ruthin	Mid Glamorgan	111
Salterwath	Cumbria	174
Sandfold	West Yorkshire	117
Sandside	Cumbria	576
Savoch	Grampian	187
Scabba Wood	South Yorkshire	374
Sconser	Highland	290
Scord	Shetland	504
Scout Moor	Greater Manchester	351
Scratchmill Scar	Cumbria	175
Setters	Shetland	482
Settrington	North Yorkshire	233
Shadforth	Durham	502
Shadwell	Shropshire	521

Shap	Cumbria	31
Shap Beck	Cumbria	31
Shap Blue	Cumbria	496
Shap Pink	Cumbria	497
Shavers End	Hereford and Worcester	208
Shawk	Cumbria	323
Shellingford	Oxfordshire	386
Shennington	Oxfordshire	93
Sherburn	North Yorkshire	284
Shierglas	Tayside	601
Shining Bank	Derbyshire	668
Shipham	Somerset	614
Shipley Bank	Durham	505
Shire Hill	Derbyshire	353
Shoreham	West Sussex	112
Shotley	Northamptonshire	94
Silverdale	Lancashire	32
Skelbrooke	South Yorkshire	183
Skelpie Mine	Fife	167
Skiddaw	Cumbria	146
Skipton Rock	North Yorkshire	640
Smaws	North Yorkshire	530
Soil Hill	West Yorkshire	266
Southard	Dorset	287
South Barrule	Isle of Man	303
Southorpe	Lincolnshire	461
South Witham	Lincolnshire	517
Sovereign	West Yorkshire	334
Spaunton	North Yorkshire	283
Spikers Hill	North Yorkshire	352
Spittal	Highland	546
Spittlegate Level	Lincolnshire	333
Spout Cragg	Cumbria	147
Spring Lodge	South Yorkshire	633
Springwell	Tyne and Wear	398
Spynie	Grampian	380
Squire Hill	West Yorkshire	404
St Aldhelm's	Dorset	286
Stainton	Cumbria	641
Stamford	Cambridgeshire	446
Stancliffe	Derbyshire	452
Stancombe	Avon	695
Stanton Moor	Derbyshire	453
Staple Farm	Nottinghamshire	124
Station	Somerset	76
Steangabhal	Western Isles	309
Stoke Hall	South Yorkshire	538
Stone	West Yorkshire	445
Stonegrave	North Yorkshire	234
Stoneraise	Cumbria	454
Stoney Brow	Lancashire	276
Stoneycombe	Devon	211
Stowe Hill	Gloucestershire	540
Stowey	Avon	541
Stowfield	Gloucestershire	615
Stranraer	Dumfries and Galloway	33
Sutton	North Yorkshire	548

Swanage	Dorset	547
Swanworth	Dorset	602
Swinburne	Northumberland	34
Swinden	North Yorkshire	634
Sycamore	Cheshire	198
Syke	Dyfed	440
Syreford	Gloucestershire	255
Taffs Well	South Glamorgan	522
Tam's Loup	Strathclyde	441
Tan-y-Foel	Powys	119
Tearne	Staffordshire	407
Tendley	Cumbria	577
Ten Yards Lane	North Yorkshire	71
Thornaugh	Lincolnshire	427
Threshfield	North Yorkshire	562
Thrislington	Durham	531
Thumpas	West Yorkshire	166
Tincornhill	Strathclyde	544
Tippetcraig	Central	536
Tom's Forest	Grampian	658
Tonfanau	Gwynedd	391
Tongland	Strathclyde	81
Topley Pike	Derbyshire	578
Torcoed	Dyfed	696
Torcoedfawr	Dyfed	359
Tor Down	Cornwall	388
Torlundy	Highland	289
Tormitchell	Strathclyde	82
Torrin	Highland	330
Town	Oxfordshire	176
Trearne	Strathclyde	465
Trebarwith	Cornwall	660
Tredinnick	Cornwall	661
Tredennick Downs	Cornwall	314
Trefil	Gwent	272
Trenoweth	Cornwall	662
Trevassack	Cornwall	162
Trevillet	Cornwall	74
Trevone	Cornwall	664
Trimm Rock	Clwyd	635
Triscombe	Somerset	616
Trusham	Devon	63
Tuckingmill	Devon	481
Tunstead	Derbyshire	301
Turkeyland	Isle of Man	321
Twll Llwyd	Gwynedd	297
Tyrebagger	Grampian	659
Tytherington	Gloucestershire	55
Underheugh	Strathclyde	364
Upper Lawn	Avon	275
Vale Road	Nottinghamshire	154
Vatseter	Shetland	244
Vaynor	Mid Glamorgan	46
Veizey's	Gloucestershire	619
Venn	Devon	216
Vronlog	Gwynedd	315
Waddington Fell	Lancashire	665

Wardlow	Staffordshire	462
Water Swallows	Derbyshire	579
Wath	North Yorkshire	532
Watson	West Yorkshire	97
Watts Cliff	Derbyshire	455
Waulkmill	Tayside	648
Webscott	Shropshire	271
Wenvoe	South Glamorgan	702
Westbury	Wiltshire	113
Westbury	Somerset	442
Western New Forres	Grampian	261
Westleigh	Devon	217
Westwood Ground	Wiltshire	85
Whatley	Somerset	56
Whitecleaves	Devon	64
White Hill	Oxfordshire	57
Whittle Hill	Lancashire	676
Whitwell	Nottinghamshire	533
Whitwick	Leicestershire	17
Whitworth	Lancashire	229
Wick	Avon	674
Wickwar	Gloucestershire	675
Wigtown	Dumfries and Galloway	35
Wilderness	Gloucestershire	448
Wimberry Moss	Cheshire	199
Windy Hill	Durham	698
Witch Hill	Durham	503
Woodbury	Hereford and Worcester	523
Woodeaton	Oxfordshire	78
Woodside	West Yorkshire	122
Woodside	Grampian	414
Wooton	Derbyshire	408
Wormersley	South Yorkshire	534
Worsham	Oxfordshire	518
Wredon	Staffordshire	360
Yalberton Tor	Devon	65

Index 3

Quarries in England, Scotland and Wales

County by county within each country; and by name within counties

Entries in this index refer to the
unique quarry numbers (1 – 705)
in the main Directory

Crushing problem?

Call Babbitless

There are no new crushing problems . . . only old ones.
So a company like Babbitless, with more than sixty years'
experience in designing and building crushers, will know all the
problems . . . and all the answers.
The well-proven range of Babbitless machines is truly
comprehensive and their international team of agents and
engineers are all experts.
Whatever the material, whatever the problem, if you're looking for
machine reliability, good product shape and economic crushing . . .
call Babbitless.

**Babbitless Company
(Gt. Britain) Ltd.,**
Stonefield Way,
Ruislip,
Middlesex, HA4 0JT
Tel: 081-841 4221   Telex: 8811853   Fax: 081-842 1254

DRAGON BABBITLESS

362

England

AVON

Backwell	672
Chipping Sodbury	49
Flax Bourton	686
Gurney Slade	382
Harnhill	688
Hayes Wood Mine	86
Longwood	439
Moons Hill	666
Mount Pleasant	447
Stancombe	695
Stowey	541
Upper Lawn	275
Wick	674

BEDFORDSHIRE

Kensworth	479

CAMBRIDGESHIRE

Barrington	476
Norman	108
Stamford	446

CHESHIRE

Afonwen	511
Borrow Pit	245
Bridge	341
Sycamore	198
Wimberry Moss	199

CLEVELAND

Hart	501

CORNWALL

Antron Hill	8
Bearrah Tor	443
Black Hill	419
Bosahan	663
Bowithic	163
Burnthouse	9
Callywith	537
Carnsew	420
Castle-an-Dinas	421
Chywoon	326
Dairy	422
Dean	457
De Lank	469
Delabole	474
Greystone	212
Hantergantick	395
Hertbury	417
Hingston Down	60
Kessel Downs	423
Laneast	475
Lean	473

Luxulyan	213
Penlee	424
Penryn	420
Pigsden	418
Pilsamoor	549
Prince of Wales	451
Tor Down	388
Trebarwith	660
Tredinnick	661
Tredennick Downs	314
Trenoweth	662
Trevassack	162
Trevillet	74
Trevone	664

CUMBRIA

Bankend	168
Baycliff	137
Birkhams	169
Blencowe	223
Brandy Crag	138
Brathay	139
Broad Oak	170
Burlington	141
Bursting Stone	142
Elterwater	143
Eskett	225
Ghyll Scaur	226
Goldmire	22
Hardendale	126
Hartley	279
Holmescales	472
Honister	357
Kendal Fell	574
Kirkby	144
Lazonby Fell	171
Leipsic	498
Moota	118
Moss Rigg	145
Parkhead	28
Pets	320
Pickerings	172
Rooks	173
Salterwath	174
Sandside	576
Scratchmill Scar	175
Shap	31
Shap Beck	31
Shap Blue	496
Shap Pink	497
Shawk	323
Skiddaw	146
Spout Cragg	147
Stainton	641
Stoneraise	454

Tendley	577	Plaistow	403
		Plymstock	110
DERBYSHIRE		Stoneycombe	211
Ann Twyford	468	Trusham	63
Ashwood Dale	563	Tuckingmill	481
Ballidon	624	Venn	216
Bolehill	449	Westleigh	217
Bolsover Moor	552	Whitecleaves	64
Brierlow	470	Yalberton Tor	65
Cowdor and Hall Dale	553		
Dene	554	**DORSET**	
Dove Holes	471	Admiralty	2
Dowlow	525	Bowyers	1
Dukes	450	Downs	338
Eldon Hill	399	Horn Park	247
Goddards	667	Independent	2
Grange Mill	89	Keates	317
Hillhead	572	Landers	325
Hindlow	611	Long Bredy	61
Hopton Wood	565	Portland	62
Ivonbrook	405	Southard	287
Longcliffe	336	St Aldhelm's	286
Middle Peak	557	Swanage	547
Once-a-Week	345	Swanworth	602
Palmers	514		
Parish	358	**DURHAM**	
Ridgeway	161	Aycliffe	539
Shining Bank	668	Bishop Middleham	623
Shire Hill	353	Cat Castle	155
Stancliffe	452	Crime Rig	500
Stanton Moor	453	Dead Friars	513
Topley Pike	578	Dunhouse	196
Tunstead	301	Eastgate	102
Water Swallows	579	Force Garth	278
Watts Cliff	455	Harthorpe East	535
Wooton	408	Heights	230
		High Force	280
DEVON		Kilmond Wood	281
Bableigh Wood	157	Old Quarrington	559
Barton Wood	189	Quickburn	318
Beam	650	Raisby	560
Beer	58	Shadforth	502
Blackaller	190	Shipley Bank	505
Bray Valley	402	Thrislington	531
Chillaton	346	Windy Hill	698
Fenacre	609	Witch Hill	503
Hayne	219		
Hearson	288	**GLOUCESTERSHIRE**	
Kersdown	493	Bixhead	241
Knowle	6	Brockhill	164
Linhay Hill	254	Clearwell	625
Longford	370	Cotswold Hill	165
Meldon	125	Cromhall	50
Merivale	397	Daglingworth	51
Mill Hill	371	Drybrook	52
Moorcroft	214	Farmington	231
New England	215	Guiting	53

Huntsmans	299	Back Lane	203
Mine Train	375	Bankfield	568
Quartzite	673	Bold Venture	569
Rogers	629	Broughton Moor	140
Stowe Hill	540	Dunald Mill	571
Stowfield	615	East	21
Syreford	255	Ellel	304
Tytherington	55	Fords	240
Veizey's	619	Highmoor	227
Wickwar	675	Holme Park	205
Wilderness	448	Hutchbank	228
		Ingleton	24
GREATER MANCHESTER		Jamestone	305
Buckton Vale	354	Leapers Wood	690
Chinley Moor	369	Leeming	132
Fletcher Bank	349	Little	67
Hayfield	324	Montcliffe	27
Pilkington	483	Ribblesdale Works	152
Scout Moor	351	Round 'O'	30
		Silverdale	32
HEREFORD AND WORCESTER		Stoney Brow	276
Fish Hill	508	Waddington Fell	665
Happylands	411	Whittle Hill	676
Hornsleasow	298	Whitworth	229
Leinthall	313		
Perton	300	**LEICESTERSHIRE**	
Shavers End	208	Bardon	75
Woodbury	523	Big Pits	135
		Breedon Hill	121
HERTFORDSHIRE		Buddon Wood	456
Flaunden	159	Charnwood	11
		Cliffe Hill	564
HUMBERSIDE		Croft	204
Brantingham	670	Groby	15
Lowthorpe	529	Holleywell	368
Melton	107	Huncote	5
Middlesgate	480	Mountsorrel	456
Queensgate	201	Old Cliffe Hill	566
Ruston Parva	529	Whitwick	17
ISLE OF MAN		**LINCOLNSHIRE**	
Balladoole	96	Ancaster	267
Granite Mountain	509	Clipsham	135
Poortown	302	Collyweston	136
South Barrule	303	Creeton	444
Turkeyland	321	Crossleys	425
		Glebe	268
KENT		Grange Top	151
Allington	47	Great Ponton	516
Greys Pit	478	Greetham	426
Holborough	104	Greetwell	149
Northfleet	109	Ketton	151
Offham	54	Kirton Lindsey	150
		Melton Ross	506
LANCASHIRE		South Witham	517
Altham	131	Southorpe	461
Appley Bridge	21	Spittlegate Level	333

Thornaugh	427	Doddington Hill	394
		Harden	638
NORFOLK		High Nick	396
Mount Pleasant	158	Howick	556
		Ladycross	322
NORTH YORKSHIRE		Mootlaw	558
Arcow	567	Prudham	128
Barton	499	Swinburne	34
Black	277		
Blue Bank	69	**NOTTINGHAMSHIRE**	
Coldstones	434	Abbey	3
Cool Scar	224	Bantycock	123
Darrington	178	Mansfield	269
Dry Rigg	458	Staple Farm	124
Forcett	637	Vale Road	154
Foxcliffe	179	Whitwell	533
Gebdykes	526		
Giggleswick	632	**OXFORDSHIRE**	
Highmoor	550	Alkerton	90
Horton	573	Ardley	507
Hovingham	66	Burford	87
Jackdaw Crag	180	Chinnor	477
Newbridge	282	Filkins	239
Potgate	528	Shellingford	386
Redmire	561	Shennington	93
Rock Cottage	377	Town	176
Settrington	233	White Hill	57
Sherburn	284	Woodeaton	78
Skipton Rock	640	Worsham	518
Smaws	530		
Spaunton	283	**SHROPSHIRE**	
Spikers Hill	352	Bayston Hill	603
Stonegrave	234	Blodwell	10
Sutton	548	Callow Hill	605
Swinden	634	Clee Hill	12
Ten Yards Lane	71	Criggion	13
Threshfield	562	Grinshill	270
Wath	532	Leaton	312
		Lea	206
NORTHAMPTONSHIRE		Lillishall	207
Boughton	91	Llynclys	520
Camp Hill	98	Shadwell	521
Corby	515	Webscott	271
Cowltick	515		
Pitsford	92	**SOMERSET**	
Pury End	98	Batts Combe	48
Shotley	94	Callow Rock	209
		Castle Hill	153
NORTHUMBERLAND		Colemans	210
Barrasford	551	Doulting	83
Belford	636	Dulcote	703
Black Pasture	487	Halecombe	687
Blaxter	392	Ham Hill	221
Copp Grag	127	Ham Hill	378
Cragmill	366	Lime Kiln Hill	332
Darney	393	Merehead	704
Divethill	367	Peddles Lane	416

Shipham	614	Hailstone		610
Station	76			
Triscombe	616	**WEST SUSSEX**		
Westbury	442	Bognor Common		335
Whatley	56	Philpots		428
		Shoreham		112
SOUTH YORKSHIRE				
Blackmoor	285	**WEST YORKSHIRE**		
Brodsworth	631	Apex		306
Cadeby	524	Appleton		293
Darlton	685	Bank Top		671
(The) Dolomite	432	Barnsdale Bar		191
Glen	350	Barnshaw		192
Harrycroft	527	Beacon Lodge		185
Holme Hall	555	Blackhill		376
Hope	105	Bolton Woods		429
Park Nook	148	Brittania		415
Scabba Wood	374	Chellow Grange		412
Skelbrooke	183	Clock Face		347
Spring Lodge	633	Cromwell		348
Stoke Hall	538	Crosland Moor		310
Wormersley	534	Deep Lane		430
		Fagley		431
STAFFORDSHIRE		Fly Delph		342
Caldon Low	604	Fly Flatts		129
Cauldon	100	Friendly		70
Croxden	608	Hainworth Shaw		72
Great Gate	512	Harden Moor		307
Kevin	627	Hillhouse Edge		484
Red Hole	512	Honley Wood		311
Tearne	407	Horn Crag		294
Wardlow	462	Hunters Hill		130
Wredon	360	Long Lane		181
		Naylor Hill Quarries		248
SUFFOLK		Newthorpe		182
Great Blakenham	103	Northowram Hills		232
Ingham	7	Peckfield		406
Masons	106	Rawdon		328
		Sandfold		117
TYNE AND WEAR		Soil Hill		266
Houghton	23	Sovereign		334
Marsden	639	Squire Hill		404
Springwell	398	Stone		445
		Thumpas		166
WARWICKSHIRE		Watson		97
Avon Dassett	222	Woodside		122
Griff	437			
Hartshill	612	**WILTSHIRE**		
Horton	295	Charnage		59
Judkins	16	East Grinstead		200
Mancetter	628	Monks Park Mine		84
Nuneaton	612	Quidhampton		202
		Westbury		113
WEST MIDLANDS		Westwood Ground		85
Edwin Richards	14			

Scotland

BORDERS	
Blynlee	242
Borthwick	678
Craighouse	584
Dunion	590
Edstone	116
Hazelbank	337

CENTRAL	
Boards	677
Dunmore	488
Murrayshall	599
Northfield	156
Tippetcraig	536

DUMFRIES AND GALLOWAY	
Barlockhart	79
Boreland Fell	193
Cairnryan	80
Coatsgate	582
Corncockle	194
Corsehill	195
Craignair	585
Dumfries	20
Gatelawbridge	489
Kelhead	25
Kirkmabreck	595
Locharbriggs	73
Morrinton	598
Stranraer	33
Wigtown	35

FIFE	
Balmullo	372
Belliston	88
Clatchard Craig	581
Craigfoot	649
Craigs	587
Cruicks	645
Goat	592
Kingoodie	490
Langside	373
Newbiggin	491
Orrock	682
Skelpie Mine	167

GRAMPIAN	
Balmedie	257
Black Hills	133
Bluehill	258
Boyne Bay	120
Cairds Hill	383
Clashach	379
Concraigs	669
Corrennie	654

Cottonhill	339
Craigenlow	583
Craiglash	259
Gedloch	220
Greenbrey	274
Kemnay	656
Longside	134
North Lasts	329
North Mains	186
Parkmore	177
Pitcaple	260
Rothes Glen	510
Savoch	187
Spynie	380
Tom's Forest	658
Tyrebagger	659
Western New Forres	261
Woodside	414

HIGHLAND	
Achnagart	651
Ardchronie	652
Banavie	642
Borrowston	344
Bower	653
Daviot	679
Glensanda	705
Morefield	597
Sconser	290
Spittal	546
Torlundy	289
Torrin	330

LOTHIAN	
Bangley	362
Blairhill	580
Broadlaw	363
Craigpark	586
Dunbar Northwest	101
Hillwood	681
Kaimes	594
Middleton Mine	464
Ravelrig	600

ORKNEY	
Curister	410
Heddle Hill	409

SHETLAND	
Brindister	545
Scord	504
Setters	482
Vatseter	244

STRATHCLYDE	
Airdriehill	361

Auchinleck	18		Tam's Loup	441
Ballygrant	197		Tincornhill	544
Bannerbank	643		Tongland	81
Bonawe	243		Tormitchell	82
Cairneyhill	644		Trearne	465
Cairngryffe	542		Underheugh	364
Calliburn	319			
Craigie Hill	655		TAYSIDE	
Croy	588		Ardownie	246
Dumbuckhill	646		Boysack	617
Dunduff	413		Collace	618
Duntilland	543		Cunmont	589
Furnace	647		Ethiebeaton	591
Hallyards	327		Friarton	680
Hessilhead	463		Shierglas	601
High Craig	593		Waulkmill	648
Hillend	485			
Hillhouse	292		WESTERN ISLES	
Kilbarchan	486		Ardhasaig	340
Loanhead	596		Askernish	68
Miltonhill	621		Cleat and Lower Grean	343
Old Mill	218		Crogarry Beag	308
Pettinain	160		Lic	331
Riganagower	622		Marybank	657
Riskend	467		Steangabhal	309
Ross of Mull	492			

Wales

CLWYD			Gellihalog	700
Aberduna	433		Kiln	495
Abergele	630		Middle Mill	253
Burley Hill	114		Mynydd y Garreg	316
Clogau	95		Rhyndaston	381
Denbigh	570		Syke	440
Graig	459		Torcoed	696
Halkyn	400		Torcoedfawr	359
Hendre	356			
Llanddulas	26		GWENT	
Minera	575		Callow	184
Pant (Halkyn)	693		Dayhouse	620
Pant-y-Pwll Dwr	400		Hafod	460
Plas Gwilym	296		Ifton	39
Raynes	401		Livox	41
Trimm Rock	635		Machen	42
			Penhow	44
DYFED			Risca	45
Alltgoch	249		Trefil	272
Ammanford	519			
Blaenyfan	699		GWYNEDD	
Bolton Hill	250		Aberstrecht	4
Carew Quarries	494		Braich Ddu	262
Cerrigyrwyn	251		Bryn Engan	384
Coygen	252		Caer Glaw	19
Dinas	355		Cwt-y-Bugail	235
Glogue	188		Diphwys Casson	263

Gloddfa Ganol	236
Gwyndy	273
Hengae	385
Llechedd Slate Mine	264
Maenofferen	265
Minffordd	691
Nanhoron	387
Nantlle	237
Nant Newydd	389
Oakeley	238
Penmaenmawr	29
Penrhyn	365
Rhuddlan Bach	466
Tonfanau	391
Twll Llwyd	297
Vronlog	315

MID GLAMORGAN

Blaengwynlais	683
Cornelly	684
Craig-yr-Hesg	37
Creigiau	607
Cwmleyshon	38
Forest Wood	435
Gelligaer	436
Grove	438
Hendy	689
Lithalun	40

Pant	692
Pencaemawr	436
Penderyn	43
Ruthin	111
Vaynor	46

POWYS

Aberllefeni	697
Builth Wells	36
Carrigwynion	613
Dolyhir and Strinds	390
Gore	626
Middletown	115
Penstrowed	256
Penwyllt	694
Rhyader	613
Tan-y-Foel	119

SOUTH GLAMORGAN

Aberthaw	99
Cefn Gawr	606
Pantyffynnon	291
Taffs Well	522
Wenvoe	702

WEST GLAMORGAN

Barland	77
Gilfach	701

Index 4

Rock types in Britain
by quarry name and location

Entries in this index refer to the
unique quarry numbers (1 – 705)
in the main Directory

Do you gamble with the steel you specify? Do you settle for less? For the toughest jobs there is only one real alternative – Hardox.

If you are really serious about producing tough products, then you have to specify abrasion resistant Hardox, the long life, guaranteed performance steel.

Don't risk breakdowns, unhappy clients or lost production – whatever the challenge, *Hardox can take it!*

Swedish Steel Limited De Salis Court, De Salis Drive, Hampton Lovett, Droitwich, Worcestershire WR9 0QE. Tel: 0905 795794. Telex: 335982. Fax: 0905 794736.

ABRASION RESISTANT STEEL PLATE
AVAILABLE
from samples to truck loads
WITHIN HOURS!

GET TOUGH, GET-

HARDOX

SWEDISH STEEL

ANDESITE

Balmullo	Fife	372
Bardon	Leicestershire	75
Ghyll Scaur	Cumbria	226
Harden	Northumberland	638
Moons Hill	Avon	666
Waulkmill	Tayside	648
Whitwick	Leicestershire	17

BASALT

Airdriehill	Strathclyde	361
Ardhasaig	Western Isles	340
Askernish	Western Isles	68
Bannerbank	Strathclyde	643
Blairhill	Lothian	580
Blodwell	Shropshire	10
Boysack	Tayside	617
Broadlaw	Lothian	363
Cairds Hill	Grampian	383
Clatchard Craig	Fife	581
Cleat and Lower Grean	Western Isles	343
Clee Hill	Shropshire	12
Craighouse	Borders	584
Craigie Hill	Strathclyde	655
Craigpark	Lothian	586
Craigs	Fife	587
Crogarry Beag	Western Isles	308
Croy	Strathclyde	588
Dumbuckhill	Strathclyde	646
Duntilland	Strathclyde	543
Edwin Richards	West Midlands	14
Friarton	Tayside	680
Hallyards	Strathclyde	327
High Craig	Strathclyde	593
Hillend	Strathclyde	485
Howick	Northumberland	556
Kilbarchan	Strathclyde	486
Laneast	Cornwall	475
Lean	Cornwall	473
Leaton	Shropshire	312
Loanhead	Strathclyde	596
Minffordd	Gwynedd	691
Murrayshall	Central	599
Orrock	Fife	682
Ravelrig	Lothian	600
Riganagower	Strathclyde	622
Steangabhal	Western Isles	309
Underheugh	Strathclyde	364
Vatseter	Shetland	244
Water Swallows	Derbyshire	579
Whitecleaves	Devon	64

CHALK

Barrington	Cambridgeshire	476
Beer	Devon	58
Charnage	Wiltshire	59
Chinnor	Oxfordshire	477

East Grinstead	Wiltshire	200
Flaunden	Hertfordshire	159
Great Blakenham	Suffolk	103
Greys Pit	Kent	478
Holborough	Kent	104
Kensworth	Bedfordshire	479
Long Bredy	Dorset	61
Lowthorpe	Humberside	529
Masons	Suffolk	106
Melton	Humberside	107
Melton Ross	Lincolnshire	506
Middlesgate	Humberside	480
Mount Pleasant	Norfolk	158
Norman	Cambridgeshire	108
Northfleet	Kent	109
Queensgate	Humberside	201
Quidhampton	Wiltshire	202
Ruston Parva	Humberside	529
Shoreham	West Sussex	112
Westbury	Wiltshire	113

DIORITE

Bolton Hill	Dyfed	250
Charnwood	Leicestershire	11
Cliffe Hill	Leicestershire	564
Croft	Leicestershire	204
Griff	Warwickshire	437
Groby	Leicestershire	15
Old Cliffe Hill	Leicestershire	566
Penmaenmawr	Gwynedd	29

DOLERITE

Ardownie	Tayside	246
Barrasford	Northumberland	551
Belford	Northumberland	636
Belliston	Fife	88
Black Hill	Cornwall	419
Boards	Central	677
Borthwick	Borders	678
Builth Wells	Powys	36
Cairneyhill	Strathclyde	644
Callow Hill	Shropshire	605
Cerrigyrwyn	Dyfed	251
Collace	Tayside	618
Coygen	Dyfed	252
Cragmill	Northumberland	366
Craigfoot	Fife	649
Criggion	Shropshire	13
Cruicks	Fife	645
Cunmont	Tayside	589
Divethill	Northumberland	367
Dunion	Borders	590
Ethiebeaton	Tayside	591
Force Garth	Durham	278
Goat	Fife	592
Greystone	Cornwall	212
Hailstone	West Midlands	610

Hillhouse	Strathclyde	292
Hillwood	Lothian	681
Kaimes	Lothian	594
Leaton	Shropshire	312
Mancetter	Warwickshire	628
Middletown	Powys	115
Miltonhill	Strathclyde	621
New England	Devon	215
Northfield	Central	156
Riskend	Strathclyde	467
Swinburne	Northumberland	34
Tam's Loup	Strathclyde	441
Trusham	Devon	63

FLAGSTONE

Borrowston	Highland	344
Cromwell	West Yorkshire	348
Friendly	West Yorkshire	70
Jamestone	Lancashire	305
Thumpas	West Yorkshire	166

GABBRO

Balmedie	Grampian	257
Corrennie	Grampian	654
Dean	Cornwall	457
North Lasts	Grampian	329
North Mains	Grampian	186
Pitcaple	Grampian	260
Tyrebagger	Grampian	659

GNEISS

Lic	Western Isles	331

GRANITE

Antron Hill	Cornwall	8
Banavie	Highland	642
Bearrah Tor	Cornwall	443
Black Hills	Grampian	133
Bonawe	Strathclyde	243
Bosahan	Cornwall	663
Buddon Wood	Leicestershire	456
Burnthouse	Cornwall	9
Caer Glaw	Gwynedd	19
Cairngryffe	Strathclyde	542
Carnsew	Cornwall	420
Castle-an-Dinas	Cornwall	421
Chywoon	Cornwall	326
Craigenlow	Grampian	583
Craignair	Dumfries and Galloway	585
De Lank	Cornwall	469
Furnace	Strathclyde	647
Glensanda	Highland	705
Granite Mountain	Isle of Man	509
Gwyndy	Gwynedd	273
Hantergantick	Cornwall	395
Hengae	Gwynedd	385
Hingston Down	Cornwall	60

Kemnay	Grampian	656
Kessel Downs	Cornwall	423
Kirkmabreck	Dumfries and Galloway	595
Longside	Grampian	134
Luxulyan	Cornwall	213
Merivale	Devon	397
Middle Mill	Dyfed	253
Mountsorrel	Leicestershire	456
Nanhoron	Gwynedd	387
Penryn	Cornwall	420
Pettinain	Strathclyde	160
Poortown	Isle of Man	302
Ross of Mull	Strathclyde	492
Savoch	Grampian	187
Shap	Cumbria	31
Shap Beck	Cumbria	31
Shap Blue	Cumbria	496
Tom's Forest	Grampian	658
Tor Down	Cornwall	388
Trenoweth	Cornwall	662
Trevone	Cornwall	664

GRANODIORITE

Broad Oak	Cumbria	170
Tincornhill	Strathclyde	544

GREYWACKE

Barlockhart	Dumfries and Galloway	79
Bayston Hill	Shropshire	603
Boreland Fell	Dumfries and Galloway	193
Carrigwynion	Powys	613
Coatsgate	Dumfries and Galloway	582
Dolyhir and Strinds	Powys	390
Dunduff	Strathclyde	413
Edstone	Borders	116
Hazelbank	Borders	337
Morrinton	Dumfries and Galloway	598
Rhyader	Powys	613
Tongland	Strathclyde	81

GRITSTONE

Ann Twyford	Derbyshire	468
Bank Top	West Yorkshire	671
Blackhill	West Yorkshire	376
Bolton Woods	West Yorkshire	429
Buckton Vale	Greater Manchester	354
Chinley Moor	Greater Manchester	369
Clock Face	West Yorkshire	347
Crosland Moor	West Yorkshire	310
Dairy	Cornwall	422
Dead Friars	Durham	513
Deep Lane	West Yorkshire	430
Ellel	Lancashire	304
Fagley	West Yorkshire	431
Fletcher Bank	Greater Manchester	349
Fly Delph	West Yorkshire	342
Fords	Lancashire	240

Great Gate	Staffordshire	512
Hainworth Shaw	West Yorkshire	72
Harden Moor	West Yorkshire	307
Harthorpe East	Durham	535
Hertbury	Cornwall	417
Honley Wood	West Yorkshire	311
Horn Crag	West Yorkshire	294
Ingleton	Lancashire	24
Ladycross	Northumberland	322
Leeming	Lancashire	132
Montcliffe	Lancashire	27
Mynydd y Garreg	Dyfed	316
Palmers	Derbyshire	514
Pigsden	Cornwall	418
Pilsamoor	Cornwall	549
Quickburn	Durham	318
Rawdon	West Yorkshire	328
Red Hole	Staffordshire	512
Round 'O'	Lancashire	30
Shawk	Cumbria	323
Shipley Bank	Durham	505
Sovereign	West Yorkshire	334
Stone	West Yorkshire	445
Tormitchell	Strathclyde	82
Tredinnick	Cornwall	661
Waddington Fell	Lancashire	665
Watson	West Yorkshire	97
Whittle Hill	Lancashire	676
Wimberry Moss	Cheshire	199
Woodside	West Yorkshire	122

HORNFELS

Meldon	Devon	125
Penlee	Cornwall	424
Shap Pink	Cumbria	497

IRONSTONE

Alkerton	Oxfordshire	90
Shennington	Oxfordshire	93

LIMESTONE

Abbey	Nottinghamshire	3
Aberduna	Clwyd	433
Abergele	Clwyd	630
Aberstrecht	Gwynedd	4
Aberthaw	South Glamorgan	99
Admiralty	Dorset	2
Afonwen	Cheshire	511
Ammanford	Dyfed	519
Ancaster	Lincolnshire	267
Ardley	Oxfordshire	507
Ashwood Dale	Derbyshire	563
Auchinleck	Strathclyde	18
Avon Dassett	Warwickshire	222
Aycliffe	Durham	539
Back Lane	Lancashire	203
Backwell	Avon	672

Balladoole	Isle of Man	96
Ballidon	Derbyshire	624
Ballygrant	Strathclyde	197
Bankfield	Lancashire	568
Barland	West Glamorgan	77
Barnsdale Bar	West Yorkshire	191
Barton	North Yorkshire	499
Batts Combe	Somerset	48
Big Pits	Lincolnshire	135
Bishop Middleham	Durham	623
Black	North Yorkshire	277
Blackaller	Devon	190
Blaengwynlais	Mid Glamorgan	683
Blaenyfan	Dyfed	699
Blencowe	Cumbria	223
Bognor Common	West Sussex	335
Bold Venture	Lancashire	569
Bolsover Moor	Derbyshire	552
Bowyers	Dorset	1
Boyne Bay	Grampian	120
Brantingham	Humberside	670
Breedon Hill	Leicestershire	121
Brierlow	Derbyshire	470
Brockhill	Gloucestershire	164
Brodsworth	South Yorkshire	631
Bryn Engan	Gwynedd	384
Burford	Oxfordshire	87
Burley Hill	Clwyd	114
Cadeby	South Yorkshire	524
Caldon Low	Staffordshire	604
Calliburn	Strathclyde	319
Callow Rock	Somerset	209
Camp Hill	Northamptonshire	98
Carew Quarries	Dyfed	494
Castle Hill	Somerset	153
Cauldon	Staffordshire	100
Cefn Gawr	South Glamorgan	606
Chipping Sodbury	Avon	49
Clearwell	Gloucestershire	625
Clipsham	Lincolnshire	135
Coldstones	North Yorkshire	434
Colemans	Somerset	210
Collyweston	Lincolnshire	136
Concraigs	Grampian	669
Cool Scar	North Yorkshire	224
Copp Grag	Northumberland	127
Corby	Northamptonshire	515
Cornelly	Mid Glamorgan	684
Cotswold Hill	Gloucestershire	165
Cottonhill	Grampian	339
Cowdor and Hall Dale	Derbyshire	553
Cowltick	Northamptonshire	515
Craiglash	Grampian	259
Creeton	Lincolnshire	444
Creigiau	Mid Glamorgan	607
Crime Rig	Durham	500
Cromhall	Gloucestershire	50

Crossleys	Lincolnshire	425
Cwmleyshon	Mid Glamorgan	38
Daglingworth	Gloucestershire	51
Darlton	South Yorkshire	685
Darrington	North Yorkshire	178
Dayhouse	Gwent	620
Denbigh	Clwyd	570
Dene	Derbyshire	554
(The) Dolomite	South Yorkshire	432
Dolyhir and Strinds	Powys	390
Doulting	Somerset	83
Dove Holes	Derbyshire	471
Dowlow	Derbyshire	525
Downs	Dorset	338
Drybrook	Gloucestershire	52
Dulcote	Somerset	703
Dumfries	Dumfries and Galloway	20
Dunald Mill	Lancashire	571
Dunbar Northwest	Lothian	101
Eastgate	Durham	102
Eldon Hill	Derbyshire	399
Eskett	Cumbria	225
Farmington	Gloucestershire	231
Fenacre	Devon	609
Filkins	Oxfordshire	239
Fish Hill	Hereford and Worcester	508
Flax Bourton	Avon	686
Forcett	North Yorkshire	637
Forest Wood	Mid Glamorgan	435
Foxcliffe	North Yorkshire	179
Gebdykes	North Yorkshire	526
Gellihalog	Dyfed	700
Giggleswick	North Yorkshire	632
Glebe	Lincolnshire	268
Glen	South Yorkshire	350
Goddards	Derbyshire	667
Goldmire	Cumbria	22
Graig	Clwyd	459
Grange Mill	Derbyshire	89
Grange Top	Lincolnshire	151
Great Ponton	Lincolnshire	516
Greetham	Lincolnshire	426
Greetwell	Lincolnshire	149
Grove	Mid Glamorgan	438
Guiting	Gloucestershire	53
Gurney Slade	Avon	382
Halecombe	Somerset	687
Halkyn	Clwyd	400
Ham Hill	Somerset	221
Ham Hill	Somerset	378
Happylands	Hereford and Worcester	411
Hardendale	Cumbria	126
Harnhill	Avon	688
Harrycroft	South Yorkshire	527
Hart	Cleveland	501
Hartley	Cumbria	279
Hayes Wood Mine	Avon	86

Heights	Durham	230
Hendre	Clwyd	356
Hendy	Mid Glamorgan	689
Hessilhead	Strathclyde	463
High Force	Durham	280
Highmoor	North Yorkshire	550
Hillhead	Derbyshire	572
Hindlow	Derbyshire	611
Holleywell	Leicestershire	368
Holme Hall	South Yorkshire	555
Holme Park	Lancashire	205
Hope	South Yorkshire	105
Hopton Wood	Derbyshire	565
Horn Park	Dorset	247
Hornsleasow	Hereford and Worcester	298
Horton	North Yorkshire	573
Horton	Warwickshire	295
Houghton	Tyne and Wear	23
Hovingham	North Yorkshire	66
Huntsmans	Gloucestershire	299
Ifton	Gwent	39
Independent	Dorset	2
Ingham	Suffolk	7
Ivonbrook	Derbyshire	405
Jackdaw Crag	North Yorkshire	180
Keates	Dorset	317
Kelhead	Dumfries and Galloway	25
Kendal Fell	Cumbria	574
Kersdown	Devon	493
Ketton	Lincolnshire	151
Kevin	Staffordshire	627
Kilmond Wood	Durham	281
Kiln	Dyfed	495
Kirton Lindsey	Lincolnshire	150
Landers	Dorset	325
Lea	Shropshire	206
Leapers Wood	Lancashire	690
Leinthall	Hereford and Worcester	313
Lillishall	Shropshire	207
Lime Kiln Hill	Somerset	332
Linhay Hill	Devon	254
Lithalun	Mid Glamorgan	40
Livox	Gwent	41
Llanddulas	Clwyd	26
Llynclys	Shropshire	520
Longcliffe	Derbyshire	336
Long Lane	West Yorkshire	181
Longwood	Avon	439
Machen	Gwent	42
Mansfield	Nottinghamshire	269
Marsden	Tyne and Wear	639
Merehead	Somerset	704
Middle Peak	Derbyshire	557
Middleton Mine	Lothian	464
Minera	Clwyd	575
Monks Park Mine	Wiltshire	84
Moorcroft	Devon	214

Moota	Cumbria	118
Mootlaw	Northumberland	558
Morefield	Highland	597
Mount Pleasant	Avon	447
Nant Newydd	Gwynedd	389
Newbridge	North Yorkshire	282
Newthorpe	West Yorkshire	182
Old Mill	Strathclyde	218
Old Quarrington	Durham	559
Once-a-Week	Derbyshire	345
Pant	Mid Glamorgan	692
Pant (Halkyn)	Clwyd	693
Pantyffynnon	South Glamorgan	291
Pant-y-Pwll Dwr	Clwyd	400
Parish	Derbyshire	358
Park Nook	South Yorkshire	148
Parkhead	Cumbria	28
Parkmore	Grampian	177
Peckfield	West Yorkshire	406
Peddles Lane	Somerset	416
Penderyn	Mid Glamorgan	43
Penhow	Gwent	44
Penwyllt	Powys	694
Perton	Hereford and Worcester	300
Pickerings	Cumbria	172
Pitsford	Northamptonshire	92
Plas Gwilym	Clwyd	296
Plymstock	Devon	110
Portland	Dorset	62
Potgate	North Yorkshire	528
Prudham	Northumberland	128
Pury End	Northamptonshire	98
Raisby	Durham	560
Raynes	Clwyd	401
Redmire	North Yorkshire	561
Rhuddlan Bach	Gwynedd	466
Ribblesdale Works	Lancashire	152
Risca	Gwent	45
Rock Cottage	North Yorkshire	377
Rogers	Gloucestershire	629
Rooks	Cumbria	173
Ruthin	Mid Glamorgan	111
Salterwath	Cumbria	174
Sandside	Cumbria	576
Scabba Wood	South Yorkshire	374
Settrington	North Yorkshire	233
Shadforth	Durham	502
Shadwell	Shropshire	521
Shavers End	Hereford and Worcester	208
Shellingford	Oxfordshire	386
Sherburn	North Yorkshire	284
Shierglas	Tayside	601
Shining Bank	Derbyshire	668
Shipham	Somerset	614
Shotley	Northamptonshire	94
Silverdale	Lancashire	32
Skelbrooke	South Yorkshire	183

Skelpie Mine	Fife	167
Skipton Rock	North Yorkshire	640
Smaws	North Yorkshire	530
Southard	Dorset	287
Southorpe	Lincolnshire	461
South Witham	Lincolnshire	517
Spaunton	North Yorkshire	283
Spikers Hill	North Yorkshire	352
Spittlegate Level	Lincolnshire	333
Spring Lodge	South Yorkshire	633
St Aldhelm's	Dorset	286
Stainton	Cumbria	641
Stamford	Cambridgeshire	446
Stancombe	Avon	695
Station	Somerset	76
Stonegrave	North Yorkshire	234
Stoneycombe	Devon	211
Stowe Hill	Gloucestershire	540
Stowey	Avon	541
Stowfield	Gloucestershire	615
Stranraer	Dumfries and Galloway	33
Sutton	North Yorkshire	548
Swanage	Dorset	547
Swanworth	Dorset	602
Swinden	North Yorkshire	634
Syreford	Gloucestershire	255
Taffs Well	South Glamorgan	522
Tendley	Cumbria	577
Thornaugh	Lincolnshire	427
Threshfield	North Yorkshire	562
Thrislington	Durham	531
Tonfanau	Gwynedd	391
Topley Pike	Derbyshire	578
Torcoed	Dyfed	696
Torcoedfawr	Dyfed	359
Torlundy	Highland	289
Torrin	Highland	330
Town	Oxfordshire	176
Trearne	Strathclyde	465
Trefil	Gwent	272
Trimm Rock	Clwyd	635
Tunstead	Derbyshire	301
Turkeyland	Isle of Man	321
Tytherington	Gloucestershire	55
Upper Lawn	Avon	275
Vale Road	Nottinghamshire	154
Vaynor	Mid Glamorgan	46
Veizey's	Gloucestershire	619
Wardlow	Staffordshire	462
Wath	North Yorkshire	532
Wenvoe	South Glamorgan	702
Westbury	Somerset	442
Westleigh	Devon	217
Westwood Ground	Wiltshire	85
Whatley	Somerset	56
White Hill	Oxfordshire	57
Whitwell	Nottinghamshire	533

Wick	Avon	674
Wickwar	Gloucestershire	675
Wigtown	Dumfries and Galloway	35
Witch Hill	Durham	503
Woodbury	Hereford and Worcester	523
Woodeaton	Oxfordshire	78
Woodside	Grampian	414
Wormersley	South Yorkshire	534
Worsham	Oxfordshire	518
Wredon	Staffordshire	360
Yalberton Tor	Devon	65

MARL

Achnagart	Highland	651

MUDSTONE

Bantycock	Nottinghamshire	123
Staple Farm	Nottinghamshire	124

QUARTZITE

Judkins	Warwickshire	16
Hartshill	Warwickshire	612
Nuneaton	Warwickshire	612

RAGSTONE

Allington	Kent	47
Offham	Kent	54

RHYOLITE

Langside	Fife	373
Rhyndaston	Dyfed	381

SANDSTONE

Alltgoch	Dyfed	249
Altham	Lancashire	131
Apex	West Yorkshire	306
Appleton	West Yorkshire	293
Appley Bridge	Lancashire	21
Arcow	North Yorkshire	567
Bableigh Wood	Devon	157
Bankend	Cumbria	168
Barnshaw	West Yorkshire	192
Barton Wood	Devon	189
Beacon Lodge	West Yorkshire	185
Beam	Devon	650
Birkhams	Cumbria	169
Bixhead	Gloucestershire	241
Blackmoor	South Yorkshire	285
Black Pasture	Northumberland	487
Blaxter	Northumberland	392
Blue Bank	North Yorkshire	69
Blynlee	Borders	242
Bolehill	Derbyshire	449
Bolton Woods	West Yorkshire	429
Borrow Pit	Cheshire	245
Boughton	Northamptonshire	91
Bray Valley	Devon	402

Bridge	Cheshire	341
Brindister	Shetland	545
Brittania	West Yorkshire	415
Cairnryan	Dumfries and Galloway	80
Callow	Gwent	184
Cat Castle	Durham	155
Chellow Grange	West Yorkshire	412
Clashach	Grampian	379
Corncockle	Dumfries and Galloway	194
Corsehill	Dumfries and Galloway	195
Craig-yr-Hesg	Mid Glamorgan	37
Cromwell	West Yorkshire	348
Croxden	Staffordshire	608
Curister	Orkney	410
Dairy	Cornwall	422
Darney	Northumberland	393
Deep Lane	West Yorkshire	430
Dinas	Dyfed	355
Doddington Hill	Northumberland	394
Dry Rigg	North Yorkshire	458
Dukes	Derbyshire	450
Dunduff	Strathclyde	413
Dunhouse	Durham	196
Dunmore	Central	488
East	Lancashire	21
Fagley	West Yorkshire	431
Fly Flatts	West Yorkshire	129
Fords	Lancashire	240
Friendly	West Yorkshire	70
Gatelawbridge	Dumfries and Galloway	489
Gelligaer	Mid Glamorgan	436
Gilfach	West Glamorgan	701
Gore	Powys	626
Great Gate	Staffordshire	512
Greenbrey	Grampian	274
Grinshill	Shropshire	270
Hafod	Gwent	460
Harthorpe East	Durham	535
Hayfield	Greater Manchester	324
Hayne	Devon	219
Hearson	Devon	288
Heddle Hill	Orkney	409
Hertbury	Cornwall	417
Highmoor	Lancashire	227
High Nick	Northumberland	396
Hillhouse Edge	West Yorkshire	484
Holmescales	Cumbria	472
Huncote	Leicestershire	5
Hunters Hill	West Yorkshire	130
Hutchbank	Lancashire	228
Ingleton	Lancashire	24
Judkins	Warwickshire	16
Kingoodie	Fife	490
Knowle	Devon	6
Lazonby Fell	Cumbria	171
Leipsic	Cumbria	498
Little	Lancashire	67

Locharbriggs	Dumfries and Galloway	73
Mine Train	Gloucestershire	375
Montcliffe	Lancashire	27
Naylor Hill Quarries	West Yorkshire	248
Newbiggin	Fife	491
Northowram Hills	West Yorkshire	232
Palmers	Derbyshire	514
Pencaemawr	Mid Glamorgan	436
Penstrowed	Powys	256
Philpots	West Sussex	428
Pigsden	Cornwall	418
Pilkington	Greater Manchester	483
Pilsamoor	Cornwall	549
Plaistow	Devon	403
Quartzite	Gloucestershire	673
Red Hole	Staffordshire	512
Ridgeway	Derbyshire	161
Round 'O'	Lancashire	30
Sandfold	West Yorkshire	117
Sconser	Highland	290
Scout Moor	Greater Manchester	351
Scratchmill Scar	Cumbria	175
Shipley Bank	Durham	505
Shire Hill	Derbyshire	353
Soil Hill	West Yorkshire	266
Spittal	Highland	546
Springwell	Tyne and Wear	398
Spynie	Grampian	380
Squire Hill	West Yorkshire	404
Stancliffe	Derbyshire	452
Stanton Moor	Derbyshire	453
Stoke Hall	South Yorkshire	538
Stoneraise	Cumbria	454
Stoney Brow	Lancashire	276
Sycamore	Cheshire	198
Syke	Dyfed	440
Tan-y-Foel	Powys	119
Tearne	Staffordshire	407
Ten Yards Lane	North Yorkshire	71
Thumpas	West Yorkshire	166
Triscombe	Somerset	616
Tuckingmill	Devon	481
Venn	Devon	216
Watson	West Yorkshire	97
Watts Cliff	Derbyshire	455
Webscott	Shropshire	271
Whitworth	Lancashire	229
Wilderness	Gloucestershire	448
Windy Hill	Durham	698
Wooton	Derbyshire	408

SCHIST

Ardchronie	Highland	652
Bluehill	Grampian	258
Bower	Highland	653
Daviot	Highland	679
Gedloch	Grampian	220

Marybank	Western Isles	657
Meldon	Devon	125
North Mains	Grampian	186
Rothes Glen	Grampian	510
Scord	Shetland	504
Setters	Shetland	482
Western New Forres	Grampian	261

SERPENTINE

Trevassack	Cornwall	162

SHALE

Tippetcraig	Central	536

SLATE

Aberllefeni	Powys	697
Baycliff	Cumbria	137
Bowithic	Cornwall	163
Braich Ddu	Gwynedd	262
Brandy Crag	Cumbria	138
Brathay	Cumbria	139
Broughton Moor	Lancashire	140
Burlington	Cumbria	141
Bursting Stone	Cumbria	142
Callywith	Cornwall	537
Chillaton	Devon	346
Clogau	Clwyd	95
Cwt-y-Bugail	Gwynedd	235
Delabole	Cornwall	474
Diphwys Casson	Gwynedd	263
Elterwater	Cumbria	143
Gloddfa Ganol	Gwynedd	236
Glogue	Dyfed	188
Honister	Cumbria	357
Kirkby	Cumbria	144
Llechedd Slate Mine	Gwynedd	264
Longford	Devon	370
Maenofferen	Gwynedd	265
Mill Hill	Devon	371
Moss Rigg	Cumbria	145
Nantlle	Gwynedd	237
Oakeley	Gwynedd	238
Penrhyn	Gwynedd	365
Pets	Cumbria	320
Prince of Wales	Cornwall	451
Skiddaw	Cumbria	146
South Barrule	Isle of Man	303
Spout Cragg	Cumbria	147
Trebarwith	Cornwall	660
Tredennick Downs	Cornwall	314
Trevillet	Cornwall	74
Twll Llwyd	Gwynedd	297
Vronlog	Gwynedd	315

TRACHYTE

Bangley	Lothian	362

Index 5

Rock types in England, Scotland and Wales

County by county within each country;
and by quarry name within counties

Entries in this index refer to the
unique quarry numbers (1 – 705)
in the main Directory

England

ANDESITE
  Avon
    Moons Hill — 666
  Cumbria
    Ghyll Scaur — 226
  Leicestershire
    Bardon — 75
    Whitwick — 17
  Northumberland
    Harden — 638

BASALT
  Cornwall
    Laneast — 475
    Lean — 473
  Derbyshire
    Water Swallows — 579
  Devon
    Whitecleaves — 64
  Northumberland
    Howick — 556
  Shropshire
    Blodwell — 10
    Clee Hill — 12
    Leaton — 312
  West Midlands
    Edwin Richards — 14

CHALK
  Bedfordshire
    Kensworth — 479
  Cambridgeshire
    Barrington — 476
    Norman — 108
  Devon
    Beer — 58
  Dorset
    Long Bredy — 61
  Hertfordshire
    Flaunden — 159
  Humberside
    Lowthorpe — 529
    Melton — 107
    Middlesgate — 480
    Queensgate — 201
    Ruston Parva — 529
  Kent
    Greys Pit — 478
    Holborough — 104
    Northfleet — 109
  Lincolnshire
    Melton Ross — 506
  Norfolk
    Mount Pleasant — 158

Oxfordshire
    Chinnor — 477
  Suffolk
    Great Blakenham — 103
    Masons — 106
  West Sussex
    Shoreham — 112
  Wiltshire
    Charnage — 59
    East Grinstead — 200
    Quidhampton — 202
    Westbury — 113

DIORITE
  Leicestershire
    Charnwood — 11
    Cliffe Hill — 564
    Croft — 204
    Groby — 15
    Old Cliffe Hill — 566
  Warwickshire
    Griff — 437

DOLERITE
  Cornwall
    Black Hill — 419
    Greystone — 212
  Devon
    New England — 215
    Trusham — 63
  Durham
    Force Garth — 278
  Northumberland
    Barrasford — 551
    Belford — 636
    Cragmill — 366
    Divethill — 367
    Swinburne — 34
  Shropshire
    Callow Hill — 605
    Criggion — 13
    Leaton — 312
  Warwickshire
    Mancetter — 628
  West Midlands
    Hailstone — 610

FLAGSTONE
  Lancashire
    Jamestone — 305
  West Yorkshire
    Cromwell — 348
    Friendly — 70
    Thumpas — 166

GABBRO
    Cornwall
        Dean — 457

GRANITE
    Cornwall
        Antron Hill — 8
        Bearrah Tor — 443
        Bosahan — 663
        Burnthouse — 9
        Carnsew — 420
        Castle-an-Dinas — 421
        Chywoon — 326
        De Lank — 469
        Hantergantick — 395
        Hingston Down — 60
        Kessel Downs — 423
        Luxulyan — 213
        Penryn — 420
        Tor Down — 388
        Trenoweth — 662
        Trevone — 664
    Cumbria
        Shap — 31
        Shap Beck — 31
        Shap Blue — 496
    Devon
        Merivale — 397
    Isle of Man
        Granite Mountain — 509
        Poortown — 302
    Leicestershire
        Buddon Wood — 456
        Mountsorrel — 456

GRANODIORITE
    Cumbria
        Broad Oak — 170

GREYWACKE
    Shropshire
        Bayston Hill — 603

GRITSTONE
    Cheshire
        Wimberry Moss — 199
    Cornwall
        Dairy — 422
        Hertbury — 417
        Pigsden — 418
        Pilsamoor — 549
        Tredinnick — 661
    Cumbria
        Shawk — 323
    Derbyshire
        Ann Twyford — 468
        Palmers — 514

Durham
    Dead Friars — 513
    Harthorpe East — 535
    Quickburn — 318
    Shipley Bank — 505
Greater Manchester
    Buckton Vale — 354
    Chinley Moor — 369
    Fletcher Bank — 349
Lancashire
    Ellel — 304
    Fords — 240
    Ingleton — 24
    Leeming — 132
    Montcliffe — 27
    Round 'O' — 30
    Waddington Fell — 665
    Whittle Hill — 676
Northumberland
    Ladycross — 322
Staffordshire
    Great Gate — 512
    Red Hole — 512
West Yorkshire
    Bank Top — 671
    Blackhill — 376
    Bolton Woods — 429
    Clock Face — 347
    Crosland Moor — 310
    Deep Lane — 430
    Fagley — 431
    Fly Delph — 342
    Hainworth Shaw — 72
    Harden Moor — 307
    Honley Wood — 311
    Horn Crag — 294
    Rawdon — 328
    Sovereign — 334
    Stone — 445
    Watson — 97
    Woodside — 122

HORNFELS
    Cornwall
        Penlee — 424
    Cumbria
        Shap Pink — 497
    Devon
        Meldon — 125

IRONSTONE
    Oxfordshire
        Alkerton — 90
        Shennington — 93

LIMESTONE
    Avon
        Backwell — 672

Chipping Sodbury	49
Flax Bourton	686
Gurney Slade	382
Harnhill	688
Hayes Wood Mine	86
Longwood	439
Mount Pleasant	447
Stancombe	695
Stowey	541
Upper Lawn	275
Wick	674
Cambridgeshire	
Stamford	446
Cheshire	
Afonwen	511
Cleveland	
Hart	501
Cumbria	
Blencowe	223
Eskett	225
Goldmire	22
Hardendale	126
Hartley	279
Kendal Fell	574
Moota	118
Parkhead	28
Pickerings	172
Rooks	173
Salterwath	174
Sandside	576
Stainton	641
Tendley	577
Derbyshire	
Ashwood Dale	563
Ballidon	624
Bolsover Moor	552
Brierlow	470
Cowdor and Hall Dale	553
Dene	554
Dove Holes	471
Dowlow	525
Eldon Hill	399
Goddards	667
Grange Mill	89
Hillhead	572
Hindlow	611
Hopton Wood	565
Ivonbrook	405
Longcliffe	336
Middle Peak	557
Once-a-Week	345
Parish	358
Shining Bank	668
Topley Pike	578
Tunstead	301
Devon	
Blackaller	190

Fenacre	609
Kersdown	493
Linhay Hill	254
Moorcroft	214
Plymstock	110
Stoneycombe	211
Westleigh	217
Yalberton Tor	65
Dorset	
Admiralty	2
Bowyers	1
Downs	338
Horn Park	247
Independent	2
Keates	317
Landers	325
Portland	62
Southard	287
St Aldhelm's	286
Swanage	547
Swanworth	602
Durham	
Aycliffe	539
Bishop Middleham	623
Crime Rig	500
Eastgate	102
Heights	230
High Force	280
Kilmond Wood	281
Old Quarrington	559
Raisby	560
Shadforth	502
Thrislington	531
Witch Hill	503
Gloucestershire	
Brockhill	164
Clearwell	625
Cotswold Hill	165
Cromhall	50
Daglingworth	51
Drybrook	52
Farmington	231
Guiting	53
Huntsmans	299
Rogers	629
Stowe Hill	540
Stowfield	615
Syreford	255
Tytherington	55
Veizey's	619
Wickwar	675
Hereford and Worcester	
Fish Hill	508
Happylands	411
Hornsleasow	298
Leinthall	313
Perton	300

Shavers End	208	Skipton Rock	640
Woodbury	523	Smaws	530
Humberside		Spaunton	283
Brantingham	670	Spikers Hill	352
Isle of Man		Stonegrave	234
Balladoole	96	Sutton	548
Turkeyland	321	Swinden	634
Lancashire		Threshfield	562
Back Lane	203	Wath	532
Bankfield	568	Northamptonshire	
Bold Venture	569	Camp Hill	98
Dunald Mill	571	Corby	515
Holme Park	205	Cowltick	515
Leapers Wood	690	Pitsford	92
Ribblesdale Works	152	Pury End	98
Silverdale	32	Shotley	94
Leicestershire		Northumberland	
Breedon Hill	121	Copp Grag	127
Holleywell	368	Mootlaw	558
Lincolnshire		Prudham	128
Ancaster	267	Nottinghamshire	
Big Pits	135	Abbey	3
Clipsham	135	Mansfield	269
Collyweston	136	Vale Road	154
Creeton	444	Whitwell	533
Crossleys	425	Oxfordshire	
Glebe	268	Ardley	507
Grange Top	151	Burford	87
Great Ponton	516	Filkins	239
Greetham	426	Shellingford	386
Greetwell	149	Town	176
Ketton	151	White Hill	57
Kirton Lindsey	150	Woodeaton	78
South Witham	517	Worsham	518
Southorpe	461	Shropshire	
Spittlegate Level	333	Lea	206
Thornaugh	427	Lillishall	207
North Yorkshire		Llynclys	520
Barton	499	Shadwell	521
Black	277	Somerset	
Coldstones	434	Batts Combe	48
Cool Scar	224	Callow Rock	209
Darrington	178	Castle Hill	153
Forcett	637	Colemans	210
Foxcliffe	179	Doulting	83
Gebdykes	526	Dulcote	703
Giggleswick	632	Halecombe	687
Highmoor	550	Ham Hill	221
Horton	573	Ham Hill	378
Hovingham	66	Lime Kiln Hill	332
Jackdaw Crag	180	Merehead	704
Newbridge	282	Peddles Lane	416
Potgate	528	Shipham	614
Redmire	561	Station	76
Rock Cottage	377	Westbury	442
Settrington	233	Whatley	56
Sherburn	284		

South Yorkshire	
Brodsworth	631
Cadeby	524
Darlton	685
(The) Dolomite	432
Glen	350
Harrycroft	527
Holme Hall	555
Hope	105
Park Nook	148
Scabba Wood	374
Skelbrooke	183
Spring Lodge	633
Wormersley	534
Staffordshire	
Caldon Low	604
Cauldon	100
Kevin	627
Wardlow	462
Wredon	360
Suffolk	
Ingham	7
Tyne and Wear	
Houghton	23
Marsden	639
Warwickshire	
Avon Dassett	222
Horton	295
West Sussex	
Bognor Common	335
West Yorkshire	
Barnsdale Bar	191
Long Lane	181
Newthorpe	182
Peckfield	406
Wiltshire	
Monks Park Mine	84
Westwood Ground	85

MUDSTONE	
Nottinghamshire	
Bantycock	123
Staple Farm	124

QUARTZITE	
Northumberland	
Swinburne	34
Warwickshire	
Judkins	16
Nuneaton	612
Hartshill	612

RAGSTONE	
Kent	
Allington	47
Offham	54

SANDSTONE	
Cheshire	
Borrow Pit	245
Bridge	341
Sycamore	198
Cornwall	
Dairy	422
Hertbury	417
Pigsden	418
Pilsamoor	549
Cumbria	
Bankend	168
Birkhams	169
Holmescales	472
Lazonby Fell	171
Leipsic	498
Scratchmill Scar	175
Stoneraise	454
Derbyshire	
Bolehill	449
Dukes	450
Palmers	514
Ridgeway	161
Shire Hill	353
Stancliffe	452
Stanton Moor	453
Watts Cliff	455
Wooton	408
Devon	
Bableigh Wood	157
Barton Wood	189
Beam	650
Bray Valley	402
Hayne	219
Hearson	288
Knowle	6
Plaistow	403
Tuckingmill	481
Venn	216
Durham	
Cat Castle	155
Dunhouse	196
Harthorpe East	535
Shipley Bank	505
Windy Hill	698
Gloucestershire	
Bixhead	241
Mine Train	375
Quartzite	673
Wilderness	448
Greater Manchester	
Hayfield	324
Pilkington	483
Scout Moor	351
Lancashire	
Altham	131
Appley Bridge	21

East	21	Deep Lane	430
Fords	240	Fagley	431
Highmoor	227	Fly Flatts	129
Hutchbank	228	Friendly	70
Ingleton	24	Hillhouse Edge	484
Little	67	Hunters Hill	130
Montcliffe	27	Naylor Hill Quarries	248
Round 'O'	30	Northowram Hills	232
Stoney Brow	276	Sandfold	117
Whitworth	229	Soil Hill	266
Leicestershire		Squire Hill	404
Huncote	5	Thumpas	166
North Yorkshire		Watson	97
Arcow	567		
Blue Bank	69	**SCHIST**	
Dry Rigg	458	Devon	
Ten Yards Lane	71	Meldon	125
Northamptonshire			
Boughton	91	**SERPENTINE**	
Northumberland		Cornwall	
Black Pasture	487	Trevassack	162
Blaxter	392		
Darney	393	**SLATE**	
Doddington Hill	394	Cornwall	
High Nick	396	Bowithic	163
Shropshire		Callywith	537
Grinshill	270	Delabole	474
Webscott	271	Prince of Wales	451
Somerset		Trebarwith	660
Triscombe	616	Tredennick Downs	314
South Yorkshire		Trevillet	74
Blackmoor	285	Cumbria	
Halltoke Hall	538	Baycliff	137
Staffordshire		Brandy Crag	138
Croxden	608	Brathay	139
Great Gate	512	Burlington	141
Red Hole	512	Bursting Stone	142
Tearne	407	Elterwater	143
Tyne and Wear		Honister	357
Springwell	398	Kirkby	144
Warwickshire		Moss Rigg	145
Judkins	16	Pets	320
West Sussex		Skiddaw	146
Philpots	428	Spout Cragg	147
West Yorkshire		Devon	
Apex	306	Chillaton	346
Appleton	293	Longford	370
Barnshaw	192	Mill Hill	371
Beacon Lodge	185	Isle of Man	
Bolton Woods	429	South Barrule	303
Brittania	415	Lancashire	
Chellow Grange	412	Broughton Moor	140
Cromwell	348		

Scotland

ANDESITE
    Fife
        Balmullo   372
    Tayside
        Waulkmill   648

BASALT
    Borders
        Craighouse   584
    Central
        Murrayshall   599
    Fife
        Clatchard Craig   581
        Craigs   587
        Orrock   682
    Grampian
        Cairds Hill   383
    Lothian
        Blairhill   580
        Broadlaw   363
        Craigpark   586
        Ravelrig   600
    Shetland
        Vatseter   244
    Strathclyde
        Airdriehill   361
        Bannerbank   643
        Craigie Hill   655
        Croy   588
        Dumbuckhill   646
        Duntilland   543
        Hallyards   327
        High Craig   593
        Hillend   485
        Kilbarchan   486
        Loanhead   596
        Riganagower   622
        Underheugh   364
    Tayside
        Boysack   617
        Friarton   680
    Western Isles
        Ardhasaig   340
        Askernish   68
        Cleat and Lower Grean   343
        Crogarry Beag   308
        Steangabhal   309

DOLERITE
    Borders
        Borthwick   678
        Dunion   590
    Central
        Boards   677
        Northfield   156

Fife
    Belliston   88
    Craigfoot   649
    Cruicks   645
    Goat   592
Lothian
    Hillwood   681
    Kaimes   594
Strathclyde
    Cairneyhill   644
    Hillhouse   292
    Miltonhill   621
    Riskend   467
    Tam's Loup   441
Tayside
    Ardownie   246
    Collace   618
    Cunmont   589
    Ethiebeaton   591

FLAGSTONE
    Highland
        Borrowston   344

GABBRO
    Grampian
        Balmedie   257
        Corrennie   654
        North Lasts   329
        North Mains   186
        Pitcaple   260
        Tyrebagger   659

GNEISS
    Western Isles
        Lic   331

GRANITE
    Dumfries and Galloway
        Craignair   585
        Kirkmabreck   595
    Grampian
        Black Hills   133
        Craigenlow   583
        Kemnay   656
        Longside   134
        Savoch   187
        Tom's Forest   658
    Highland
        Banavie   642
        Glensanda   705
    Strathclyde
        Bonawe   243
        Cairngryffe   542
        Furnace   647

Pettinain	160	**RHYOLITE**		
Ross of Mull	492	Fife		
		Langside	373	
GRANODIORITE				
Strathclyde		**SANDSTONE**		
Tincornhill	544	Borders		
		Blynlee	242	
GREYWACKE		Central		
Borders		Dunmore	488	
Edstone	116	Dumfries and Galloway		
Hazelbank	337	Cairnryan	80	
Dumfries and Galloway		Corncockle	194	
Barlockhart	79	Corsehill	195	
Boreland Fell	193	Gatelawbridge	489	
Coatsgate	582	Locharbriggs	73	
Morrinton	598	Fife		
Strathclyde		Kingoodie	490	
Dunduff	413	Newbiggin	491	
Tongland	81	Grampian		
		Clashach	379	
GRITSTONE		Greenbrey	274	
Strathclyde		Spynie	380	
Tormitchell	82	Highland		
		Sconser	290	
LIMESTONE		Spittal	546	
Dumfries and Galloway		Orkney		
Dumfries	20	Curister	410	
Kelhead	25	Heddle Hill	409	
Stranraer	33	Shetland		
Wigtown	35	Brindister	545	
Fife		Strathclyde		
Skelpie Mine	167	Dunduff	413	
Grampian				
Boyne Bay	120	**SCHIST**		
Concraigs	669	Grampian		
Cottonhill	339	Bluehill	258	
Craiglash	259	Gedloch	220	
Parkmore	177	North Mains	186	
Woodside	414	Rothes Glen	510	
Highland		Western New Forres	261	
Morefield	597	Highland		
Torlundy	289	Ardchronie	652	
Torrin	330	Bower	653	
Lothian		Daviot	679	
Dunbar Northwest	101	Shetland		
Middleton Mine	464	Scord	504	
Strathclyde		Setters	482	
Auchinleck	18	Western Isles		
Ballygrant	197	Marybank	657	
Calliburn	319			
Hessilhead	463	**SHALE**		
Old Mill	218	Central		
Trearne	465	Tippetcraig	536	
Tayside				
Shierglas	601	**TRACHYTE**		
		Lothian		
MARL		Bangley	362	
Highland				
Achnagart	651			

Wales

Gwent			Dyfed	
Callow	184		Glogue	188
Hafod	460		Gwynedd	
Mid Glamorgan			Braich Ddu	262
Craig-yr-Hesg	37		Cwt-y-Bugail	235
Gelligaer	436		Diphwys Casson	263
Pencaemawr	436		Gloddfa Ganol	236
Powys			Llechedd Slate Mine	264
Gore	626		Maenofferen	265
Penstrowed	256		Nantlle	237
Tan-y-Foel	119		Oakeley	238
West Glamorgan			Penrhyn	365
Gilfach	701		Twll Llwyd	297
			Vronlog	315
SLATE			Powys	
Clwyd			Aberllefeni	697
Clogau	95			

Index 6

Rock colours in Britain
by quarry name and location

Entries in this index refer to the
unique quarry numbers (1 – 705)
in the main Directory

BLACK

Ardhasaig	Western Isles	340
Ardownie	Tayside	246
Askernish	Western Isles	68
Balmullo	Fife	372
Barlockhart	Dumfries and Galloway	79
Belliston	Fife	88
Blairhill	Lothian	580
Bolton Hill	Dyfed	250
Boysack	Tayside	617
Broadlaw	Lothian	363
Cairds Hill	Grampian	383
Carrigwynion	Powys	613
Cerrigyrwyn	Dyfed	251
Charnwood	Leicestershire	11
Clee Hill	Shropshire	12
Collace	Tayside	618
Coygen	Dyfed	252
Craigfoot	Fife	649
Craigie Hill	Strathclyde	655
Craigpark	Lothian	586
Duntilland	Strathclyde	543
Edstone	Borders	116
Ethiebeaton	Tayside	591
Goat	Fife	592
Groby	Leicestershire	15
Hainworth Shaw	West Yorkshire	72
Hallyards	Strathclyde	327
Howick	Northumberland	556
North Mains	Grampian	186
Ravelrig	Lothian	600
Rhyader	Powys	613
Steangabhal	Western Isles	309
Tongland	Strathclyde	81
Underheugh	Strathclyde	364
Vatseter	Shetland	244

BLUE/LILAC/MAUVE/PURPLE

Alkerton	Oxfordshire	90
Appleton	West Yorkshire	293
Bixhead	Gloucestershire	241
Blackmoor	South Yorkshire	285
Clogau	Clwyd	95
Dukes	Derbyshire	450
Dunion	Borders	590
Gelligaer	Mid Glamorgan	436
Glogue	Dyfed	188
Hafod	Gwent	460
Hearson	Devon	288
Horton	Warwickshire	295
Keates	Dorset	317
Little	Lancashire	67
Longford	Devon	370
Middleton Mine	Lothian	464
Mill Hill	Devon	371
Mine Train	Gloucestershire	375
Mount Pleasant	Norfolk	158

Nantlle	Gwynedd	237
North Mains	Grampian	186
Once-a-Week	Derbyshire	345
Pencaemawr	Mid Glamorgan	436
Penrhyn	Gwynedd	365
Shennington	Oxfordshire	93
Sovereign	West Yorkshire	334
Station	Somerset	76
Stowey	Avon	541
Syke	Dyfed	440
Trebarwith	Cornwall	660
Watts Cliff	Derbyshire	455
Wimberry Moss	Cheshire	199

BROWNS

Brown

Afonwen	Cheshire	511
Alkerton	Oxfordshire	90
Apex	West Yorkshire	306
Blue Bank	North Yorkshire	69
Boughton	Northamptonshire	91
Callow	Gwent	184
Cat Castle	Durham	155
Copp Crag	Northumberland	127
Fly Delph	West Yorkshire	342
Horn Crag	West Yorkshire	294
Horton	Warwickshire	295
Little	Lancashire	67
Longwood	Avon	439
Miltonhill	Strathclyde	621
Nanhoron	Gwynedd	387
Naylor Hill Quarries	West Yorkshire	248
Ribblesdale Works	Lancashire	152
Shennington	Oxfordshire	93
Stone	West Yorkshire	445
Ten Yards Lane	North Yorkshire	71
Trevillet	Cornwall	74
Twll Llwyd	Gwynedd	297
Windy Hill	Durham	698

Dark Brown

Bableigh Wood	Devon	157
Gelligaer	Mid Glamorgan	436
Huntsmans	Gloucestershire	299
Pencaemawr	Mid Glamorgan	436
Pilsamoor	Cornwall	549
Shap Blue	Cumbria	496
Syke	Dyfed	440

Medium Brown/Rustic

Beam	Devon	650
Blackhill	West Yorkshire	376
Chillaton	Devon	346
Gelligaer	Mid Glamorgan	436
Lean	Cornwall	473
Pencaemawr	Mid Glamorgan	436
Prince of Wales	Cornwall	451

Light Brown/Buff/Fawn/Honey

Abbey	Nottinghamshire	3

Abergele	Clwyd	630
Ann Twyford	Derbyshire	468
Apex	West Yorkshire	306
Backwell	Avon	672
Bankfield	Lancashire	568
Beer	Devon	58
Big Pits	Lincolnshire	135
Black Pasture	Northumberland	487
Bognor Common	West Sussex	335
Bolehill	Derbyshire	449
Bolton Woods	West Yorkshire	429
Bridge	Cheshire	341
Brittania	West Yorkshire	415
Brodsworth	South Yorkshire	631
Burford	Oxfordshire	87
Caldon Low	Staffordshire	604
Callywith	Cornwall	537
Castle-an-Dinas	Cornwall	421
Chellow Grange	West Yorkshire	412
Chinley Moor	Greater Manchester	369
Clashach	Grampian	379
Clearwell	Gloucestershire	625
Clipsham	Lincolnshire	135
Colemans	Somerset	210
Cromwell	West Yorkshire	348
Crosland Moor	West Yorkshire	310
Deep Lane	West Yorkshire	430
Doulting	Somerset	83
Drybrook	Gloucestershire	52
Dukes	Derbyshire	450
Dunhouse	Durham	196
Ellel	Lancashire	304
Fagley	West Yorkshire	431
Fly Delph	West Yorkshire	342
Fords	Lancashire	240
Friendly	West Yorkshire	70
Gebdykes	North Yorkshire	526
Grange Top	Lincolnshire	151
Greenbrey	Grampian	274
Grinshill	Shropshire	270
Ham Hill	Somerset	221
Ham Hill	Somerset	378
Happylands	Hereford and Worcester	411
Hayes Wood Mine	Avon	86
Hayne	Devon	219
Hillhouse Edge	West Yorkshire	484
Holleywell	Leicestershire	368
Honley Wood	West Yorkshire	311
Jamestone	Lancashire	305
Ketton	Lincolnshire	151
Ladycross	Northumberland	322
Long Lane	West Yorkshire	181
Luxulyan	Cornwall	213
Mansfield	Nottinghamshire	269
Marsden	Tyne and Wear	639
Naylor Hill Quarries	West Yorkshire	248
Newbiggin	Fife	491

Northowram Hills	West Yorkshire	232
Palmers	Derbyshire	514
Park Nook	South Yorkshire	148
Philpots	West Sussex	428
Pilsamoor	Cornwall	549
Portland	Dorset	62
Prudham	Northumberland	128
Rawdon	West Yorkshire	328
Rhyndaston	Dyfed	381
Ribblesdale Works	Lancashire	152
Ridgeway	Derbyshire	161
Rogers	Gloucestershire	629
Shipley Bank	Durham	505
Skiddaw	Cumbria	146
Soil Hill	West Yorkshire	266
Southorpe	Lincolnshire	461
Springwell	Tyne and Wear	398
Squire Hill	West Yorkshire	404
Stamford	Cambridgeshire	446
Stancliffe	Derbyshire	452
Stanton Moor	Derbyshire	453
Station	Somerset	76
Stoke Hall	South Yorkshire	538
Stone	West Yorkshire	445
Sycamore	Cheshire	198
Thumpas	West Yorkshire	166
Tredennick Downs	Cornwall	314
Tredinnick	Cornwall	661
Veizey's	Gloucestershire	619
Waddington Fell	Lancashire	665
Watson	West Yorkshire	97
White Hill	Oxfordshire	57
Whittle Hill	Lancashire	676

CREAM

Blaxter	Northumberland	392
Blue Bank	North Yorkshire	69
Bolsover Moor	Derbyshire	552
Cadeby	South Yorkshire	524
Creeton	Lincolnshire	444
Darney	Northumberland	393
Doddington Hill	Northumberland	394
(The) Dolomite	South Yorkshire	432
Dunmore	Central	488
Fish Hill	Hereford and Worcester	508
Friendly	West Yorkshire	70
Grange Top	Lincolnshire	151
Guiting	Gloucestershire	53
High Nick	Northumberland	396
Holme Hall	South Yorkshire	555
Houghton	Tyne and Wear	23
Jackdaw Crag	North Yorkshire	180
Ketton	Lincolnshire	151
Kirton Lindsey	Lincolnshire	150
Middle Peak	Derbyshire	557
Monks Park Mine	Wiltshire	84
Mount Pleasant	Avon	447

Newbiggin	Fife	491
Prudham	Northumberland	128
Scabba Wood	South Yorkshire	374
Sherburn	North Yorkshire	284
Skelbrooke	South Yorkshire	183
Skiddaw	Cumbria	146
Spynie	Grampian	380
Squire Hill	West Yorkshire	404
Swanworth	Dorset	602
Ten Yards Lane	North Yorkshire	71
Thornaugh	Lincolnshire	427
Upper Lawn	Avon	275
Vale Road	Nottinghamshire	154
Veizey's	Gloucestershire	619

GREEN

Bardon	Leicestershire	75
Bixhead	Gloucestershire	241
Bognor Common	West Sussex	335
Bolehill	Derbyshire	449
Borthwick	Borders	678
Broughton Moor	Lancashire	140
Bursting Stone	Cumbria	142
Criggion	Shropshire	13
Elterwater	Cumbria	143
Ghyll Scaur	Cumbria	226
Gilfach	West Glamorgan	701
Gore	Powys	626
Honister	Cumbria	357
Leaton	Shropshire	312
Minffordd	Gwynedd	691
Moss Rigg	Cumbria	145
Penrhyn	Gwynedd	365
Pets	Cumbria	320
Spout Cragg	Cumbria	147
Trevassack	Cornwall	162
Vronlog	Gwynedd	315
Water Swallows	Derbyshire	579

GREYS
Grey

Abergele	Clwyd	630
Aberthaw	South Glamorgan	99
Aycliffe	Durham	539
Bangley	Lothian	362
Barlockhart	Dumfries and Galloway	79
Barnsdale Bar	West Yorkshire	191
Bishop Middleham	Durham	623
Blackaller	Devon	190
Blaenyfan	Dyfed	699
Blairhill	Lothian	580
Bognor Common	West Sussex	335
Bosahan	Cornwall	663
Bowyers	Dorset	1
Bridge	Cheshire	341
Callywith	Cornwall	537
Carrigwynion	Powys	613

Chellow Grange	West Yorkshire	412
Chywoon	Cornwall	326
Craighouse	Borders	584
Dene	Derbyshire	554
Eastgate	Durham	102
Edstone	Borders	116
Farmington	Gloucestershire	231
Gebdykes	North Yorkshire	526
Giggleswick	North Yorkshire	632
Goddards	Derbyshire	667
Hainworth Shaw	West Yorkshire	72
Hardendale	Cumbria	126
Hartshill	Warwickshire	612
Hopton Wood	Derbyshire	565
Houghton	Tyne and Wear	23
Hutchbank	Lancashire	228
Keates	Dorset	317
Ladycross	Northumberland	322
Lean	Cornwall	473
Lic	Western Isles	331
Little	Lancashire	67
Llanddulas	Clwyd	26
Marsden	Tyne and Wear	639
Middle Peak	Derbyshire	557
Moorcroft	Devon	214
Morefield	Highland	597
Northowram Hills	West Yorkshire	232
Nuneaton	Warwickshire	612
Pantyffynnon	South Glamorgan	291
Penmaenmawr	Gwynedd	29
Rhyader	Powys	613
Sandside	Cumbria	576
Scord	Shetland	504
Setters	Shetland	482
Sovereign	West Yorkshire	334
Swinden	North Yorkshire	634
Sycamore	Cheshire	198
Tongland	Strathclyde	81
Trebarwith	Cornwall	660
Trevone	Cornwall	664
Tunstead	Derbyshire	301
Water Swallows	Derbyshire	579
Whitwick	Leicestershire	17
Whitworth	Lancashire	229
Dark Grey		
Balmedie	Grampian	257
Bannerbank	Strathclyde	643
Barrasford	Northumberland	551
Barton Wood	Devon	189
Belford	Northumberland	636
Black	North Yorkshire	277
Bolehill	Derbyshire	449
Borthwick	Borders	678
Cairneyhill	Strathclyde	644
Callow Hill	Shropshire	605
Croy	Strathclyde	588
Cruicks	Fife	645

Dolyhir and Strinds	Powys	390
Dry Rigg	North Yorkshire	458
Dumbuckhill	Strathclyde	646
Edwin Richards	West Midlands	14
Force Garth	Durham	278
Friarton	Tayside	680
Gelligaer	Mid Glamorgan	436
Gellihalog	Dyfed	700
Goldmire	Cumbria	22
Hailstone	West Midlands	610
High Craig	Strathclyde	593
High Force	Durham	280
Hillend	Strathclyde	485
Hillhouse	Strathclyde	292
Hillwood	Lothian	681
Huntsmans	Gloucestershire	299
Kaimes	Lothian	594
Kilbarchan	Strathclyde	486
Kilmond Wood	Durham	281
Knowle	Devon	6
Leaton	Shropshire	312
Linhay Hill	Devon	254
Loanhead	Strathclyde	596
Northfield	Central	156
North Lasts	Grampian	329
North Mains	Grampian	186
Orrock	Fife	682
Parkhead	Cumbria	28
Pencaemawr	Mid Glamorgan	436
Penmaenmawr	Gwynedd	29
Pitcaple	Grampian	260
Riganagower	Strathclyde	622
Riskend	Strathclyde	467
Round 'O'	Lancashire	30
Skipton Rock	North Yorkshire	640
Spittal	Highland	546
Swinburne	Northumberland	34
Syke	Dyfed	440
Tam's Loup	Strathclyde	441
Trefil	Gwent	272
Venn	Devon	216
Whitecleaves	Devon	64
Medium Grey		
Barton Wood	Devon	189
Bayston Hill	Shropshire	603
Blodwell	Shropshire	10
Boards	Central	677
Bold Venture	Lancashire	569
Callow Rock	Somerset	209
Coldstones	North Yorkshire	434
Cragmill	Northumberland	366
Craig-yr-Hesg	Mid Glamorgan	37
Croft	Leicestershire	204
Divethill	Northumberland	367
Dunald Mill	Lancashire	571
Fenacre	Devon	609
Graig	Clwyd	459

Griff	Warwickshire	437
Hailstone	West Midlands	610
Knowle	Devon	6
Mootlaw	Northumberland	558
Murrayshall	Central	599
Plymstock	Devon	110
Ruthin	Mid Glamorgan	111
Silverdale	Lancashire	32
Threshfield	North Yorkshire	562
Venn	Devon	216
Westbury	Wiltshire	113
Westbury	Somerset	442
Westleigh	Devon	217

Light Grey

Aberstrecht	Gwynedd	4
Antron Hill	Cornwall	8
Back Lane	Lancashire	203
Backwell	Avon	672
Banavie	Highland	642
Batts Combe	Somerset	48
Bearrah Tor	Cornwall	443
Brandy Crag	Cumbria	138
Burnthouse	Cornwall	9
Caer Glaw	Gwynedd	19
Callow Rock	Somerset	209
Carnsew	Cornwall	420
Castle-an-Dinas	Cornwall	421
Coldstones	North Yorkshire	434
Colemans	Somerset	210
Cowdor and Hall Dale	Derbyshire	553
Craig-yr-Hesg	Mid Glamorgan	37
Cwmleyshon	Mid Glamorgan	38
De Lank	Cornwall	469
Dolyhir and Strinds	Powys	390
Dove Holes	Derbyshire	471
Eldon Hill	Derbyshire	399
Gilfach	West Glamorgan	701
Gore	Powys	626
Griff	Warwickshire	437
Gwyndy	Gwynedd	273
Hantergantick	Cornwall	395
Hartley	Cumbria	279
Hindlow	Derbyshire	611
Hingston Down	Cornwall	60
Holme Park	Lancashire	205
Ifton	Gwent	39
Kemnay	Grampian	656
Kendal Fell	Cumbria	574
Kirkmabreck	Dumfries and Galloway	595
Livox	Gwent	41
Mancetter	Warwickshire	628
Meldon	Devon	125
Merehead	Somerset	704
Merivale	Devon	397
Once-a-Week	Derbyshire	345
Penhow	Gwent	44
Penlee	Cornwall	424

Penryn	Cornwall	420
Pickerings	Cumbria	172
Rhuddlan Bach	Gwynedd	466
Rogers	Gloucestershire	629
Rooks	Cumbria	173
Scratchmill Scar	Cumbria	175
Shap Pink	Cumbria	497
Spaunton	North Yorkshire	283
Sutton	North Yorkshire	548
Tendley	Cumbria	577
Tor Down	Cornwall	388
Town	Oxfordshire	176
Trenoweth	Cornwall	662
Trimm Rock	Clwyd	635
Tunstead	Derbyshire	301
Vaynor	Mid Glamorgan	46
Wenvoe	South Glamorgan	702
Wick	Avon	674
Wickwar	Gloucestershire	675
Yalberton Tor	Devon	65

GREY HUES
Bluish

Aberllefeni	Powys	697
Braich Ddu	Gwynedd	262
Broad Oak	Cumbria	170
Burlington	Cumbria	141
Creeton	Lincolnshire	444
Cunmont	Tayside	589
Cwt-y-Bugail	Gwynedd	235
Dean	Cornwall	457
Delabole	Cornwall	474
Diphwys Casson	Gwynedd	263
Downs	Dorset	338
Fly Flatts	West Yorkshire	129
Gloddfa Ganol	Gwynedd	236
Horton	North Yorkshire	573
Hunters Hill	West Yorkshire	130
Judkins	Warwickshire	16
Kirkby	Cumbria	144
Landers	Dorset	325
Lime Kiln Hill	Somerset	332
Llechedd Slate Mine	Gwynedd	264
Maenofferen	Gwynedd	265
Moota	Cumbria	118
Oakeley	Gwynedd	238
Salterwath	Cumbria	174
Shavers End	Hereford and Worcester	208
Shining Bank	Derbyshire	668
Swanage	Dorset	547
Tyrebagger	Grampian	659

Brownish

Blencowe	Cumbria	223
Chipping Sodbury	Avon	49
Cornelly	Mid Glamorgan	684
Cromhall	Gloucestershire	50
Dayhouse	Gwent	620

Denbigh	Clwyd	570
Forcett	North Yorkshire	637
Forest Wood	Mid Glamorgan	435
Grove	Mid Glamorgan	438
Hendy	Mid Glamorgan	689
Lithalun	Mid Glamorgan	40
Machen	Gwent	42
Minera	Clwyd	575
Moons Hill	Avon	666
Pant	Mid Glamorgan	692
Penderyn	Mid Glamorgan	43
Risca	Gwent	45
Spring Lodge	South Yorkshire	633
Tytherington	Gloucestershire	55
Waulkmill	Tayside	648
Whatley	Somerset	56
Buffish		
Barton	North Yorkshire	499
Downs	Dorset	338
Landers	Dorset	325
Stancombe	Avon	695
Creamy		
Aberduna	Clwyd	433
Ballidon	Derbyshire	624
Cauldon	Staffordshire	100
Collyweston	Lincolnshire	136
Cool Scar	North Yorkshire	224
Darlton	South Yorkshire	685
Halkyn	Clwyd	400
Kevin	Staffordshire	627
Leapers Wood	Lancashire	690
Newbridge	North Yorkshire	282
Pant (Halkyn)	Clwyd	693
Pant-y-Pwll Dwr	Clwyd	400
Wardlow	Staffordshire	462
Greenish		
Allington	Kent	47
Appley Bridge	Lancashire	21
Builth Wells	Powys	36
East	Lancashire	21
Greystone	Cornwall	212
Ingleton	Lancashire	24
New England	Devon	215
Offham	Kent	54
Trusham	Devon	63
Twll Llwyd	Gwynedd	297
Pinkish		
Blaengwynlais	Mid Glamorgan	683
Bonawe	Strathclyde	243
Cefn Gawr	South Glamorgan	606
Creigiau	Mid Glamorgan	607
Daglingworth	Gloucestershire	51
Daviot	Highland	679
Eskett	Cumbria	225
Plas Gwilym	Clwyd	296
Shap	Cumbria	31
Shap Beck	Cumbria	31

Stoneycombe	Devon	211
Taffs Well	South Glamorgan	522
Reddish		
Breedon Hill	Leicestershire	121
Stowfield	Gloucestershire	615

PINK

Ann Twyford	Derbyshire	468
Bardon	Leicestershire	75
Black Hills	Grampian	133
Buddon Wood	Leicestershire	456
Croft	Leicestershire	204
Doddington Hill	Northumberland	394
Lazonby Fell	Cumbria	171
Lic	Western Isles	331
Locharbriggs	Dumfries and Galloway	73
Longside	Grampian	134
Mountsorrel	Leicestershire	456
Palmers	Derbyshire	514
Quartzite	Gloucestershire	673
Raynes	Clwyd	401
Ridgeway	Derbyshire	161
Ross of Mull	Strathclyde	492
Shap Blue	Cumbria	496
Shap Pink	Cumbria	497
Stainton	Cumbria	641
Stanton Moor	Derbyshire	453
Stoneraise	Cumbria	454
Waddington Fell	Lancashire	665
Wimberry Moss	Cheshire	199

RED

Abbey	Nottinghamshire	3
Bankend	Cumbria	168
Birkhams	Cumbria	169
Blackmoor	South Yorkshire	285
Gore	Powys	626
Great Gate	Staffordshire	512
Huncote	Leicestershire	5
Lazonby Fell	Cumbria	171
Linhay Hill	Devon	254
Pettinain	Strathclyde	160
Red Hole	Staffordshire	512
Ross of Mull	Strathclyde	492
Shawk	Cumbria	323
Tredennick Downs	Cornwall	314
Tredinnick	Cornwall	661
Trevassack	Cornwall	162
Triscombe	Somerset	616
Twll Llwyd	Gwynedd	297
Webscott	Shropshire	271
Wilderness	Gloucestershire	448

WHITE

Admiralty	Dorset	2
Ancaster	Lincolnshire	267
Barrington	Cambridgeshire	476

Beer	Devon	58
Bolton Hill	Dyfed	250
Caer Glaw	Gwynedd	19
Cat Castle	Durham	155
Charnage	Wiltshire	59
Charnwood	Leicestershire	11
Chinnor	Oxfordshire	477
Copp Grag	Northumberland	127
Darney	Northumberland	393
East Grinstead	Wiltshire	200
Farmington	Gloucestershire	231
Flaunden	Hertfordshire	159
Glebe	Lincolnshire	268
Grange Mill	Derbyshire	89
Great Blakenham	Suffolk	103
Greys Pit	Kent	478
Groby	Leicestershire	15
Highmoor	North Yorkshire	550
Holborough	Kent	104
Holme Hall	South Yorkshire	555
Horn Park	Dorset	247
Hovingham	North Yorkshire	66
Independent	Dorset	2
Ingham	Suffolk	7
Keates	Dorset	317
Kensworth	Bedfordshire	479
Llanddulas	Clwyd	26
Long Bredy	Dorset	61
Masons	Suffolk	106
Melton	Humberside	107
Melton Ross	Lincolnshire	506
Middlegate	Humberside	480
Newbiggin	Fife	491
Norman	Cambridgeshire	108
Northfleet	Kent	109
Queensgate	Humberside	201
Quidhampton	Wiltshire	202
Rock Cottage	North Yorkshire	377
Scabba Wood	South Yorkshire	374
Shoreham	West Sussex	112
Shotley	Northamptonshire	94
Skelbrooke	South Yorkshire	183
Southard	Dorset	287
St Aldhelm's	Dorset	286
Stowey	Avon	541
Trevassack	Cornwall	162
Wath	North Yorkshire	532
Westwood Ground	Wiltshire	85
Woodeaton	Oxfordshire	78
Woodside	West Yorkshire	122

YELLOW/YELLOW-ORANGE

Beacon Lodge	West Yorkshire	185
Brockhill	Gloucestershire	164
Clashach	Grampian	379
Cotswold Hill	Gloucestershire	165
Dukes	Derbyshire	450

Fish Hill	Hereford and Worcester	508
Happylands	Hereford and Worcester	411
Huntsmans	Gloucestershire	299
Trevassack	Cornwall	162
Waddington Fell	Lancashire	665
Woodside	West Yorkshire	122

MOTTLED/SPECKLED COLOURS

MOTTLED
Grey

Cliffe Hill	Leicestershire	564
De Lank	Cornwall	469
Dolyhir and Strinds	Powys	390
Lea	Shropshire	206
Old Cliffe Hill	Leicestershire	566

Pink

Old Cliffe Hill	Leicestershire	566
Tearne	Staffordshire	407

Red

Tearne	Staffordshire	407

SPECKLED
Black

Bolton Hill	Dyfed	250
Charnwood	Leicestershire	11
Groby	Leicestershire	15

Brown

Hillhouse Edge	West Yorkshire	484

Grey

Banavie	Highland	642
Broad Oak	Cumbria	170
Hingston Down	Cornwall	60
Meldon	Devon	125
Penlee	Cornwall	424

Pink

Buddon Wood	Leicestershire	456
Mountsorrel	Leicestershire	456

White

Bolton Hill	Dyfed	250
Charnwood	Leicestershire	11
Groby	Leicestershire	15

Geotechnical
and Geological Engineering

Editor, **D G Toll**, School of Engineering and Applied Science, University of Durham
Co Editor, **Prof J M Kemeny**, Dept., of Mining and Geological Engineering, University of Arizona, Tucson, Arizona

Geotechnical and Geological Engineering is a new quarterly international journal. Formerly International Journal of Mining and Geological Engineering, the journal's title has been changed to reflect a widening of its scope.

With a new Editor, Co-Editor, *Geotechnical and Geological Engineering* will emphasize the practical aspects of geotechnical engineering and engineering geology. Papers on theoretical and experimental advances in soil and rock mechanics will also be welcomed. Particular priority will be given to the following areas:

* Case histories describing ground engineering projects, including contractual and logistical aspects.
* Novel geotechnical construction techniques
* Pollution and environmental problems
* Tropical soil and rock engineering
* Ground investigation; engineering geological and hydrological appraisals
* Computer-aided geotechnical and geological engineering including the application of information technology

The high technical and publication standards of *International Journal of Mining and Geological Engineering* will be maintained by *Geotechnical and Geological Engineering.*

Subscription Information

ISSN: 0960-3182 Published quarterly Volume 9 will published in 1991
European Community: £120
USA/Canada: $215
Rest of World: £132

For more information about this journal and others published by us, please contact:
The Journals Promotion Dept., Chapman and Hall, 2-6 Boundary Row,
***London SE1 8HN** (Telephone 071 865 0066)*

Index 7

Rock colours in England, Scotland and Wales

by quarry name, rock type and location

Entries in this index refer to the
unique quarry numbers (1 – 705)
in the main Directory

England

BLACK

Charnwood	Diorite, Coarse	Leicestershire	11
Clee Hill	Basalt, Fine	Shropshire	12
Groby	Diorite, Coarse	Leicestershire	15
Hainworth Shaw	Gritstone, Fine	West Yorkshire	72
Howick	Basalt, Fine	Northumberland	556

BLUE/LILAC/MAUVE/PURPLE

Alkerton	Ironstone, Fine	Oxfordshire	90
Appleton	Sandstone, Fine	West Yorkshire	293
Bixhead	Sandstone, Fine to medium	Gloucestershire	241
Blackmoor	Sandstone, Fine	South Yorkshire	285
Dukes	Sandstone, Medium	Derbyshire	450
Hearson	Sandstone, Medium	Devon	288
Horton	Limestone, Fine	Warwickshire	295
Keates	Limestone, Fine	Dorset	317
Little	Sandstone, Coarse	Lancashire	67
Longford	Slate, Fine	Devon	370
Mill Hill	Slate, Fine	Devon	371
Mine Train	Sandstone, Medium	Gloucestershire	375
Mount Pleasant	Chalk, Round	Norfolk	158
Once-a-Week	Limestone, Fine	Derbyshire	345
Shennington	Ironstone, Fine	Oxfordshire	93
Sovereign	Gritstone, Coarse	West Yorkshire	334
Station	Limestone	Somerset	76
Stowey	Limestone, Fine	Avon	541
Trebarwith	Slate, Fine	Cornwall	660
Watts Cliff	Sandstone, Medium	Derbyshire	455
Wimberry Moss	Gritstone, Fine	Cheshire	199

BROWNS

Brown

Afonwen	Limestone, Coarse	Cheshire	511
Alkerton	Ironstone, Fine	Oxfordshire	90
Apex	Sandstone, Medium	West Yorkshire	306
Blue Bank	Sandstone, Coarse	North Yorkshire	69
Boughton	Sandstone, Fine	Northamptonshire	91
Cat Castle	Sandstone, Coarse	Durham	155
Copp Grag	Limestone	Northumberland	127
Fly Delph	Gritstone, Medium	West Yorkshire	342
Horn Crag	Gritstone, Fine	West Yorkshire	294
Horton	Limestone, Fine	Warwickshire	295
Little	Sandstone, Coarse	Lancashire	67
Longwood	Limestone	Avon	439
Naylor Hill Quarries	Sandstone, Fine	West Yorkshire	248
Ribblesdale Works	Limestone, Fine	Lancashire	152
Shennington	Ironstone, Fine	Oxfordshire	93
Stone	Gritstone, Medium	West Yorkshire	445
Ten Yards Lane	Sandstone, Fine	North Yorkshire	71
Trevillet	Slate, Medium	Cornwall	74
Windy Hill	Sandstone, Medium	Durham	698

Dark Brown

Bableigh Wood	Sandstone, Medium	Devon	157
Huntsmans	Limestone, Fine	Gloucestershire	299

Pilsamoor	Sandstone, Medium	Cornwall	549
Shap Blue	Hornfels	Cumbria	496
Medium Brown/Rustic			
Beam	Sandstone, Medium	Devon	650
Blackhill	Gritstone, Medium to coarse	West Yorkshire	376
Chillaton	Slate, Fine	Devon	346
Lean	Basalt, Fine	Cornwall	473
Prince of Wales	Slate, Fine	Cornwall	451
Light Brown/Buff/Fawn/Honey			
Abbey	Limestone, Fine to medium	Nottinghamshire	3
Ann Twyford	Gritstone, Medium to coarse	Derbyshire	468
Apex	Sandstone, Medium	West Yorkshire	306
Backwell	Limestone	Avon	672
Bankfield	Limestone	Lancashire	568
Beer	Chalk, Fine	Devon	58
Big Pits	Limestone, Fine	Lincolnshire	135
Black Pasture	Sandstone, Medium	Northumberland	487
Bognor Common	Limestone, Fine to medium	West Sussex	335
Bolehill	Sandstone, Fine	Derbyshire	449
Bolton Woods	Sandstone, Fine	West Yorkshire	429
Bridge	Sandstone, Fine	Cheshire	341
Brittania	Sandstone, Fine	West Yorkshire	415
Brodsworth	Limestone	South Yorkshire	631
Burford	Limestone, Fossiliferous	Oxfordshire	87
Caldon Low	Limestone	Staffordshire	604
Callywith	Slate, Fine	Cornwall	537
Castle-an-Dinas	Granite, Coarse	Cornwall	421
Chellow Grange	Sandstone, Fine	West Yorkshire	412
Chinley Moor	Gritstone, Fine to medium	Greater Manchester	369
Clearwell	Limestone	Gloucestershire	625
Clipsham	Limestone, Fine	Lincolnshire	135
Colemans	Limestone	Somerset	210
Cromwell	Sandstone, Fine	West Yorkshire	348
Crosland Moor	Gritstone, Fine	West Yorkshire	310
Deep Lane	Sandstone, Fine	West Yorkshire	430
Doulting	Limestone, Fine to coarse	Somerset	83
Drybrook	Limestone, Coarse	Gloucestershire	52
Dukes	Sandstone, Medium	Derbyshire	450
Dunhouse	Sandstone, Medium	Durham	196
Ellel	Gritstone, Medium	Lancashire	304
Fagley	Sandstone, Fine	West Yorkshire	431
Fly Delph	Gritstone, Medium	West Yorkshire	342
Fords	Sandstone, Fine	Lancashire	240
Friendly	Sandstone, Fine	West Yorkshire	70
Gebdykes	Limestone	North Yorkshire	526
Grange Top	Limestone, Fine	Lincolnshire	151
Grinshill	Sandstone, Fine	Shropshire	270
Ham Hill	Limestone, Shelly	Somerset	221
Ham Hill	Limestone, Fine to coarse	Somerset	378
Happylands	Limestone, Fine	Hereford and Worcester	411
Hayes Wood Mine	Limestone, Fine, fossiliferous	Avon	86
Hayne	Sandstone, Medium	Devon	219
Hillhouse Edge	Sandstone, Medium	West Yorkshire	484
Holleywell	Limestone, Shelly	Leicestershire	368
Honley Wood	Gritstone, Fine	West Yorkshire	311
Jamestone	Flagstone, Fine	Lancashire	305
Ketton	Limestone, Fine	Lincolnshire	151

Ladycross	Gritstone, Fine	Northumberland	322
Long Lane	Limestone	West Yorkshire	181
Luxulyan	Granite, Coarse	Cornwall	213
Mansfield	Limestone, Fine	Nottinghamshire	269
Marsden	Limestone, Coarse	Tyne and Wear	639
Naylor Hill Quarries	Sandstone, Fine	West Yorkshire	248
Northowram Hills	Sandstone, Fine	West Yorkshire	232
Palmers	Sandstone, Fine to medium	Derbyshire	514
Park Nook	Limestone, Fine	South Yorkshire	148
Philpots	Sandstone, Medium	West Sussex	428
Pilsamoor	Sandstone, Medium	Cornwall	549
Portland	Limestone, Shelly	Dorset	62
Prudham	Limestone, Fine	Northumberland	128
Rawdon	Gritstone, Fine	West Yorkshire	328
Ribblesdale Works	Limestone, Fine	Lancashire	152
Ridgeway	Sandstone, Coarse	Derbyshire	161
Rogers	Limestone	Gloucestershire	629
Shipley Bank	Sandstone, Fine	Durham	505
Skiddaw	Slate, Fine	Cumbria	146
Soil Hill	Sandstone, Fine	West Yorkshire	266
Southorpe	Limestone, Fine	Lincolnshire	461
Springwell	Sandstone, Fine to medium	Tyne and Wear	398
Squire Hill	Sandstone, Medium	West Yorkshire	404
Stamford	Limestone, Fine to medium	Cambridgeshire	446
Stancliffe	Sandstone, Fine	Derbyshire	452
Stanton Moor	Sandstone, Fine to medium	Derbyshire	453
Station	Limestone	Somerset	76
Stoke Hall	Sandstone, Fine	South Yorkshire	538
Stone	Gritstone, Medium	West Yorkshire	445
Sycamore	Sandstone, Fine	Cheshire	198
Thumpas	Sandstone, Fine	West Yorkshire	166
Tredennick Downs	Slate, Fine	Cornwall	314
Tredinnick	Gritstone, Medium	Cornwall	661
Veizey's	Limestone, Coarse	Gloucestershire	619
Waddington Fell	Gritstone, Fine	Lancashire	665
Watson	Sandstone, Fine	West Yorkshire	97
White Hill	Limestone, Fine	Oxfordshire	57
Whittle Hill	Gritstone, Coarse	Lancashire	676

CREAM

Blaxter	Sandstone, Fine to medium	Northumberland	392
Blue Bank	Sandstone, Coarse	North Yorkshire	69
Bolsover Moor	Limestone, Coarse	Derbyshire	552
Cadeby	Limestone	South Yorkshire	524
Creeton	Limestone, Fine	Lincolnshire	444
Darney	Sandstone, Fine	Northumberland	393
Doddington Hill	Sandstone, Fine to medium	Northumberland	394
(The) Dolomite	Limestone, Coarse	South Yorkshire	432
Fish Hill	Limestone, Fine	Hereford and Worcester	508
Friendly	Sandstone, Fine	West Yorkshire	70
Grange Top	Limestone, Fine	Lincolnshire	151
Guiting	Limestone, Fine	Gloucestershire	53
High Nick	Sandstone, Medium	Northumberland	396
Holme Hall	Limestone	South Yorkshire	555
Houghton	Limestone, Coarse	Tyne and Wear	23
Jackdaw Crag	Limestone	North Yorkshire	180
Ketton	Limestone, Fine	Lincolnshire	151

Kirton Lindsey	Limestone, Fine	Lincolnshire	150
Middle Peak	Limestone	Derbyshire	557
Monks Park Mine	Limestone, Fine	Wiltshire	84
Mount Pleasant	Limestone, Medium	Avon	447
Prudham	Limestone, Fine	Northumberland	128
Scabba Wood	Limestone, Coarse	South Yorkshire	374
Sherburn	Limestone, Fine	North Yorkshire	284
Skelbrooke	Limestone	South Yorkshire	183
Skiddaw	Slate, Fine	Cumbria	146
Squire Hill	Sandstone, Medium	West Yorkshire	404
Swanworth	Limestone, Fine	Dorset	602
Ten Yards Lane	Sandstone, Fine	North Yorkshire	71
Thornaugh	Limestone, Fine	Lincolnshire	427
Upper Lawn	Limestone, Medium	Avon	275
Vale Road	Limestone, Fine	Nottinghamshire	154
Veizey's	Limestone, Coarse	Gloucestershire	619

GREEN

Bardon	Andesite, Fine	Leicestershire	75
Bixhead	Sandstone, Fine to medium	Gloucestershire	241
Bognor Common	Limestone, Fine to medium	West Sussex	335
Bolehill	Sandstone, Fine	Derbyshire	449
Broughton Moor	Slate, Fine	Lancashire	140
Bursting Stone	Slate, Fine	Cumbria	142
Criggion	Dolerite, Medium	Shropshire	13
Elterwater	Slate, Fine	Cumbria	143
Ghyll Scaur	Andesite, Fine	Cumbria	226
Honister	Slate, Fine	Cumbria	357
Leaton	Dolerite, Medium	Shropshire	312
Moss Rigg	Slate, Fine	Cumbria	145
Pets	Slate, Fine	Cumbria	320
Spout Cragg	Slate, Fine	Cumbria	147
Trevassack	Serpentine, Medium to coarse	Cornwall	162
Water Swallows	Basalt, Fine	Derbyshire	579

GREYS

Grey

Aycliffe	Limestone, Coarse	Durham	539
Barnsdale Bar	Limestone	West Yorkshire	191
Bishop Middleham	Limestone, Coarse	Durham	623
Blackaller	Limestone	Devon	190
Bognor Common	Limestone, Fine to medium	West Sussex	335
Bosahan	Granite, Medium	Cornwall	663
Bowyers	Limestone, Fine	Dorset	1
Bridge	Sandstone, Fine	Cheshire	341
Callywith	Slate, Fine	Cornwall	537
Chellow Grange	Sandstone, Fine	West Yorkshire	412
Chywoon	Granite, Coarse	Cornwall	326
Dene	Limestone	Derbyshire	554
Eastgate	Limestone	Durham	102
Farmington	Limestone, Fine	Gloucestershire	231
Gebdykes	Limestone	North Yorkshire	526
Giggleswick	Limestone	North Yorkshire	632
Goddards	Limestone	Derbyshire	667
Hainworth Shaw	Gritstone, Fine	West Yorkshire	72
Hardendale	Limestone	Cumbria	126
Hartshill	Sandstone, Medium	Warwickshire	612

Hopton Wood	Limestone	Derbyshire	565
Houghton	Limestone, Coarse	Tyne and Wear	23
Hutchbank	Sandstone, Medium	Lancashire	228
Keates	Limestone, Fine	Dorset	317
Ladycross	Gritstone, Fine	Northumberland	322
Lean	Basalt, Fine	Cornwall	473
Little	Sandstone, Coarse	Lancashire	67
Marsden	Limestone, Coarse	Tyne and Wear	639
Middle Peak	Limestone	Derbyshire	557
Moorcroft	Limestone	Devon	214
Northowram Hills	Sandstone, Fine	West Yorkshire	232
Nuneaton	Sandstone, Medium	Warwickshire	612
Sandside	Limestone	Cumbria	576
Sovereign	Gritstone, Coarse	West Yorkshire	334
Swinden	Limestone	North Yorkshire	634
Sycamore	Sandstone, Fine	Cheshire	198
Trebarwith	Slate, Fine	Cornwall	660
Trevone	Granite, Medium	Cornwall	664
Tunstead	Limestone	Derbyshire	301
Water Swallows	Basalt, Fine	Derbyshire	579
Whitwick	Andesite, Fine	Leicestershire	17
Whitworth	Sandstone, Fine	Lancashire	229
Dark Grey			
Barrasford	Dolerite, Medium	Northumberland	551
Barton Wood	Sandstone, Medium	Devon	189
Belford	Dolerite, Fine	Northumberland	636
Black	Limestone, Medium	North Yorkshire	277
Bolehill	Sandstone, Fine	Derbyshire	449
Callow Hill	Dolerite, Medium	Shropshire	605
Dry Rigg	Sandstone, Medium	North Yorkshire	458
Edwin Richards	Basalt, Fine	West Midlands	14
Force Garth	Dolerite, Medium	Durham	278
Goldmire	Limestone, Medium	Cumbria	22
Hailstone	Dolerite, Medium	West Midlands	610
High Force	Limestone	Durham	280
Huntsmans	Limestone, Fine	Gloucestershire	299
Kilmond Wood	Limestone	Durham	281
Knowle	Sandstone, Medium	Devon	6
Leaton	Dolerite, Medium	Shropshire	312
Linhay Hill	Limestone	Devon	254
Parkhead	Limestone	Cumbria	28
Round 'O'	Sandstone, Medium	Lancashire	30
Skipton Rock	Limestone	North Yorkshire	640
Swinburne	Dolerite, Coarse	Northumberland	34
Venn	Sandstone, Medium	Devon	216
Whitecleaves	Basalt, Fine	Devon	64
Medium Grey			
Barton Wood	Sandstone, Medium	Devon	189
Bayston Hill	Greywacke, Angular	Shropshire	603
Blodwell	Basalt, Fine	Shropshire	10
Bold Venture	Limestone	Lancashire	569
Callow Rock	Limestone	Somerset	209
Coldstones	Limestone	North Yorkshire	434
Cragmill	Dolerite, Coarse	Northumberland	366
Croft	Diorite, Coarse	Leicestershire	204
Divethill	Dolerite, Coarse	Northumberland	367
Dunald Mill	Limestone	Lancashire	571

Fenacre	Limestone	Devon	609
Griff	Diorite, Coarse	Warwickshire	437
Hailstone	Dolerite, Medium	West Midlands	610
Knowle	Sandstone, Medium	Devon	6
Mootlaw	Limestone	Northumberland	558
Plymstock	Limestone	Devon	110
Silverdale	Limestone	Lancashire	32
Threshfield	Limestone	North Yorkshire	562
Venn	Sandstone, Medium	Devon	216
Westbury	Limestone	Somerset	442
Westbury	Chalk, Fine	Wiltshire	113
Westleigh	Limestone	Devon	217
Light Grey			
Antron Hill	Granite, Fine	Cornwall	8
Back Lane	Limestone	Lancashire	203
Backwell	Limestone	Avon	672
Batts Combe	Limestone	Somerset	48
Bearrah Tor	Granite, Coarse	Cornwall	443
Brandy Crag	Slate, Fine	Cumbria	138
Burnthouse	Granite, Fine	Cornwall	9
Callow Rock	Limestone	Somerset	209
Carnsew	Granite, Coarse	Cornwall	420
Castle-an-Dinas	Granite, Coarse	Cornwall	421
Coldstones	Limestone	North Yorkshire	434
Colemans	Limestone	Somerset	210
Cowdor and Hall Dale	Limestone	Derbyshire	553
De Lank	Granite, Fine to medium	Cornwall	469
Dove Holes	Limestone	Derbyshire	471
Eldon Hill	Limestone	Derbyshire	399
Griff	Diorite, Coarse	Warwickshire	437
Hantergantick	Granite, Fine	Cornwall	395
Hartley	Limestone	Cumbria	279
Hindlow	Limestone	Derbyshire	611
Hingston Down	Granite, Fine	Cornwall	60
Holme Park	Limestone	Lancashire	205
Kendal Fell	Limestone	Cumbria	574
Mancetter	Dolerite, Medium	Warwickshire	628
Meldon	Schist, Fine	Devon	125
Merehead	Limestone, Fine to medium	Somerset	704
Merivale	Granite, Coarse	Devon	397
Once-a-Week	Limestone, Fine	Derbyshire	345
Penlee	Hornfels, Medium to fine	Cornwall	424
Penryn	Granite, Coarse	Cornwall	420
Pickerings	Limestone	Cumbria	172
Rogers	Limestone	Gloucestershire	629
Rooks	Limestone	Cumbria	173
Scratchmill Scar	Sandstone, Fine	Cumbria	175
Shap Pink	Granite	Cumbria	497
Spaunton	Limestone, Fine	North Yorkshire	283
Sutton	Limestone	North Yorkshire	548
Tendley	Limestone	Cumbria	577
Tor Down	Granite, Coarse	Cornwall	388
Town	Limestone, Fine	Oxfordshire	176
Trenoweth	Granite, Fine	Cornwall	662
Tunstead	Limestone	Derbyshire	301
Wick	Limestone	Avon	674
Wickwar	Limestone	Gloucestershire	675

Yalberton Tor	Limestone	Devon	65

GREY HUES

Bluish

Broad Oak	Granodiorite, Fine	Cumbria	170
Burlington	Slate, Fine	Cumbria	141
Creeton	Limestone, Fine	Lincolnshire	444
Dean	Gabbro, Coarse	Cornwall	457
Delabole	Slate, Fine	Cornwall	474
Downs	Limestone, Fine	Dorset	338
Fly Flatts	Sandstone, Medium	West Yorkshire	129
Horton	Limestone	North Yorkshire	573
Hunters Hill	Sandstone, Medium	West Yorkshire	130
Judkins	Sandstone, Coarse	Warwickshire	16
Kirkby	Slate, Fine	Cumbria	144
Landers	Limestone, Fine	Dorset	325
Lime Kiln Hill	Limestone	Somerset	332
Moota	Limestone	Cumbria	118
Salterwath	Limestone	Cumbria	174
Shavers End	Limestone	Hereford and Worcester	208
Shining Bank	Limestone	Derbyshire	668
Swanage	Limestone, Shelly	Dorset	547

Brownish

Blencowe	Limestone, Fine	Cumbria	223
Chipping Sodbury	Limestone	Avon	49
Cromhall	Limestone, Coarse	Gloucestershire	50
Forcett	Limestone	North Yorkshire	637
Moons Hill	Andesite, Fine	Avon	666
Spring Lodge	Limestone, Coarse	South Yorkshire	633
Tytherington	Limestone, Medium	Gloucestershire	55
Whatley	Limestone, Medium	Somerset	56

Buffish

Barton	Limestone	North Yorkshire	499
Downs	Limestone, Fine	Dorset	338
Landers	Limestone, Fine	Dorset	325
Stancombe	Limestone	Avon	695

Creamy

Ballidon	Limestone	Derbyshire	624
Cauldon	Limestone	Staffordshire	100
Collyweston	Limestone, Fine	Lincolnshire	136
Cool Scar	Limestone	North Yorkshire	224
Darlton	Limestone	South Yorkshire	685
Kevin	Limestone	Staffordshire	627
Leapers Wood	Limestone	Lancashire	690
Newbridge	Limestone, Fine	North Yorkshire	282
Wardlow	Limestone	Staffordshire	462

Greenish

Allington	Ragstone, Fine	Kent	47
Appley Bridge	Sandstone, Medium	Lancashire	21
East	Sandstone, Medium	Lancashire	21
Greystone	Dolerite, Medium	Cornwall	212
Ingleton	Sandstone, Medium	Lancashire	24
New England	Dolerite, Medium	Devon	215
Offham	Ragstone	Kent	54
Trusham	Dolerite, Medium	Devon	63

Pinkish

Daglingworth	Limestone, Fine to medium	Gloucestershire	51

Eskett	Limestone, Medium	Cumbria	225
Shap	Granite, Coarse	Cumbria	31
Shap Beck	Granite, Coarse	Cumbria	31
Stoneycombe	Limestone	Devon	211
Reddish			
Breedon Hill	Limestone	Leicestershire	121
Stowfield	Limestone, Coarse	Gloucestershire	615

PINK

Ann Twyford	Gritstone, Medium to coarse	Derbyshire	468
Bardon	Andesite, Fine	Leicestershire	75
Buddon Wood	Granite, Coarse	Leicestershire	456
Croft	Diorite, Coarse	Leicestershire	204
Doddington Hill	Sandstone, Fine to medium	Northumberland	394
Lazonby Fell	Sandstone, Fine to medium	Cumbria	171
Mountsorrel	Granite, Coarse	Leicestershire	456
Palmers	Sandstone, Fine to medium	Derbyshire	514
Quartzite	Sandstone, Medium	Gloucestershire	673
Ridgeway	Sandstone, Coarse	Derbyshire	161
Shap Blue	Hornfels	Cumbria	496
Shap Pink	Granite	Cumbria	497
Stainton	Limestone	Cumbria	641
Stanton Moor	Sandstone, Fine to medium	Derbyshire	453
Stoneraise	Sandstone, Fine to medium	Cumbria	454
Waddington Fell	Gritstone, Fine	Lancashire	665
Wimberry Moss	Gritstone, Fine	Cheshire	199

RED

Abbey	Limestone, Fine to medium	Nottinghamshire	3
Bankend	Sandstone, Fine	Cumbria	168
Birkhams	Sandstone, Fine	Cumbria	169
Blackmoor	Sandstone, Fine	South Yorkshire	285
Great Gate	Sandstone, Fine to medium	Staffordshire	512
Huncote	Sandstone, Medium	Leicestershire	5
Lazonby Fell	Sandstone, Fine to medium	Cumbria	171
Linhay Hill	Limestone	Devon	254
Red Hole	Sandstone, Fine to medium	Staffordshire	512
Shawk	Gritstone, Fine	Cumbria	323
Tredennick Downs	Slate, Fine	Cornwall	314
Tredinnick	Gritstone, Medium	Cornwall	661
Trevassack	Serpentine, Medium to coarse	Cornwall	162
Triscombe	Sandstone, Medium	Somerset	616
Webscott	Sandstone, Fine	Shropshire	271
Wilderness	Sandstone, Fine	Gloucestershire	448

WHITE

Admiralty	Limestone, Fine	Dorset	2
Ancaster	Limestone, Fine	Lincolnshire	267
Barrington	Chalk, Fine	Cambridgeshire	476
Beer	Chalk, Fine	Devon	58
Cat Castle	Sandstone, Coarse	Durham	155
Charnage	Chalk, Fine	Wiltshire	59
Charnwood	Diorite, Coarse	Leicestershire	11
Chinnor	Chalk, Fine	Oxfordshire	477
Copp Grag	Limestone	Northumberland	127
Darney	Sandstone, Fine	Northumberland	393

East Grinstead	Chalk, Fine	Wiltshire	200
Farmington	Limestone, Fine	Gloucestershire	231
Flaunden	Chalk, Fine	Hertfordshire	159
Glebe	Limestone, Fine	Lincolnshire	268
Grange Mill	Limestone, Medium	Derbyshire	89
Great Blakenham	Chalk, Fine	Suffolk	103
Greys Pit	Chalk, Fine	Kent	478
Groby	Diorite, Coarse	Leicestershire	15
Highmoor	Limestone, Coarse	North Yorkshire	550
Holborough	Chalk, Fine	Kent	104
Holme Hall	Limestone	South Yorkshire	555
Horn Park	Limestone, Fine	Dorset	247
Hovingham	Limestone, Fine	North Yorkshire	66
Independent	Limestone, Fine	Dorset	2
Ingham	Limestone, Fine	Suffolk	7
Keates	Limestone, Fine	Dorset	317
Kensworth	Chalk, Fine	Bedfordshire	479
Long Bredy	Chalk, Fine	Dorset	61
Masons	Chalk, Fine	Suffolk	106
Melton	Chalk	Humberside	107
Melton Ross	Chalk, Fine	Lincolnshire	506
Middlegate	Chalk, Fine	Humberside	480
Norman	Chalk, Fine	Cambridgeshire	108
Northfleet	Chalk, Fine	Kent	109
Queensgate	Chalk, Fine	Humberside	201
Quidhampton	Chalk, Fine	Wiltshire	202
Rock Cottage	Limestone, Medium	North Yorkshire	377
Scabba Wood	Limestone, Coarse	South Yorkshire	374
Shoreham	Chalk, Fine	West Sussex	112
Shotley	Limestone, Medium	Northamptonshire	94
Skelbrooke	Limestone	South Yorkshire	183
Southard	Limestone, Fine	Dorset	287
St Aldhelm's	Limestone, Fine	Dorset	286
Stowey	Limestone, Fine	Avon	541
Trevassack	Serpentine, Medium to coarse	Cornwall	162
Wath	Limestone	North Yorkshire	532
Westwood Ground	Limestone, Medium	Wiltshire	85
Woodeaton	Limestone, Fine, fossiliferous	Oxfordshire	78
Woodside	Gritstone, Fine	West Yorkshire	122

YELLOW/YELLOW-ORANGE

Beacon Lodge	Sandstone, Fine	West Yorkshire	185
Brockhill	Limestone, Fine	Gloucestershire	164
Cotswold Hill	Limestone, Fine	Gloucestershire	165
Dukes	Sandstone, Medium	Derbyshire	450
Fish Hill	Limestone, Fine	Hereford and Worcester	508
Happylands	Limestone, Fine	Hereford and Worcester	411
Huntsmans	Limestone, Fine	Gloucestershire	299
Trevassack	Serpentine, Medium to coarse	Cornwall	162
Waddington Fell	Gritstone, Fine	Lancashire	665
Woodside	Gritstone, Fine	West Yorkshire	122

MOTTLED/SPECKLED COLOURS

MOTTLED
Grey

Cliffe Hill	Diorite, Coarse	Leicestershire	564

De Lank	Granite, Fine to medium	Cornwall	469
Lea	Limestone	Shropshire	206
Old Cliffe Hill	Diorite, Coarse	Leicestershire	566

Pink

Old Cliffe Hill	Diorite, Coarse	Leicestershire	566
Tearne	Sandstone, Fine to medium	Staffordshire	407

Red

Tearne	Sandstone, Fine to medium	Staffordshire	407

SPECKLED
Black

Charnwood	Diorite, Coarse	Leicestershire	11
Groby	Diorite, Coarse	Leicestershire	15

Brown

Hillhouse Edge	Sandstone, Medium	West Yorkshire	484

Grey

Broad Oak	Granodiorite, Fine	Cumbria	170
Hingston Down	Granite, Fine	Cornwall	60
Meldon	Schist, Fine	Devon	125
Penlee	Hornfels, Medium to fine	Cornwall	424

Pink

Buddon Wood	Granite, Coarse	Leicestershire	456
Mountsorrel	Granite, Coarse	Leicestershire	456

White

Charnwood	Diorite, Coarse	Leicestershire	11
Groby	Diorite, Coarse	Leicestershire	15

Scotland

BLACK

Airdriehill	Basalt, Fine	Strathclyde	361
Ardhasaig	Basalt, Fine	Western Isles	340
Ardownie	Dolerite, Medium	Tayside	246
Askernish	Basalt, Fine	Western Isles	68
Balmullo	Andesite, Fine	Fife	372
Barlockhart	Greywacke, Fine	Dumfries and Galloway	79
Belliston	Dolerite, Medium	Fife	88
Blairhill	Basalt, Fine	Lothian	580
Boysack	Basalt, Fine	Tayside	617
Broadlaw	Basalt, Fine	Lothian	363
Cairds Hill	Basalt, Medium	Grampian	383
Collace	Dolerite, Fine	Tayside	618
Craigfoot	Dolerite, Medium	Fife	649
Craigie Hill	Basalt, Fine	Strathclyde	655
Craigpark	Basalt, Fine	Lothian	586
Duntilland	Basalt, Fine	Strathclyde	543
Edstone	Greywacke, Fine	Borders	116
Ethiebeaton	Dolerite, Medium	Tayside	591
Goat	Dolerite, Medium	Fife	592
Hallyards	Basalt, Fine	Strathclyde	327
North Mains	Gabbro, Coarse	Grampian	186
Ravelrig	Basalt, Fine	Lothian	600
Steangabhal	Basalt, Fine	Western Isles	309
Tongland	Greywacke, Fine	Strathclyde	81
Underheugh	Basalt, Fine	Strathclyde	364
Vatseter	Basalt, Fine	Shetland	244

BLUE/LILAC/MAUVE/PURPLE

Dunion	Dolerite, Medium	Borders	590
Middleton Mine	Limestone	Lothian	464
North Mains	Gabbro, Coarse	Grampian	186

BROWNS

Brown

Miltonhill	Dolerite, Medium	Strathclyde	621

Light Brown/Buff/Fawn/Honey

Clashach	Sandstone, Fine	Grampian	379
Greenbrey	Sandstone, Fine	Grampian	274
Newbiggin	Sandstone, Medium	Fife	491

CREAM

Dunmore	Sandstone, Fine	Central	488
Newbiggin	Sandstone, Medium	Fife	491
Spynie	Sandstone, Fine	Grampian	380

GREEN

Borthwick	Dolerite, Medium	Borders	678

GREYS

Grey

Bangley	Trachyte, Fine	Lothian	362
Barlockhart	Greywacke, Fine	Dumfries and Galloway	79
Blairhill	Basalt, Fine	Lothian	580
Craighouse	Basalt, Fine	Borders	584
Edstone	Greywacke, Fine	Borders	116
Lic	Gneiss, Medium to coarse	Western Isles	331
Morefield	Limestone	Highland	597
Scord	Schist, Fine to medium	Shetland	504
Setters	Schist, Fine to medium	Shetland	482
Tongland	Greywacke, Fine	Strathclyde	81

Dark Grey

Balmedie	Gabbro, Coarse	Grampian	257
Bannerbank	Basalt, Fine	Strathclyde	643
Borthwick	Dolerite, Medium	Borders	678
Cairneyhill	Dolerite, Medium	Strathclyde	644
Croy	Basalt, Fine	Strathclyde	588
Cruicks	Dolerite, Medium	Fife	645
Dumbuckhill	Basalt, Fine	Strathclyde	646
Friarton	Basalt, Fine	Tayside	680
High Craig	Basalt, Fine	Strathclyde	593
Hillend	Basalt, Fine	Strathclyde	485
Hillhouse	Dolerite, Medium	Strathclyde	292
Hillwood	Dolerite, Medium	Lothian	681
Kaimes	Dolerite, Medium	Lothian	594
Kilbarchan	Basalt, Fine	Strathclyde	486
Loanhead	Basalt, Fine	Strathclyde	596
Northfield	Dolerite, Medium	Central	156
North Lasts	Gabbro, Coarse	Grampian	329
North Mains	Gabbro, Coarse	Grampian	186
Orrock	Basalt, Fine	Fife	682
Pitcaple	Gabbro, Coarse	Grampian	260
Riganagower	Basalt, Medium	Strathclyde	622
Riskend	Dolerite, Fine	Strathclyde	467

Spittal	Sandstone, Fine	Highland	546
Tam's Loup	Dolerite, Medium	Strathclyde	441
Medium Grey			
Boards	Dolerite, Medium	Central	677
Murrayshall	Basalt, Fine	Central	599
Light Grey			
Banavie	Granite, Coarse	Highland	642
Kemnay	Granite, Coarse	Grampian	656
Kirkmabreck	Granodiorite, Fine	Dumfries and Galloway	595

GREY HUES

Bluish			
Cunmont	Dolerite, Coarse	Tayside	589
Tyrebagger	Gabbro, Coarse	Grampian	659
Brownish			
Waulkmill	Andesite, Fine	Tayside	648
Pinkish			
Bonawe	Granite, Coarse	Strathclyde	243
Daviot	Schist	Highland	679

PINK

Black Hills	Granite, Coarse	Grampian	133
Lic	Gneiss, Medium to coarse	Western Isles	331
Locharbriggs	Sandstone, Medium	Dumfries and Galloway	73
Longside	Granite, Coarse	Grampian	134
Ross of Mull	Granite, Coarse	Strathclyde	492

RED

Pettinain	Granite, Fine	Strathclyde	160
Ross of Mull	Granite, Coarse	Strathclyde	492

WHITE

Newbiggin	Sandstone, Medium	Fife	491

YELLOW/YELLOW-ORANGE

Clashach	Sandstone, Fine	Grampian	379

MOTTLED/SPECKLED COLOURS

SPECKLED

Grey			
Banavie	Granite, Coarse	Highland	642

Wales

BLACK

Bolton Hill	Diorite, Coarse	Dyfed	250
Carrigwynion	Greywacke, Angular	Powys	613
Cerrigyrwyn	Dolerite, Medium	Dyfed	251
Coygen	Dolerite, Medium	Dyfed	252
Rhyader	Greywacke, Angular	Powys	613

BLUE/LILAC/MAUVE/PURPLE

Clogau	Slate, Fine	Clwyd	95
Gelligaer	Sandstone, Medium	Mid Glamorgan	436
Glogue	Slate, Fine	Dyfed	188

Hafod	Sandstone, Fine	Gwent	460
Nantlle	Slate, Fine	Gwynedd	237
Pencaemawr	Sandstone, Medium	Mid Glamorgan	436
Penrhyn	Slate, Fine	Gwynedd	365
Syke	Sandstone, Medium	Dyfed	440

BROWNS
Brown

Callow	Sandstone, Fine	Gwent	184
Nanhoron	Granite, Fine	Gwynedd	387
Twll Llwyd	Slate, Fine	Gwynedd	297

Dark Brown

Gelligaer	Sandstone, Medium	Mid Glamorgan	436
Pencaemawr	Sandstone, Medium	Mid Glamorgan	436
Syke	Sandstone, Medium	Dyfed	440

Medium Brown/Rustic

Gelligaer	Sandstone, Medium	Mid Glamorgan	436
Pencaemawr	Sandstone, Medium	Mid Glamorgan	436

Light Brown/Buff/Fawn/Honey

Abergele	Limestone	Clwyd	630
Rhyndaston	Rhyolite, Fine to very fine	Dyfed	381

GREEN

Gilfach	Sandstone, Medium to coarse	West Glamorgan	701
Gore	Sandstone, Medium to coarse	Powys	626
Minffordd	Basalt, Fine	Gwynedd	691
Penrhyn	Slate, Fine	Gwynedd	365
Vronlog	Slate, Fine	Gwynedd	315

GREYS
Grey

Abergele	Limestone	Clwyd	630
Aberthaw	Limestone, Fine	South Glamorgan	99
Blaenyfan	Limestone, Medium	Dyfed	699
Carrigwynion	Greywacke, Angular	Powys	613
Llanddulas	Limestone, Medium	Clwyd	26
Pantyffynnon	Limestone, Medium	South Glamorgan	291
Penmaenmawr	Diorite, Medium	Gwynedd	29
Rhyader	Greywacke, Angular	Powys	613

Dark Grey

Dolyhir and Strinds	Limestone, Fine	Powys	390
Gelligaer	Sandstone, Medium	Mid Glamorgan	436
Gellihalog	Limestone, Medium	Dyfed	700
Pencaemawr	Sandstone, Medium	Mid Glamorgan	436
Penmaenmawr	Diorite, Medium	Gwynedd	29
Syke	Sandstone, Medium	Dyfed	440
Trefil	Limestone, Medium	Gwent	272

Medium Grey

Craig-yr-Hesg	Sandstone, Medium	Mid Glamorgan	37
Graig	Limestone	Clwyd	459
Ruthin	Limestone, Medium	Mid Glamorgan	111

Light Grey

Aberstrecht	Limestone, Fine	Gwynedd	4
Caer Glaw	Granite, Coarse	Gwynedd	19
Craig-yr-Hesg	Sandstone, Medium	Mid Glamorgan	37
Cwmleyshon	Limestone, Coarse	Mid Glamorgan	38
Dolyhir and Strinds	Limestone, Fine	Powys	390

Gilfach	Sandstone, Medium to coarse	West Glamorgan	701
Gore	Sandstone, Medium to coarse	Powys	626
Gwyndy	Granite, Coarse	Gwynedd	273
Ifton	Limestone, Coarse	Gwent	39
Livox	Limestone, Medium to coarse	Gwent	41
Penhow	Limestone, Medium	Gwent	44
Rhuddlan Bach	Limestone, Fine	Gwynedd	466
Trimm Rock	Limestone	Clwyd	635
Vaynor	Limestone, Medium	Mid Glamorgan	46
Wenvoe	Limestone, Medium	South Glamorgan	702

GREY HUES
Bluish

Aberllefeni	Slate, Fine	Powys	697
Braich Ddu	Slate, Fine	Gwynedd	262
Cwt-y-Bugail	Slate, Fine	Gwynedd	235
Diphwys Casson	Slate, Fine	Gwynedd	263
Gloddfa Ganol	Slate, Fine	Gwynedd	236
Llechedd Slate Mine	Slate, Fine	Gwynedd	264
Maenofferen	Slate, Fine	Gwynedd	265
Oakeley	Slate, Fine	Gwynedd	238

Brownish

Cornelly	Limestone, Medium	Mid Glamorgan	684
Dayhouse	Limestone, Medium	Gwent	620
Denbigh	Limestone	Clwyd	570
Forest Wood	Limestone, Medium	Mid Glamorgan	435
Grove	Limestone, Medium	Mid Glamorgan	438
Hendy	Limestone, Medium	Mid Glamorgan	689
Lithalun	Limestone	Mid Glamorgan	40
Machen	Limestone, Coarse	Gwent	42
Minera	Limestone	Clwyd	575
Pant	Limestone, Medium	Mid Glamorgan	692
Penderyn	Limestone, Medium	Mid Glamorgan	43
Risca	Limestone, Medium to coarse	Gwent	45

Creamy

Aberduna	Limestone	Clwyd	433
Halkyn	Limestone, Fine	Clwyd	400
Pant (Halkyn)	Limestone, Fine	Clwyd	693
Pant-y-Pwll Dwr	Limestone, Fine	Clwyd	400

Greenish

Builth Wells	Dolerite, Coarse	Powys	36
Twll Llwyd	Slate, Fine	Gwynedd	297

Pinkish

Blaengwynlais	Limestone, Medium	Mid Glamorgan	683
Cefn Gawr	Limestone, Coarse	South Glamorgan	606
Creigiau	Limestone, Coarse	Mid Glamorgan	607
Plas Gwilym	Limestone	Clwyd	296
Taffs Well	Limestone	South Glamorgan	522

PINK

Raynes	Limestone, Fine	Clwyd	401

RED

Gore	Sandstone, Medium to coarse	Powys	626
Twll Llwyd	Slate, Fine	Gwynedd	297

WHITE

Bolton Hill	Diorite, Coarse	Dyfed	250

| Caer Glaw | Granite, Coarse | Gwynedd | 19 |
| Llanddulas | Limestone, Medium | Clwyd | 26 |

MOTTLED/SPECKLED COLOURS

MOTTLED
Grey

| Dolyhir and Strinds | Limestone, Fine | Powys | 390 |

SPECKLED
Black

| Bolton Hill | Diorite, Coarse | Dyfed | 250 |

White

| Bolton Hill | Diorite, Coarse | Dyfed | 250 |

Index 8

Quarry products of Britain
by quarry name and location

Entries in this index refer to the
unique quarry numbers (1 – 705)
in the main Directory

MELENCO

MANUFACTURER OF THE

POWER FORK

MULTI-PURPOSE GRAB

- Relentlessly prises out stumps, trees, rocks, boulders and concrete foundations . . .
- Riddles out earth and dirt for easy burning of roots, undergrowth etc.
- Grades rocks, stones and rubble . . .
- Carries massive debris
- Picks up, loads and stacks timber, pipes, girders, railway lines and scrap.

- *Two moving parts only*
- *Maintenance virtually nil*
- *No vulnerable hoses*
- *Fits most 360° or 180° machines*
- *Replaces bucket in 15 minutes*

MELENCO ENGINEERING LIMITED
50-52 Balena Close, Creekmoor Industrial Estate, Poole, Dorset BH17 7DY
Tel: 0202 604695 Fax: 0202 605280

AGRICULTURAL LIME

Abergele	Clwyd	630
Afonwen	Cheshire	511
Back Lane	Lancashire	203
Balladoole	Isle of Man	96
Bankfield	Lancashire	568
Batts Combe	Somerset	48
Big Pits	Lincolnshire	135
Bishop Middleham	Durham	623
Blaenyfan	Dyfed	699
Boyne Bay	Grampian	120
Brierlow	Derbyshire	470
Cadeby	South Yorkshire	524
Caldon Low	Staffordshire	604
Calliburn	Strathclyde	319
Callow Rock	Somerset	209
Cauldon	Staffordshire	100
Chipping Sodbury	Avon	49
Clearwell	Gloucestershire	625
Clipsham	Lincolnshire	135
Cool Scar	North Yorkshire	224
Cromhall	Gloucestershire	50
Dene	Derbyshire	554
The Dolomite	South Yorkshire	432
Dove Holes	Derbyshire	471
Dowlow	Derbyshire	525
Drybrook	Gloucestershire	52
Eldon Hill	Derbyshire	399
Eskett	Cumbria	225
Flaunden	Hertfordshire	159
Gebdykes	North Yorkshire	526
Grange Mill	Derbyshire	89
Great Ponton	Lincolnshire	516
Halkyn	Clwyd	400
Hart	Cleveland	501
Hartley	Cumbria	279
Hillhead	Derbyshire	572
Hindlow	Derbyshire	611
Holme Hall	South Yorkshire	555
Holme Park	Lancashire	205
Horton	North Yorkshire	573
Houghton	Tyne and Wear	23
Kevin	Staffordshire	627
Lea	Shropshire	206
Leapers Wood	Lancashire	690
Linhay Hill	Devon	254
Livox	Gwent	41
Machen	Gwent	42
Marsden	Tyne and Wear	639
Melton Ross	Lincolnshire	506
Merehead	Somerset	704
Middle Peak	Derbyshire	557
Middleton Mine	Lothian	464
Minera	Clwyd	575
Moorcroft	Devon	214
Mootlaw	Northumberland	558
Morefield	Highland	597

Mount Pleasant	Norfolk	158
Old Quarrington	Durham	559
Pant-y-Pwll Dwr	Clwyd	400
Parish	Derbyshire	358
Peckfield	West Yorkshire	406
Raisby	Durham	560
Sandside	Cumbria	576
Shadforth	Durham	502
Shierglas	Tayside	601
Skelpie Mine	Fife	167
Spikers Hill	North Yorkshire	352
Stainton	Cumbria	641
Stoneycombe	Devon	211
Swanworth	Dorset	602
Swinden	North Yorkshire	634
Taffs Well	South Glamorgan	522
Tendley	Cumbria	577
Thrislington	Durham	531
Topley Pike	Derbyshire	578
Torlundy	Highland	289
Torrin	Highland	330
Trearne	Strathclyde	465
Tunstead	Derbyshire	301
Turkeyland	Isle of Man	321
Tytherington	Gloucestershire	55
Vaynor	Mid Glamorgan	46
Veizey's	Gloucestershire	619
Wardlow	Staffordshire	462
Wath	North Yorkshire	532
Whatley	Somerset	56
Whitwell	Nottinghamshire	533
Witch Hill	Durham	503

ASPHALT OR TARMACADAM

Aberduna	Clwyd	433
Allington	Kent	47
Alltgoch	Dyfed	249
Arcow	North Yorkshire	567
Ardhasaig	Western Isles	340
Ballidon	Derbyshire	624
Banavie	Highland	642
Bangley	Lothian	362
Bankfield	Lancashire	568
Bannerbank	Strathclyde	643
Bardon	Leicestershire	75
Barrasford	Northumberland	551
Batts Combe	Somerset	48
Bayston Hill	Shropshire	603
Black	North Yorkshire	277
Blackaller	Devon	190
Blaenyfan	Dyfed	699
Blairhill	Lothian	580
Blodwell	Shropshire	10
Boards	Central	677
Bold Venture	Lancashire	569
Bolton Hill	Dyfed	250
Bonawe	Strathclyde	243

Borrowston	Highland	344
Borthwick	Borders	678
Bray Valley	Devon	402
Buddon Wood	Leicestershire	456
Builth Wells	Powys	36
Cairneyhill	Strathclyde	644
Caldon Low	Staffordshire	604
Callow Hill	Shropshire	605
Carnsew	Cornwall	420
Cauldon	Staffordshire	100
Charnwood	Leicestershire	11
Chipping Sodbury	Avon	49
Clatchard Craig	Fife	581
Clee Hill	Shropshire	12
Cliffe Hill	Leicestershire	564
Coatsgate	Dumfries and Galloway	582
Coldstones	North Yorkshire	434
Coygen	Dyfed	252
Cragmill	Northumberland	366
Craig-yr-Hesg	Mid Glamorgan	37
Craigenlow	Grampian	583
Craighouse	Borders	584
Craignair	Dumfries and Galloway	585
Craigpark	Lothian	586
Criggion	Shropshire	13
Croft	Leicestershire	204
Cromhall	Gloucestershire	50
Croy	Strathclyde	588
Cruicks	Fife	645
Darlton	South Yorkshire	685
Daviot	Highland	679
Dean	Cornwall	457
Denbigh	Clwyd	570
Dene	Derbyshire	554
The Dolomite	South Yorkshire	432
Dolyhir and Strinds	Powys	390
Dove Holes	Derbyshire	471
Dry Rigg	North Yorkshire	458
Dunald Mill	Lancashire	571
Duntilland	Strathclyde	543
Edwin Richards	West Midlands	14
Eldon Hill	Derbyshire	399
Eskett	Cumbria	225
Fenacre	Devon	609
Flax Bourton	Avon	686
Force Garth	Durham	278
Forcett	North Yorkshire	637
Forest Wood	Mid Glamorgan	435
Friarton	Tayside	680
Furnace	Strathclyde	647
Gilfach	West Glamorgan	701
Goldmire	Cumbria	22
Gore	Powys	626
Graig	Clwyd	459
Greystone	Cornwall	212
Griff	Warwickshire	437
Groby	Leicestershire	15

Halecombe	Somerset	687
Halkyn	Clwyd	400
Harnhill	Avon	688
Hartshill	Warwickshire	612
Heights	Durham	230
Hendre	Clwyd	356
High Craig	Strathclyde	593
High Force	Durham	280
Hillhouse	Strathclyde	292
Hillwood	Lothian	681
Hingston Down	Cornwall	60
Holme Hall	South Yorkshire	555
Howick	Northumberland	556
Huntsmans	Gloucestershire	299
Ivonbrook	Derbyshire	405
Judkins	Warwickshire	16
Kevin	Staffordshire	627
Kilbarchan	Strathclyde	486
Lean	Cornwall	473
Leapers Wood	Lancashire	690
Leaton	Shropshire	312
Linhay Hill	Devon	254
Livox	Gwent	41
Luxulyan	Cornwall	213
Machen	Gwent	42
Mancetter	Warwickshire	628
Marsden	Tyne and Wear	639
Merehead	Somerset	704
Middle Peak	Derbyshire	557
Minera	Clwyd	575
Minffordd	Gwynedd	691
Moons Hill	Avon	666
Moorcroft	Devon	214
Moota	Cumbria	118
Morefield	Highland	597
Morrinton	Dumfries and Galloway	598
Mountsorrel	Leicestershire	456
Murrayshall	Central	599
Newbridge	North Yorkshire	282
Northfield	Central	156
Nuneaton	Warwickshire	612
Old Cliffe Hill	Leicestershire	566
Orrock	Fife	682
Pant	Mid Glamorgan	692
Pant (Halkyn)	Clwyd	693
Pant-y-Pwll Dwr	Clwyd	400
Parkmore	Grampian	177
Penlee	Cornwall	424
Penmaenmawr	Gwynedd	29
Penryn	Cornwall	420
Raisby	Durham	560
Sandside	Cumbria	576
Sconser	Highland	290
Shap	Cumbria	31
Shap Beck	Cumbria	31
Shierglas	Tayside	601
Shining Bank	Derbyshire	668

Silverdale	Lancashire	32
Spaunton	North Yorkshire	283
Spikers Hill	North Yorkshire	352
Stoneycombe	Devon	211
Stowfield	Gloucestershire	615
Swanworth	Dorset	602
Swinburne	Northumberland	34
Tam's Loup	Strathclyde	441
Tan-y-Foel	Powys	119
Tendley	Cumbria	577
Tom's Forest	Grampian	658
Tonfanau	Gwynedd	391
Topley Pike	Derbyshire	578
Torcoed	Dyfed	696
Torcoedfawr	Dyfed	359
Triscombe	Somerset	616
Trusham	Devon	63
Tytherington	Gloucestershire	55
Vale Road	Nottinghamshire	154
Vaynor	Mid Glamorgan	46
Venn	Devon	216
Wardlow	Staffordshire	462
Water Swallows	Derbyshire	579
Waulkmill	Tayside	648
Wenvoe	South Glamorgan	702
Westbury	Somerset	442
Westleigh	Devon	217
Whatley	Somerset	56
Whitecleaves	Devon	64
Whitwick	Leicestershire	17
Wick	Avon	674
Wickwar	Gloucestershire	675
Wredon	Staffordshire	360

BURNT LIMESTONE

Abergele	Clwyd	630
Batts Combe	Somerset	48
Dove Holes	Derbyshire	471
Hartley	Cumbria	279
Hindlow	Derbyshire	611
Horton	North Yorkshire	573
Melton	Humberside	107
Melton Ross	Lincolnshire	506
Swinden	North Yorkshire	634
Tunstead	Derbyshire	301

CRUSHED STONE

Aberduna	Clwyd	433
Abergele	Clwyd	630
Aberstrecht	Gwynedd	4
Aberthaw	South Glamorgan	99
Achnagart	Highland	651
Admiralty	Dorset	2
Airdriehill	Strathclyde	361
Alkerton	Oxfordshire	90
Allington	Kent	47
Alltgoch	Dyfed	249

Altham	Lancashire	131
Ammanford	Dyfed	519
Arcow	North Yorkshire	567
Ardhasaig	Western Isles	340
Ardley	Oxfordshire	507
Ardownie	Tayside	246
Askernish	Western Isles	68
Avon Dassett	Warwickshire	222
Aycliffe	Durham	539
Bableigh Wood	Devon	157
Back Lane	Lancashire	203
Backwell	Avon	672
Balladoole	Isle of Man	96
Ballidon	Derbyshire	624
Ballygrant	Strathclyde	197
Balmedie	Grampian	257
Balmullo	Fife	372
Banavie	Highland	642
Bangley	Lothian	362
Bankfield	Lancashire	568
Bank Top	West Yorkshire	671
Bannerbank	Strathclyde	643
Bardon	Leicestershire	75
Barland	West Glamorgan	77
Barlockhart	Dumfries and Galloway	79
Barnsdale Bar	West Yorkshire	191
Barrasford	Northumberland	551
Barton	North Yorkshire	499
Barton Wood	Devon	189
Batts Combe	Somerset	48
Bayston Hill	Shropshire	603
Beer	Devon	58
Belliston	Fife	88
Big Pits	Lincolnshire	135
Bishop Middleham	Durham	623
Black	North Yorkshire	277
Blackaller	Devon	190
Blaenyfan	Dyfed	699
Blairhill	Lothian	580
Blodwell	Shropshire	10
Bluehill	Grampian	258
Blynlee	Borders	242
Boards	Central	677
Bognor Common	West Sussex	335
Bold Venture	Lancashire	569
Bolsover Moor	Derbyshire	552
Bolton Hill	Dyfed	250
Bonawe	Strathclyde	243
Borrow Pit	Cheshire	245
Borrowston	Highland	344
Borthwick	Borders	678
Bower	Highland	653
Boysack	Tayside	617
Brantingham	Humberside	670
Bray Valley	Devon	402
Breedon Hill	Leicestershire	121
Brindister	Shetland	545

Broadlaw	Lothian	363
Brodsworth	South Yorkshire	631
Bryn Engan	Gwynedd	384
Buckton Vale	Greater Manchester	354
Buddon Wood	Leicestershire	456
Builth Wells	Powys	36
Burford	Oxfordshire	87
Burley Hill	Clwyd	114
Cadeby	South Yorkshire	524
Caer Glaw	Gwynedd	19
Cairds Hill	Grampian	383
Cairneyhill	Strathclyde	644
Caldon Low	Staffordshire	604
Calliburn	Strathclyde	319
Callow Hill	Shropshire	605
Callow Rock	Somerset	209
Carew Quarries	Dyfed	494
Carnsew	Cornwall	420
Castle Hill	Somerset	153
Castle-an-Dinas	Cornwall	421
Cauldon	Staffordshire	100
Cerrigyrwyn	Dyfed	251
Charnage	Wiltshire	59
Charnwood	Leicestershire	11
Chillaton	Devon	346
Chipping Sodbury	Avon	49
Chywoon	Cornwall	326
Clatchard Craig	Fife	581
Clearwell	Gloucestershire	625
Clee Hill	Shropshire	12
Cliffe Hill	Leicestershire	564
Clipsham	Lincolnshire	135
Clock Face	West Yorkshire	347
Coatsgate	Dumfries and Galloway	582
Coldstones	North Yorkshire	434
Collace	Tayside	618
Collyweston	Lincolnshire	136
Concraigs	Grampian	669
Cool Scar	North Yorkshire	224
Corby	Northamptonshire	515
Cornelly	Mid Glamorgan	684
Cotswold Hill	Gloucestershire	165
Cottonhill	Grampian	339
Cowltick	Northamptonshire	515
Coygen	Dyfed	252
Cragmill	Northumberland	366
Craig-yr-Hesg	Mid Glamorgan	37
Craigenlow	Grampian	583
Craigfoot	Fife	649
Craighouse	Borders	584
Craigie Hill	Strathclyde	655
Craiglash	Grampian	259
Craignair	Dumfries and Galloway	585
Craigpark	Lothian	586
Creeton	Lincolnshire	444
Creigiau	Mid Glamorgan	607
Criggion	Shropshire	13

Crime Rig	Durham	500
Croft	Leicestershire	204
Cromhall	Gloucestershire	50
Croxden	Staffordshire	608
Croy	Strathclyde	588
Cruicks	Fife	645
Curister	Orkney	410
Daglingworth	Gloucestershire	51
Darlton	South Yorkshire	685
Darrington	North Yorkshire	178
Daviot	Highland	679
Dayhouse	Gwent	620
Dean	Cornwall	457
Denbigh	Clwyd	570
Dene	Derbyshire	554
Dinas	Dyfed	355
Divethill	Northumberland	367
(The) Dolomite	South Yorkshire	432
Dolyhir and Strinds	Powys	390
Dove Holes	Derbyshire	471
Dowlow	Derbyshire	525
Dry Rigg	North Yorkshire	458
Drybrook	Gloucestershire	52
Dulcote	Somerset	703
Dunald Mill	Lancashire	571
Dunbar Northwest	Lothian	101
Dunduff	Strathclyde	413
Duntilland	Strathclyde	543
Eastgate	Durham	102
Edstone	Borders	116
Edwin Richards	West Midlands	14
Eldon Hill	Derbyshire	399
Ellel	Lancashire	304
Eskett	Cumbria	225
Ethiebeaton	Tayside	591
Fenacre	Devon	609
Flax Bourton	Avon	686
Fletcher Bank	Greater Manchester	349
Force Garth	Durham	278
Forcett	North Yorkshire	637
Forest Wood	Mid Glamorgan	435
Foxcliffe	North Yorkshire	179
Friarton	Tayside	680
Furnace	Strathclyde	647
Gebdykes	North Yorkshire	526
Gedloch	Grampian	220
Gelligaer	Mid Glamorgan	436
Gellihalog	Dyfed	700
Ghyll Scaur	Cumbria	226
Giggleswick	North Yorkshire	632
Gilfach	West Glamorgan	701
Glensanda	Highland	705
Glen	South Yorkshire	350
Goat	Fife	592
Goddards	Derbyshire	667
Goldmire	Cumbria	22
Gore	Powys	626

Graig	Clwyd	459
Grange Mill	Derbyshire	89
Granite Mountain	Isle of Man	509
Great Blakenham	Suffolk	103
Great Ponton	Lincolnshire	516
Greetham	Lincolnshire	426
Greetwell	Lincolnshire	149
Greystone	Cornwall	212
Griff	Warwickshire	437
Groby	Leicestershire	15
Grove	Mid Glamorgan	438
Guiting	Gloucestershire	53
Gurney Slade	Avon	382
Gwyndy	Gwynedd	273
Hafod	Gwent	460
Halecombe	Somerset	687
Halkyn	Clwyd	400
Hallyards	Strathclyde	327
Happylands	Hereford and Worcester	411
Harden	Northumberland	638
Hardendale	Cumbria	126
Harnhill	Avon	688
Harrycroft	South Yorkshire	527
Hartley	Cumbria	279
Hart	Cleveland	501
Hartshill	Warwickshire	612
Heddle Hill	Orkney	409
Heights	Durham	230
Hendre	Clwyd	356
Hendy	Mid Glamorgan	689
Hengae	Gwynedd	385
High Craig	Strathclyde	593
High Force	Durham	280
Highmoor	Lancashire	227
Highmoor	North Yorkshire	550
Hillend	Strathclyde	485
Hillhead	Derbyshire	572
Hillhouse	Strathclyde	292
Hillwood	Lothian	681
Hindlow	Derbyshire	611
Hingston Down	Cornwall	60
Holme Hall	South Yorkshire	555
Holme Park	Lancashire	205
Hope	South Yorkshire	105
Horton	North Yorkshire	573
Houghton	Tyne and Wear	23
Howick	Northumberland	556
Huntsmans	Gloucestershire	299
Hutchbank	Lancashire	228
Independent	Dorset	2
Ingham	Suffolk	7
Ingleton	Lancashire	24
Ivonbrook	Derbyshire	405
Jackdaw Crag	North Yorkshire	180
Judkins	Warwickshire	16
Kelhead	Dumfries and Galloway	25
Kemnay	Grampian	656

Kendal Fell	Cumbria	574
Kersdown	Devon	493
Kessel Downs	Cornwall	423
Kevin	Staffordshire	627
Kilbarchan	Strathclyde	486
Kiln	Dyfed	495
Kirkmabreck	Dumfries and Galloway	595
Kirton Lindsey	Lincolnshire	150
Knowle	Devon	6
Langside	Fife	373
Lea	Shropshire	206
Lean	Cornwall	473
Leapers Wood	Lancashire	690
Leaton	Shropshire	312
Leeming	Lancashire	132
Leinthall	Hereford and Worcester	313
Lic	Western Isles	331
Lime Kiln Hill	Somerset	332
Linhay Hill	Devon	254
Lithalun	Mid Glamorgan	40
Little	Lancashire	67
Livox	Gwent	41
Llanddulas	Clwyd	26
Llynclys	Shropshire	520
Loanhead	Strathclyde	596
Longcliffe	Derbyshire	336
Long Bredy	Dorset	61
Long Lane	West Yorkshire	181
Longside	Grampian	134
Longwood	Avon	439
Lowthorpe	Humberside	529
Luxulyan	Cornwall	213
Machen	Gwent	42
Mancetter	Warwickshire	628
Marsden	Tyne and Wear	639
Marybank	Western Isles	657
Masons	Suffolk	106
Meldon	Devon	125
Melton Ross	Lincolnshire	506
Merehead	Somerset	704
Middle Peak	Derbyshire	557
Middleton Mine	Lothian	464
Middletown	Powys	115
Miltonhill	Strathclyde	621
Minera	Clwyd	575
Minffordd	Gwynedd	691
Montcliffe	Lancashire	27
Moons Hill	Avon	666
Moorcroft	Devon	214
Moota	Cumbria	118
Mootlaw	Northumberland	558
Morefield	Highland	597
Morrinton	Dumfries and Galloway	598
Mountsorrel	Leicestershire	456
Murrayshall	Central	599
Mynydd y Garreg	Dyfed	316
Nanhoron	Gwynedd	387

New England	Devon	215
Newbridge	North Yorkshire	282
Northfield	Central	156
Northfleet	Kent	109
North Lasts	Grampian	329
North Mains	Grampian	186
Nuneaton	Warwickshire	612
Offham	Kent	54
Old Cliffe Hill	Leicestershire	566
Old Quarrington	Durham	559
Orrock	Fife	682
Pant	Mid Glamorgan	692
Pant (Halkyn)	Clwyd	693
Pantyffynnon	South Glamorgan	291
Pant-y-Pwll Dwr	Clwyd	400
Parish	Derbyshire	358
Parkhead	Cumbria	28
Parkmore	Grampian	177
Peckfield	West Yorkshire	406
Pencaemawr	Mid Glamorgan	436
Penderyn	Mid Glamorgan	43
Penhow	Gwent	44
Penlee	Cornwall	424
Penmaenmawr	Gwynedd	29
Penryn	Cornwall	420
Penstrowed	Powys	256
Penwyllt	Powys	694
Perton	Hereford and Worcester	300
Pettinain	Strathclyde	160
Pigsden	Cornwall	418
Pilkington	Greater Manchester	483
Pitcaple	Grampian	260
Plaistow	Devon	403
Plymstock	Devon	110
Poortown	Isle of Man	302
Portland	Dorset	62
Potgate	North Yorkshire	528
Quartzite	Gloucestershire	673
Quickburn	Durham	318
Raisby	Durham	560
Ravelrig	Lothian	600
Raynes	Clwyd	401
Redmire	North Yorkshire	561
Rhyndaston	Dyfed	381
Riskend	Strathclyde	467
Rothes Glen	Grampian	510
Round 'O'	Lancashire	30
Ruston Parva	Humberside	529
Ruthin	Mid Glamorgan	111
Sandside	Cumbria	576
Savoch	Grampian	187
Scabba Wood	South Yorkshire	374
Sconser	Highland	290
Scord	Shetland	504
Setters	Shetland	482
Settrington	North Yorkshire	233
Shadforth	Durham	502

Shadwell	Shropshire	521
Shap	Cumbria	31
Shap Beck	Cumbria	31
Shap Blue	Cumbria	496
Shap Pink	Cumbria	497
Shavers End	Hereford and Worcester	208
Shennington	Oxfordshire	93
Shierglas	Tayside	601
Shining Bank	Derbyshire	668
Shipham	Somerset	614
Shire Hill	Derbyshire	353
Shotley	Northamptonshire	94
Silverdale	Lancashire	32
Smaws	North Yorkshire	530
South Witham	Lincolnshire	517
Southorpe	Lincolnshire	461
Spaunton	North Yorkshire	283
Spikers Hill	North Yorkshire	352
Spittlegate Level	Lincolnshire	333
Spring Lodge	South Yorkshire	633
Stainton	Cumbria	641
Steangabhal	Western Isles	309
Stonegrave	North Yorkshire	234
Stoneycombe	Devon	211
Stowe Hill	Gloucestershire	540
Stowfield	Gloucestershire	615
Sutton	North Yorkshire	548
Swanage	Dorset	547
Swanworth	Dorset	602
Swinburne	Northumberland	34
Swinden	North Yorkshire	634
Syke	Dyfed	440
Taffs Well	South Glamorgan	522
Tam's Loup	Strathclyde	441
Tan-y-Foel	Powys	119
Tendley	Cumbria	577
Thornaugh	Lincolnshire	427
Threshfield	North Yorkshire	562
Thrislington	Durham	531
Tom's Forest	Grampian	658
Tonfanau	Gwynedd	391
Tongland	Strathclyde	81
Topley Pike	Derbyshire	578
Torcoed	Dyfed	696
Torcoedfawr	Dyfed	359
Torlundy	Highland	289
Tormitchell	Strathclyde	82
Torrin	Highland	330
Town	Oxfordshire	176
Trearne	Strathclyde	465
Trefil	Gwent	272
Trevassack	Cornwall	162
Trimm Rock	Clwyd	635
Triscombe	Somerset	616
Trusham	Devon	63
Tuckingmill	Devon	481
Tunstead	Derbyshire	301

Turkeyland	Isle of Man	321
Tytherington	Gloucestershire	55
Underheugh	Strathclyde	364
Vale Road	Nottinghamshire	154
Vatseter	Shetland	244
Vaynor	Mid Glamorgan	46
Venn	Devon	216
Waddington Fell	Lancashire	665
Wardlow	Staffordshire	462
Water Swallows	Derbyshire	579
Wath	North Yorkshire	532
Waulkmill	Tayside	648
Wenvoe	South Glamorgan	702
Westbury	Wiltshire	113
Westbury	Somerset	442
Western New Forres	Grampian	261
Westleigh	Devon	217
Whatley	Somerset	56
Whitecleaves	Devon	64
Whitwell	Nottinghamshire	533
Whitwick	Leicestershire	17
Whitworth	Lancashire	229
Wick	Avon	674
Wickwar	Gloucestershire	675
Witch Hill	Durham	503
Woodbury	Hereford and Worcester	523
Woodeaton	Oxfordshire	78
Wormersley	South Yorkshire	534
Worsham	Oxfordshire	518
Wredon	Staffordshire	360
Yalberton Tor	Devon	65

DRESSED OR LUMP STONE

Abbey	Nottinghamshire	3
Aberllefeni	Powys	697
Aberstrecht	Gwynedd	4
Admiralty	Dorset	2
Afonwen	Cheshire	511
Alkerton	Oxfordshire	90
Altham	Lancashire	131
Ancaster	Lincolnshire	267
Ann Twyford	Derbyshire	468
Apex	West Yorkshire	306
Appleton	West Yorkshire	293
Bableigh Wood	Devon	157
Bankend	Cumbria	168
Bank Top	West Yorkshire	671
Barnsdale Bar	West Yorkshire	191
Barton	North Yorkshire	499
Beacon Lodge	West Yorkshire	185
Beam	Devon	650
Bearrah Tor	Cornwall	443
Big Pits	Lincolnshire	135
Birkhams	Cumbria	169
Bixhead	Gloucestershire	241
Black	North Yorkshire	277
Blackhill	West Yorkshire	376

Black Hills	Grampian	133
Blackmoor South	Yorkshire	285
Black Pasture	Northumberland	487
Blynlee	Borders	242
Bognor Common	West Sussex	335
Bolehill	Derbyshire	449
Bolton Woods	West Yorkshire	429
Bosahan	Cornwall	663
Boughton	Northamptonshire	91
Bower	Highland	653
Bowithic	Cornwall	163
Bowyers	Dorset	1
Braich Ddu	Gwynedd	262
Brandy Crag	Cumbria	138
Bridge	Cheshire	341
Brittania West	Yorkshire	415
Broad Oak	Cumbria	170
Brockhill	Gloucestershire	164
Broughton Moor	Lancashire	140
Burford	Oxfordshire	87
Burlington	Cumbria	141
Burnthouse	Cornwall	9
Bursting Stone	Cumbria	142
Cadeby	South Yorkshire	524
Callywith	Cornwall	537
Camp Hill	Northamptonshire	98
Carnsew	Cornwall	420
Cat Castle	Durham	155
Chellow Grange West	Yorkshire	412
Chillaton	Devon	346
Chinley Moor	Greater Manchester	369
Chinnor	Oxfordshire	477
Chywoon	Cornwall	326
Clashach	Grampian	379
Clipsham	Lincolnshire	135
Clogau	Clwyd	95
Corncockle	Dumfries and Galloway	194
Corsehill	Dumfries and Galloway	195
Cotswold Hill	Gloucestershire	165
Craigie Hill	Strathclyde	655
Cromwell	West Yorkshire	348
Crosland Moor	West Yorkshire	310
Cwt-y-Bugail	Gwynedd	235
Daglingworth	Gloucestershire	51
Darrington	North Yorkshire	178
Dead Friars	Durham	513
Dean	Cornwall	457
Delabole	Cornwall	474
De Lank	Cornwall	469
Diphwys Casson	Gwynedd	263
Doddington Hill	Northumberland	394
Doulting	Somerset	83
Downs	Dorset	338
Dukes	Derbyshire	450
Dunhouse	Durham	196
Dunmore	Central	488
Ellel	Lancashire	304

Elterwater	Cumbria	143
Fagley	West Yorkshire	431
Farmington	Gloucestershire	231
Filkins	Oxfordshire	239
Fish Hill	Hereford and Worcester	508
Fly Delph	West Yorkshire	342
Fly Flatts	West Yorkshire	129
Fords	Lancashire	240
Forest Wood	Mid Glamorgan	435
Foxcliffe	North Yorkshire	179
Friendly	West Yorkshire	70
Gatelawbridge	Dumfries and Galloway	489
Gebdykes	North Yorkshire	526
Gelligaer	Mid Glamorgan	436
Glebe	Lincolnshire	268
Gloddfa Ganol	Gwynedd	236
Glogue	Dyfed	188
Goddards	Derbyshire	667
Great Gate	Staffordshire	512
Grinshill	Shropshire	270
Guiting	Gloucestershire	53
Hainworth Shaw	West Yorkshire	72
Ham Hill	Somerset	221
Ham Hill	Somerset	378
Hantergantick	Cornwall	395
Happylands	Hereford and Worcester	411
Harden Moor	West Yorkshire	307
Harthorpe	East Durham	535
Hayes Wood Mine	Avon	86
Hayfield	Greater Manchester	324
Hayne	Devon	219
Hearson	Devon	288
Hessilhead	Strathclyde	463
Hillhouse Edge	West Yorkshire	484
Holleywell	Leicestershire	368
Holmescales	Cumbria	472
Honister	Cumbria	357
Honley Wood	West Yorkshire	311
Horn Crag	West Yorkshire	294
Horn Park	Dorset	247
Hornsleasow	Hereford and Worcester	298
Horton	Warwickshire	295
Hovingham	North Yorkshire	66
Huncote	Leicestershire	5
Hunters Hill	West Yorkshire	130
Independent	Dorset	2
Jackdaw Crag	North Yorkshire	180
Jamestone	Lancashire	305
Keates	Dorset	317
Kirkby	Cumbria	144
Knowle	Devon	6
Ladycross	Northumberland	322
Landers	Dorset	325
Lazonby Fell	Cumbria	171
Leeming	Lancashire	132
Leipsic	Cumbria	498
Lic	Western Isles	331

Llechedd Slate Mine	Gwynedd	264
Locharbriggs	Dumfries and Galloway	73
Long Lane	West Yorkshire	181
Longford	Devon	370
Maenofferen	Gwynedd	265
Mansfield	Nottinghamshire	269
Melton Ross	Lincolnshire	506
Merivale	Devon	397
Mill Hill	Devon	371
Mine Train	Gloucestershire	375
Monks Park Mine	Wiltshire	84
Moss Rigg	Cumbria	145
Mount Pleasant	Avon	447
Nanhoron	Gwynedd	387
Nantlle	Gwynedd	237
Nant Newydd	Gwynedd	389
Naylor Hill Quarries	West Yorkshire	248
Newthorpe	West Yorkshire	182
Northowram Hills	West Yorkshire	232
Oakeley	Gwynedd	238
Old Mill	Strathclyde	218
Once-a-Week	Derbyshire	345
Palmers	Derbyshire	514
Park Nook	South Yorkshire	148
Parkhead	Cumbria	28
Peddles Lane	Somerset	416
Pencaemawr	Mid Glamorgan	436
Penhow	Gwent	44
Penrhyn	Gwynedd	365
Penryn	Cornwall	420
Pets	Cumbria	320
Philpots	West Sussex	428
Pickerings	Cumbria	172
Pilsamoor	Cornwall	549
Plas Gwilym	Clwyd	296
Portland	Dorset	62
Prince of Wales	Cornwall	451
Pury End	Northamptonshire	98
Quickburn	Durham	318
Rawdon	West Yorkshire	328
Red Hole	Staffordshire	512
Rhuddlan Bach	Gwynedd	466
Ridgeway	Derbyshire	161
Rock Cottage	North Yorkshire	377
Rooks	Cumbria	173
Ross of Mull	Strathclyde	492
Rothes Glen	Grampian	510
Salterwath	Cumbria	174
Scout Moor	Greater Manchester	351
Scratchmill Scar	Cumbria	175
Shap Blue	Cumbria	496
Shap Pink	Cumbria	497
Shawk	Cumbria	323
Shellingford	Oxfordshire	386
Shennington	Oxfordshire	93
Sherburn	North Yorkshire	284
Shining Bank	Derbyshire	668

Shipley Bank	Durham	505
Shotley	Northamptonshire	94
Skelbrooke	South Yorkshire	183
Soil Hill	West Yorkshire	266
South Barrule	Isle of Man	303
Southard	Dorset	287
Sovereign	West Yorkshire	334
Spittal	Highland	546
Spout Cragg	Cumbria	147
Springwell	Tyne and Wear	398
Spynie	Grampian	380
Squire Hill	West Yorkshire	404
St Aldhelm's	Dorset	286
Stamford	Cambridgeshire	446
Station	Somerset	76
Stoke Hall	South Yorkshire	538
Stoneraise	Cumbria	454
Stone	West Yorkshire	445
Stoney Brow	Lancashire	276
Stowe Hill	Gloucestershire	540
Stowey	Avon	541
Swanage	Dorset	547
Sycamore	Cheshire	198
Tam's Loup	Strathclyde	441
Tearne	Staffordshire	407
Ten Yards Lane	North Yorkshire	71
Thumpas	West Yorkshire	166
Tor Down	Cornwall	388
Trebarwith	Cornwall	660
Tredinnick	Cornwall	661
Tredennick Downs	Cornwall	314
Trenoweth	Cornwall	662
Trevillet	Cornwall	74
Trevone	Cornwall	664
Tuckingmill	Devon	481
Twll Llwyd	Gwynedd	297
Upper Lawn	Avon	275
Vale Road	Nottinghamshire	154
Veizey's	Gloucestershire	619
Vronlog	Gwynedd	315
Waddington Fell	Lancashire	665
Wath	North Yorkshire	532
Watson	West Yorkshire	97
Watts Cliff	Derbyshire	455
Webscott	Shropshire	271
Westwood Ground	Wiltshire	85
Whittle Hill	Lancashire	676
Wilderness	Gloucestershire	448
Wimberry Moss	Cheshire	199
Windy Hill	Durham	698
Woodside	West Yorkshire	122
Woodside	Grampian	414
Wooton	Derbyshire	408

INDUSTRIAL LIMESTONE

Aberthaw	South Glamorgan	99
Aycliffe	Durham	539

Back Lane	Lancashire	203
Ballidon	Derbyshire	624
Bankfield	Lancashire	568
Barrington	Cambridgeshire	476
Batts Combe	Somerset	48
Beer	Devon	58
Bold Venture	Lancashire	569
Brierlow	Derbyshire	470
Cadeby	South Yorkshire	524
Caldon Low	Staffordshire	604
Callow Rock	Somerset	209
Cauldon	Staffordshire	100
Charnage	Wiltshire	59
Chipping Sodbury	Avon	49
Clearwell	Gloucestershire	625
Cool Scar	North Yorkshire	224
Cornelly	Mid Glamorgan	684
Cromhall	Gloucestershire	50
Daglingworth	Gloucestershire	51
Dene	Derbyshire	554
Dolomite (The)	South Yorkshire	432
Dolyhir and Strinds	Powys	390
Dove Holes	Derbyshire	471
Dunald Mill	Lancashire	571
Dunbar Northwest	Lothian	101
Eastgate	Durham	102
East Grinstead	Wiltshire	200
Eldon Hill	Derbyshire	399
Eskett	Cumbria	225
Fenacre	Devon	609
Forcett	North Yorkshire	637
Gebdykes	North Yorkshire	526
Goddards	Derbyshire	667
Grange Mill	Derbyshire	89
Grange Top	Lincolnshire	151
Greys Pit	Kent	478
Guiting	Gloucestershire	53
Halkyn	Clwyd	400
Hardendale	Cumbria	126
Hartley	Cumbria	279
Hindlow	Derbyshire	611
Holme Hall	South Yorkshire	555
Holme Park	Lancashire	205
Hope	South Yorkshire	105
Horton	North Yorkshire	573
Kensworth	Bedfordshire	479
Ketton	Lincolnshire	151
Kevin	Staffordshire	627
Lea	Shropshire	206
Llynclys	Shropshire	520
Long Bredy	Dorset	61
Marsden	Tyne and Wear	639
Melton	Humberside	107
Melton Ross	Lincolnshire	506
Merehead	Somerset	704
Middle Peak	Derbyshire	557
Middlegate	Humberside	480

Middleton Mine	Lothian	464
Minera	Clwyd	575
Moorcroft	Devon	214
Mootlaw	Northumberland	558
Pant-y-Pwll Dwr	Clwyd	400
Penhow	Gwent	44
Plymstock	Devon	110
Queensgate	Humberside	201
Quidhampton	Wiltshire	202
Raisby	Durham	560
Ribblesdale Works	Lancashire	152
Ruthin	Mid Glamorgan	111
Sandside	Cumbria	576
Shining Bank	Derbyshire	668
Shoreham	West Sussex	112
Spring Lodge	South Yorkshire	633
Swanworth	Dorset	602
Swinden	North Yorkshire	634
Thrislington	Durham	531
Topley Pike	Derbyshire	578
Trefil	Gwent	272
Trimm Rock	Clwyd	635
Tunstead	Derbyshire	301
Tytherington	Gloucestershire	55
Veizey's	Gloucestershire	619
Wardlow	Staffordshire	462
Westbury	Wiltshire	113
Whatley	Somerset	56
Whitwell	Nottinghamshire	533
Yalberton Tor	Devon	65

INDUSTRIAL USE

Bantycock	Nottinghamshire	123
Staple Farm	Nottinghamshire	124

PRE-CAST CONCRETE PRODUCTS

Ardley	Oxfordshire	507
Bardon	Leicestershire	75
Barrasford	Northumberland	551
Blaenyfan	Dyfed	699
Bonawe	Strathclyde	243
Borrowston	Highland	344
Brindister	Shetland	545
Buddon Wood	Leicestershire	456
Callow Rock	Somerset	209
Chipping Sodbury	Avon	49
Craig-yr-Hesg	Mid Glamorgan	37
Croft	Leicestershire	204
Dove Holes	Derbyshire	471
Flax Bourton	Avon	686
Goldmire	Cumbria	22
Gwyndy	Gwynedd	273
Highmoor	North Yorkshire	550
Hingston Down	Cornwall	60
Judkins	Warwickshire	16
Kendal Fell	Cumbria	574
Moons Hill	Avon	666

Moorcroft	Devon	214
Mountsorrel	Leicestershire	456
Old Quarrington	Durham	559
Stoneycombe	Devon	211
Swinburne	Northumberland	34
Whitwick	Leicestershire	17
Wick	Avon	674

READY-MIXED CONCRETE

Abergele	Clwyd	630
Allington	Kent	47
Alltgoch	Dyfed	249
Backwell	Avon	672
Ballidon	Derbyshire	624
Bayston Hill	Shropshire	603
Blaenyfan	Dyfed	699
Bolton Hill	Dyfed	250
Bonawe	Strathclyde	243
Borrowston	Highland	344
Brindister	Shetland	545
Buddon Wood	Leicestershire	456
Charnwood	Leicestershire	11
Chipping Sodbury	Avon	49
Cornelly	Mid Glamorgan	684
Cragmill	Northumberland	366
Craig-yr-Hesg	Mid Glamorgan	37
Daglingworth	Gloucestershire	51
Denbigh	Clwyd	570
Divethill	Northumberland	367
Dolyhir and Strinds	Powys	390
Eldon Hill	Derbyshire	399
Flax Bourton	Avon	686
Gilfach	West Glamorgan	701
Goldmire	Cumbria	22
Griff	Warwickshire	437
Groby	Leicestershire	15
Harnhill	Avon	688
Hillhouse	Strathclyde	292
Hingston Down	Cornwall	60
Holme Hall	South Yorkshire	555
Houghton	Tyne and Wear	23
Kendal Fell	Cumbria	574
Kessel Downs	Cornwall	423
Lean	Cornwall	473
Linhay Hill	Devon	254
Longwood	Avon	439
Machen	Gwent	42
Minffordd	Gwynedd	691
Mountsorrel	Leicestershire	456
Old Cliffe Hill	Leicestershire	566
Penderyn	Mid Glamorgan	43
Penlee	Cornwall	424
Penmaenmawr	Gwynedd	29
Riskend	Strathclyde	467
Spring Lodge	South Yorkshire	633
Taffs Well	South Glamorgan	522
Tan-y-Foel	Powys	119

Thrislington	Durham	531
Tom's Forest	Grampian	658
Tonfanau	Gwynedd	391
Turkeyland	Isle of Man	321
Tytherington	Gloucestershire	55
Vaynor	Mid Glamorgan	46
Wenvoe	South Glamorgan	702
Whitwell	Nottinghamshire	533
Wickwar	Gloucestershire	675

Index 9

Quarry products in England, Scotland and Wales

County by county within each country;
and by quarry name within counties

Entries in this index refer to the
unique quarry numbers (1 – 705)
in the main Directory

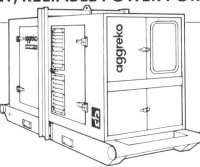

458

England

AGRICULTURAL LIME		Lincolnshire	
Avon		Big Pits	135
Chipping Sodbury	49	Clipsham	135
Cheshire		Great Ponton	516
Afonwen	511	Melton Ross	506
Cleveland		Norfolk	
Hart	501	Mount Pleasant	158
Cumbria		North Yorkshire	
Eskett	225	Cool Scar	224
Hartley	279	Gebdykes	526
Sandside	576	Horton	573
Stainton	641	Spikers Hill	352
Tendley	577	Swinden	634
Derbyshire		Wath	532
Brierlow	470	Northumberland	
Dene	554	Mootlaw	558
Dove Holes	471	Nottinghamshire	
Dowlow	525	Whitwell	533
Eldon Hill	399	Shropshire	
Grange Mill	89	Lea	206
Hillhead	572	Somerset	
Hindlow	611	Batts Combe	48
Middle Peak	557	Callow Rock	209
Parish	358	Merehead	704
Topley Pike	578	Whatley	56
Tunstead	301	South Yorkshire	
Devon		Cadeby	524
Linhay Hill	254	The Dolomite	432
Moorcroft	214	Holme Hall	555
Stoneycombe	211	Staffordshire	
Dorset		Caldon Low	604
Swanworth	602	Cauldon	100
Durham		Kevin	627
Bishop Middleham	623	Wardlow	462
Old Quarrington	559	Tyne and Wear	
Raisby	560	Houghton	23
Shadforth	502	Marsden	639
Thrislington	531	West Yorkshire	
Witch Hill	503	Peckfield	406
Gloucestershire			
Clearwell	625	ASPHALT OR TARMACADAM	
Cromhall	50	Avon	
Drybrook	52	Chipping Sodbury	49
Tytherington	55	Flax Bourton	686
Veizey's	619	Harnhill	688
Hertfordshire		Moons Hill	666
Flaunden	159	Wick	674
Isle of Man		Cornwall	
Balladoole	96	Carnsew	420
Turkeyland	321	Dean	457
Lancashire		Greystone	212
Back Lane	203	Hingston Down	60
Bankfield	568	Lean	473
Holme Park	205	Luxulyan	213
Leapers Wood	690	Penlee	424

Penryn	420	Groby	15
Cumbria		Mountsorrel	456
Eskett	225	Old Cliffe Hill	566
Goldmire	22	Whitwick	17
Moota	118	**North Yorkshire**	
Sandside	576	Arcow	567
Shap Beck	31	Black	277
Shap	31	Coldstones	434
Tendley	577	Dry Rigg	458
Derbyshire		Forcett	637
Ballidon	624	Newbridge	282
Dene	554	Spaunton	283
Dove Holes	471	Spikers Hill	352
Eldon Hill	399	**Northumberland**	
Ivonbrook	405	Barrasford	551
Middle Peak	557	Cragmill	366
Shining Bank	668	Howick	556
Topley Pike	578	Swinburne	34
Water Swallows	579	**Nottinghamshire**	
Devon		Vale Road	154
Blackaller	190	**Shropshire**	
Bray Valley	402	Bayston Hill	603
Fenacre	609	Blodwell	10
Linhay Hill	254	Callow Hill	605
Moorcroft	214	Clee Hill	12
Stoneycombe	211	Criggion	13
Trusham	63	Leaton	312
Venn	216	**Somerset**	
Westleigh	217	Batts Combe	48
Whitecleaves	64	Halecombe	687
Dorset		Merehead	704
Swanworth	602	Triscombe	616
Durham		Westbury	442
Force Garth	278	Whatley	56
Heights	230	**South Yorkshire**	
High Force	280	Darlton	685
Raisby	560	(The) Dolomite	432
Gloucestershire		Holme Hall	555
Cromhall	50	**Staffordshire**	
Huntsmans	299	Caldon Low	604
Stowfield	615	Cauldon	100
Tytherington	55	Kevin	627
Wickwar	675	Wardlow	462
Kent		Wredon	360
Allington	47	**Tyne and Wear**	
Lancashire		Marsden	639
Bankfield	568	**Warwickshire**	
Bold Venture	569	Griff	437
Dunald Mill	571	Hartshill	612
Leapers Wood	690	Judkins	16
Silverdale	32	Mancetter	628
Leicestershire		Nuneaton	612
Bardon	75	**West Midlands**	
Buddon Wood	456	Edwin Richards	14
Charnwood	11		
Cliffe Hill	564		
Croft	204		

BURNT LIMESTONE		Shap Pink	497
Cumbria		Shap	31
Hartley	279	Stainton	641
Derbyshire		Tendley	577
Dove Holes	471	Derbyshire	
Hindlow	611	Ballidon	624
Tunstead	301	Bolsover Moor	552
Humberside		Dene	554
Melton	107	Dove Holes	471
Lincolnshire		Dowlow	525
Melton Ross	506	Eldon Hill	399
North Yorkshire		Goddards	667
Horton	573	Grange Mill	89
Swinden	634	Hillhead	572
Somerset		Hindlow	611
Batts Combe	48	Ivonbrook	405
		Longcliffe	336
CRUSHED STONE		Middle Peak	557
Avon		Parish	358
Backwell	672	Shining Bank	668
Chipping Sodbury	49	Shire Hill	353
Flax Bourton	686	Topley Pike	578
Gurney Slade	382	Tunstead	301
Harnhill	688	Water Swallows	579
Longwood	439	Devon	
Moons Hill	666	Bableigh Wood	157
Wick	674	Barton Wood	189
Cheshire		Beer	58
Borrow Pit	245	Blackaller	190
Cleveland		Bray Valley	402
Hart	501	Chillaton	346
Cornwall		Fenacre	609
Carnsew	420	Kersdown	493
Castle-an-Dinas	421	Knowle	6
Chywoon	326	Linhay Hill	254
Dean	457	Meldon	125
Greystone	212	Moorcroft	214
Hingston Down	60	New England	215
Kessel Downs	423	Plaistow	403
Lean	473	Plymstock	110
Luxulyan	213	Stoneycombe	211
Penlee	424	Trusham	63
Penryn	420	Tuckingmill	481
Pigsden	418	Venn	216
Trevassack	162	Westleigh	217
Cumbria		Whitecleaves	64
Eskett	225	Yalberton Tor	65
Ghyll Scaur	226	Dorset	
Goldmire	22	Admiralty	2
Hardendale	126	Independent	2
Hartley	279	Long Bredy	61
Kendal Fell	574	Portland	62
Moota	118	Swanage	547
Parkhead	28	Swanworth	602
Sandside	576	Durham	
Shap Beck	31	Aycliffe	539
Shap Blue	496	Bishop Middleham	623

Crime Rig	500	Ingleton	24
Eastgate	102	Leapers Wood	690
Force Garth	278	Leeming	132
Heights	230	Little	67
High Force	280	Montcliffe	27
Old Quarrington	559	Round 'O'	30
Quickburn	318	Silverdale	32
Raisby	560	Waddington Fell	665
Shadforth	502	Whitworth	229
Thrislington	531	Leicestershire	
Witch Hill	503	Bardon	75
Gloucestershire		Breedon Hill	121
Clearwell	625	Buddon Wood	456
Cotswold Hill	165	Charnwood	11
Cromhall	50	Cliffe Hill	564
Daglingworth	51	Croft	204
Drybrook	52	Groby	15
Guiting	53	Mountsorrel	456
Huntsmans	299	Old Cliffe Hill	566
Quartzite	673	Whitwick	17
Stowe Hill	540	Lincolnshire	
Stowfield	615	Big Pits	135
Tytherington	55	Clipsham	135
Wickwar	675	Collyweston	136
Greater Manchester		Creeton	444
Buckton Vale	354	Great Ponton	516
Fletcher Bank	349	Greetham	426
Pilkington	483	Greetwell	149
Hereford and Worcester		Kirton Lindsey	150
Happylands	411	Melton Ross	506
Leinthall	313	South Witham	517
Perton	300	Southorpe	461
Shavers End	208	Spittlegate Level	333
Woodbury	523	Thornaugh	427
Humberside		North Yorkshire	
Brantingham	670	Arcow	567
Lowthorpe	529	Barton	499
Ruston Parva	529	Black	277
Isle of Man		Coldstones	434
Balladoole	96	Cool Scar	224
Granite Mountain	509	Darrington	178
Poortown	302	Dry Rigg	458
Turkeyland	321	Forcett	637
Kent		Foxcliffe	179
Allington	47	Gebdykes	526
Northfleet	109	Giggleswick	632
Offham	54	Highmoor	550
Lancashire		Horton	573
Altham	131	Jackdaw Crag	180
Back Lane	203	Newbridge	282
Bankfield	568	Potgate	528
Bold Venture	569	Redmire	561
Dunald Mill	571	Settrington	233
Ellel	304	Smaws	530
Highmoor	227	Spaunton	283
Holme Park	205	Spikers Hill	352
Hutchbank	228	Stonegrave	234

Sutton	548	Hope	105
Swinden	634	Scabba Wood	374
Threshfield	562	Spring Lodge	633
Wath	532	Wormersley	534
Northamptonshire		Staffordshire	
Corby	515	Caldon Low	604
Cowltick	515	Cauldon	100
Shotley	94	Croxden	608
Northumberland		Kevin	627
Barrasford	551	Wardlow	462
Cragmill	366	Wredon	360
Divethill	367	Suffolk	
Harden	638	Great Blakenham	103
Howick	556	Ingham	7
Mootlaw	558	Masons	106
Swinburne	34	Tyne and Wear	
Nottinghamshire		Houghton	23
Vale Road	154	Marsden	639
Whitwell	533	Warwickshire	
Oxfordshire		Avon Dassett	222
Alkerton	90	Griff	437
Ardley	507	Hartshill	612
Burford	87	Judkins	16
Shennington	93	Mancetter	628
Town	176	Nuneaton	612
Woodeaton	78	West Midlands	
Worsham	518	Edwin Richards	14
Shropshire		West Sussex	
Bayston Hill	603	Bognor Common	335
Blodwell	10	West Yorkshire	
Callow Hill	605	Bank Top	671
Clee Hill	12	Barnsdale Bar	191
Criggion	13	Clock Face	347
Leaton	312	Long Lane	181
Lea	206	Peckfield	406
Llynclys	520	Wiltshire	
Shadwell	521	Charnage	59
Somerset		Westbury	113
Batts Combe	48		
Callow Rock	209	**DRESSED OR LUMP STONE**	
Castle Hill	153	Avon	
Dulcote	703	Hayes Wood Mine	86
Halecombe	687	Mount Pleasant	447
Lime Kiln Hill	332	Stowey	541
Merehead	704	Upper Lawn	275
Shipham	614	Cambridgeshire	
Triscombe	616	Stamford	446
Westbury	442	Cheshire	
Whatley	56	Afonwen	511
South Yorkshire		Bridge	341
Brodsworth	631	Sycamore	198
Cadeby	524	Wimberry Moss	199
Darlton	685	Cornwall	
(The) Dolomite	432	Bearrah Tor	443
Glen	350	Bosahan	663
Harrycroft	527	Bowithic	163
Holme Hall	555	Burnthouse	9

Callywith	537	Hayne	219
Carnsew	420	Hearson	288
Chywoon	326	Knowle	6
De Lank	469	Longford	370
Dean	457	Merivale	397
Delabole	474	Mill Hill	371
Hantergantick	395	Tuckingmill	481
Penryn	420	Dorset	
Pilsamoor	549	Admiralty	2
Prince of Wales	451	Bowyers	1
Tor Down	388	Downs	338
Trebarwith	660	Horn Park	247
Tredennick Downs	314	Independent	2
Tredinnick	661	Keates	317
Trenoweth	662	Landers	325
Trevillet	74	Portland	62
Trevone	664	Southard	287
Cumbria		St Aldhelm's	286
Bankend	168	Swanage	547
Birkhams	169	Durham	
Brandy Crag	138	Cat Castle	155
Broad Oak	170	Dead Friars	513
Burlington	141	Dunhouse	196
Bursting Stone	142	Harthorpe East	535
Elterwater	143	Quickburn	318
Holmescales	472	Shipley Bank	505
Honister	357	Windy Hill	698
Kirkby	144	Gloucestershire	
Lazonby Fell	171	Bixhead	241
Leipsic	498	Brockhill	164
Moss Rigg	145	Cotswold Hill	165
Parkhead	28	Daglingworth	51
Pets	320	Farmington	231
Pickerings	172	Guiting	53
Rooks	173	Mine Train	375
Salterwath	174	Stowe Hill	540
Scratchmill Scar	175	Veizey's	619
Shap Blue	496	Wilderness	448
Shap Pink	497	Greater Manchester	
Shawk	323	Chinley Moor	369
Spout Cragg	147	Hayfield	324
Stoneraise	454	Scout Moor	351
Derbyshire		Hereford and Worcester	
Ann Twyford	468	Fish Hill	508
Bolehill	449	Happylands	411
Dukes	450	Hornsleasow	298
Goddards	667	Isle of Man	
Once-a-Week	345	South Barrule	303
Palmers	514	Lancashire	
Ridgeway	161	Altham	131
Shining Bank	668	Broughton Moor	140
Watts Cliff	455	Ellel	304
Wooton	408	Fords	240
Devon		Jamestone	305
Bableigh Wood	157	Leeming	132
Beam	650	Stoney Brow	276
Chillaton	346	Waddington Fell	665

Whittle Hill	676	Staffordshire	
Leicestershire		Great Gate	512
Holleywell	368	Red Hole	512
Huncote	5	Tearne	407
Lincolnshire		Tyne and Wear	
Ancaster	267	Springwell	398
Big Pits	135	Warwickshire	
Clipsham	135	Horton	295
Glebe	268	West Sussex	
Melton Ross	506	Bognor Common	335
North Yorkshire		Philpots	428
Barton	499	West Yorkshire	
Black	277	Apex	306
Darrington	178	Appleton	293
Foxcliffe	179	Bank Top	671
Gebdykes	526	Barnsdale Bar	191
Hovingham	66	Beacon Lodge	185
Jackdaw Crag	180	Blackhill	376
Rock Cottage	377	Bolton Woods	429
Sherburn	284	Brittania	415
Ten Yards Lane	71	Chellow Grange	412
Wath	532	Cromwell	348
Northamptonshire		Crosland Moor	310
Boughton	91	Fagley	431
Camp Hill	98	Fly Delph	342
Pury End	98	Fly Flatts	129
Shotley	94	Friendly	70
Northumberland		Hainworth Shaw	72
Black Pasture	487	Harden Moor	307
Doddington Hill	394	Hillhouse Edge	484
Ladycross	322	Honley Wood	311
Nottinghamshire		Horn Crag	294
Abbey	3	Hunters Hill	130
Mansfield	269	Long Lane	181
Vale Road	154	Naylor Hill Quarries	248
Oxfordshire		Newthorpe	182
Alkerton	90	Northowram Hills	232
Burford	87	Rawdon	328
Chinnor	477	Soil Hill	266
Filkins	239	Sovereign	334
Shellingford	386	Squire Hill	404
Shennington	93	Stone	445
Shropshire		Thumpas	166
Grinshill	270	Watson	97
Webscott	271	Woodside	122
Somerset		Wiltshire	
Doulting	83	Monks Park Mine	84
Ham Hill	221	Westwood Ground	85
Ham Hill	378		
Peddles Lane	416		
Station	76	**INDUSTRIAL LIMESTONE**	
South Yorkshire		Avon	
Blackmoor	285	Chipping Sodbury	49
Cadeby	524	Bedfordshire	
Park Nook	148	Kensworth	479
Skelbrooke	183	Cambridgeshire	
Stoke Hall	538	Barrington	476

Cumbria		Cool Scar	224
Eskett	225	Forcett	637
Hardendale	126	Gebdykes	526
Hartley	279	Horton	573
Sandside	576	Swinden	634
Derbyshire		Northumberland	
Ballidon	624	Mootlaw	558
Brierlow	470	Nottinghamshire	
Dene	554	Whitwell	533
Dove Holes	471	Shropshire	
Eldon Hill	399	Lea	206
Goddards	667	Llynclys	520
Grange Mill	89	Somerset	
Hindlow	611	Batts Combe	48
Middle Peak	557	Callow Rock	209
Shining Bank	668	Merehead	704
Topley Pike	578	Whatley	56
Tunstead	301	South Yorkshire	
Devon		Cadeby	524
Beer	58	(The) Dolomite	432
Fenacre	609	Holme Hall	555
Moorcroft	214	Hope	105
Plymstock	110	Spring Lodge	633
Yalberton Tor	65	Staffordshire	
Dorset		Caldon Low	604
Long Bredy	61	Cauldon	100
Swanworth	602	Kevin	627
Durham		Wardlow	462
Aycliffe	539	Tyne and Wear	
Eastgate	102	Marsden	639
Raisby	560	West Sussex	
Thrislington	531	Shoreham	112
Gloucestershire		Wiltshire	
Clearwell	625	Charnage	59
Cromhall	50	East Grinstead	200
Daglingworth	51	Quidhampton	202
Guiting	53	Westbury	113
Tytherington	55		
Veizey's	619	INDUSTRIAL USE	
Humberside		Nottinghamshire	
Melton	107	Bantycock	123
Middlegate	480	Staple Farm	124
Queensgate	201		
Kent		PRE-CAST CONCRETE PRODUCTS	
Greys Pit	478	Avon	
Lancashire		Chipping Sodbury	49
Back Lane	203	Flax Bourton	686
Bankfield	568	Moons Hill	666
Bold Venture	569	Wick	674
Dunald Mill	571	Cornwall	
Holme Park	205	Hingston Down	60
Ribblesdale Works	152	Cumbria	
Lincolnshire		Goldmire	22
Grange Top	151	Kendal Fell	574
Ketton	151	Derbyshire	
Melton Ross	506	Dove Holes	471
North Yorkshire			

Devon		Kendal Fell	574
Moorcroft	214	Derbyshire	
Stoneycombe	211	Ballidon	624
Durham		Eldon Hill	399
Old Quarrington	559	Devon	
Leicestershire		Linhay Hill	254
Bardon	75	Durham	
Buddon Wood	456	Thrislington	531
Croft	204	Gloucestershire	
Mountsorrel	456	Daglingworth	51
Whitwick	17	Tytherington	55
North Yorkshire		Wickwar	675
Highmoor	550	Isle of Man	
Northumberland		Turkeyland	321
Barrasford	551	Kent	
Swinburne	34	Allington	47
Oxfordshire		Leicestershire	
Ardley	507	Buddon Wood	456
Somerset		Charnwood	11
Callow Rock	209	Groby	15
Warwickshire		Mountsorrel	456
Judkins	16	Old Cliffe Hill	566
		Northumberland	
READY-MIXED CONCRETE		Cragmill	366
Avon		Divethill	367
Backwell	672	Nottinghamshire	
Chipping Sodbury	49	Whitwell	533
Flax Bourton	686	Shropshire	
Harnhill	688	Bayston Hill	603
Longwood	439	South Yorkshire	
Cornwall		Holme Hall	555
Hingston Down	60	Spring Lodge	633
Kessel Downs	423	Tyne and Wear	
Lean	473	Houghton	23
Penlee	424	Warwickshire	
Cumbria		Griff	437
Goldmire	22		

Scotland

AGRICULTURAL LIME		ASPHALT OR TARMACADAM	
Fife		Borders	
Skelpie Mine	167	Borthwick	678
Grampian		Craighouse	584
Boyne Bay	120	Central	
Highland		Boards	677
Morefield	597	Murrayshall	599
Torlundy	289	Northfield	156
Torrin	330	Dumfries and Galloway	
Lothian		Coatsgate	582
Middleton Mine	464	Craignair	585
Strathclyde		Morrinton	598
Calliburn	319	Fife	
Trearne	465	Clatchard Craig	581
Tayside		Cruicks	645
Shierglas	601	Orrock	682

Grampian			Langside	373
Craigenlow	583		Orrock	682
Parkmore	177		Grampian	
Tom's Forest	658		Balmedie	257
Highland			Bluehill	258
Banavie	642		Cairds Hill	383
Borrowston	344		Concraigs	669
Daviot	679		Cottonhill	339
Morefield	597		Craigenlow	583
Sconser	290		Craiglash	259
Lothian			Gedloch	220
Bangley	362		Kemnay	656
Blairhill	580		Longside	134
Craigpark	586		North Lasts	329
Hillwood	681		North Mains	186
Strathclyde			Parkmore	177
Bannerbank	643		Pitcaple	260
Bonawe	243		Rothes Glen	510
Cairneyhill	644		Savoch	187
Croy	588		Tom's Forest	658
Duntilland	543		Western New Forres	261
Furnace	647		Highland	
High Craig	593		Achnagart	651
Hillhouse	292		Banavie	642
Kilbarchan	486		Borrowston	344
Tam's Loup	441		Bower	653
Tayside			Daviot	679
Friarton	680		Glensanda	705
Shierglas	601		Morefield	597
Waulkmill	648		Sconser	290
Western Isles			Torlundy	289
Ardhasaig	340		Torrin	330
			Lothian	
CRUSHED STONE			Bangley	362
Borders			Blairhill	580
Blynlee	242		Broadlaw	363
Borthwick	678		Craigpark	586
Craighouse	584		Dunbar Northwest	101
Edstone	116		Hillwood	681
Central			Middleton Mine	464
Boards	677		Ravelrig	600
Murrayshall	599		Orkney	
Northfield	156		Curister	410
Dumfries and Galloway			Heddle Hill	409
Barlockhart	79		Shetland	
Coatsgate	582		Brindister	545
Craignair	585		Scord	504
Kelhead	25		Setters	482
Kirkmabreck	595		Vatseter	244
Morrinton	598		Strathclyde	
Fife			Airdriehill	361
Balmullo	372		Ballygrant	197
Belliston	88		Bannerbank	643
Clatchard Craig	581		Bonawe	243
Craigfoot	649		Cairneyhill	644
Cruicks	645		Calliburn	319
Goat	592		Craigie Hill	655

Croy	588	Grampian		
Dunduff	413	Black Hills	133	
Duntilland	543	Clashach	379	
Furnace	647	Rothes Glen	510	
Hallyards	327	Spynie	380	
High Craig	593	Woodside	414	
Hillend	485	Highland		
Hillhouse	292	Bower	653	
Kilbarchan	486	Spittal	546	
Loanhead	596	Strathclyde		
Miltonhill	621	Craigie Hill	655	
Pettinain	160	Hessilhead	463	
Riskend	467	Old Mill	218	
Tam's Loup	441	Ross of Mull	492	
Tongland	81	Tam's Loup	441	
Tormitchell	82	Western Isles		
Trearne	465	Lic	331	
Underheugh	364			
Tayside		**INDUSTRIAL LIMESTONE**		
Ardownie	246	Lothian		
Boysack	617	Dunbar Northwest	101	
Collace	618	Middleton Mine	464	
Ethiebeaton	591			
Friarton	680	**PRE-CAST CONCRETE PRODUCTS**		
Shierglas	601	Highland		
Waulkmill	648	Borrowston	344	
Western Isles		Shetland		
Ardhasaig	340	Brindister	545	
Askernish	68	Strathclyde		
Lic	331	Bonawe	243	
Marybank	657			
Steangabhal	309	**READY-MIXED CONCRETE**		
		Grampian		
DRESSED OR LUMP STONE		Tom's Forest	658	
Borders		Highland		
Blynlee	242	Borrowston	344	
Central		Shetland		
Dunmore	488	Brindister	545	
Dumfries and Galloway		Strathclyde		
Corncockle	194	Bonawe	243	
Corsehill	195	Hillhouse	292	
Gatelawbridge	489	Riskend	467	
Locharbriggs	73			

Wales

AGRICULTURAL LIME		Machen	42
Clwyd		Mid Glamorgan	
Abergele	630	Vaynor	46
Halkyn	400	South Glamorgan	
Minera	575	Taffs Well	522
Pant-y-Pwll Dwr	400		
Dyfed		**ASPHALT OR TARMACADAM**	
Blaenyfan	699	Clwyd	
Gwent		Aberduna	433
Livox	41	Denbigh	570

Graig	459	Blaenyfan	699
Halkyn	400	Bolton Hill	250
Hendre	356	Carew Quarries	494
Minera	575	Cerrigyrwyn	251
Pant (Halkyn)	693	Coygen	252
Pant-y-Pwll Dwr	400	Dinas	355
Dyfed		Gellihalog	700
Alltgoch	249	Kiln	495
Blaenyfan	699	Mynydd y Garreg	316
Bolton Hill	250	Rhyndaston	381
Coygen	252	Syke	440
Torcoed	696	Torcoed	696
Torcoedfawr	359	Torcoedfawr	359
Gwent		Gwent	
Livox	41	Dayhouse	620
Machen	42	Hafod	460
Gwynedd		Livox	41
Minffordd	691	Machen	42
Penmaenmawr	29	Penhow	44
Tonfanau	391	Trefil	272
Mid Glamorgan		Gwynedd	
Craig-yr-Hesg	37	Aberstrecht	4
Forest Wood	435	Bryn Engan	384
Pant	692	Caer Glaw	19
Vaynor	46	Gwyndy	273
Powys		Hengae	385
Builth Wells	36	Minffordd	691
Dolyhir and Strinds	390	Nanhoron	387
Gore	626	Penmaenmawr	29
Tan-y-Foel	119	Tonfanau	391
South Glamorgan		Mid Glamorgan	
Wenvoe	702	Cornelly	684
West Glamorgan		Craig-yr-Hesg	37
Gilfach	701	Creigiau	607
		Forest Wood	435
BURNT LIMESTONE		Gelligaer	436
Clwyd		Grove	438
Abergele	630	Hendy	689
		Lithalun	40
CRUSHED STONE		Pant	692
Clwyd		Pencaemawr	436
Aberduna	433	Penderyn	43
Abergele	630	Ruthin	111
Burley Hill	114	Vaynor	46
Denbigh	570	Powys	
Graig	459	Builth Wells	36
Halkyn	400	Dolyhir and Strinds	390
Hendre	356	Gore	626
Llanddulas	26	Middletown	115
Minera	575	Penstrowed	256
Pant (Halkyn)	693	Penwyllt	694
Pant-y-Pwll Dwr	400	Tan-y-Foel	119
Raynes	401	South Glamorgan	
Trimm Rock	635	Aberthaw	99
Dyfed		Pantyffynnon	291
Alltgoch	249	Taffs Well	522
Ammanford	519	Wenvoe	702

West Glamorgan
    Barland   77
    Gilfach   701

DRESSED OR LUMP STONE
  Clwyd
    Clogau   95
    Plas Gwilym   296
  Dyfed
    Glogue   188
  Gwent
    Penhow   44
  Gwynedd
    Aberstrecht   4
    Braich Ddu   262
    Cwt-y-Bugail   235
    Diphwys Casson   263
    Gloddfa Ganol   236
    Llechedd Slate Mine   264
    Maenofferen   265
    Nanhoron   387
    Nantlle   237
    Nant Newydd   389
    Oakeley   238
    Penrhyn   365
    Rhuddlan Bach   466
    Twll Llwyd   297
    Vronlog   315
  Mid Glamorgan
    Forest Wood   435
    Gelligaer   436
    Pencaemawr   436
  Powys
    Aberllefeni   697

INDUSTRIAL LIMESTONE
  Clwyd
    Halkyn   400
    Minera   575
    Pant-y-Pwll Dwr   400
    Trimm Rock   635
  Gwent
    Penhow   44
    Trefil   272

Mid Glamorgan
    Cornelly   684
    Ruthin   111
  Powys
    Dolyhir and Strinds   390
  South Glamorgan
    Aberthaw   99

PRE-CAST CONCRETE PRODUCTS
  Dyfed
    Blaenyfan   699
  Gwynedd
    Gwyndy   273
  Mid Glamorgan
    Craig-yr-Hesg   37

READY-MIXED CONCRETE
  Clwyd
    Abergele   630
    Denbigh   570
  Dyfed
    Alltgoch   249
    Blaenyfan   699
    Bolton Hill   250
  Gwent
    Machen   42
  Gwynedd
    Minffordd   691
    Penmaenmawr   29
    Tonfanau   391
  Mid Glamorgan
    Cornelly   684
    Craig-yr-Hesg   37
    Penderyn   43
    Vaynor   46
  Powys
    Dolyhir and Strinds   390
    Tan-y-Foel   119
  South Glamorgan
    Taffs Well   522
    Wenvoe   702
  West Glamorgan
    Gilfach   701

Index 10

Personnel
in quarry companies and their quarries

Entries in this index refer to the
page numbers (in *italics*) for personnel based at
company head offices, or to the unique
quarry numbers (1 – 705 in ordinary typeface)
for personnel based at quarries.
Both page *and* quarry numbers are used for
personnel who are to be found at either address.

Sand & gravel in Holland
Manganese in Gabon
Copper in Portugal
Chromium in Madagascar
Phosphates in Tunisia
Gold in Ghana
Diamonds in Sierra-Leone
Uranium in France
Iron ore in Sweden
Talc in Norway
Coal in Germany
Tin in England
Limestone in Wales
Bauxite in Greece
Fluorspar in Russia

The world of Hewitt-Robins

In a world where experience and engineering excellence
really count, talk to Hewitt-Robins International.
We've probably solved your production problems before.

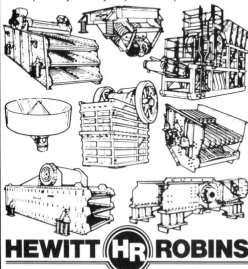

HEWITT (HR) ROBINS

Hewitt-Robins International Ltd.
Orchard House, Tebbutts Road, St. Neots, Cambs PE19 1AW
Tel: 44 (0) 480-404041   Fax: 44 (0) 480 403413

SITE INVESTIGATION

EXPLORATION

HARD-ROCK DRILLING EQUIPMENT

**H & F Drilling Supplies Ltd
Millhall, Stirling FK7 7LT**

Tel: 0786 79575   *Fax:* 0786 65803   *Telex:* 778583

Index 11

Advertisers